MILITARY SPACE ETHICS

Issues in Military Ethics

Edited by Don Carrick, King's College London, Michael Skerker, United States Naval Academy, and David Whetham, King's College London at the Joint Services Command and Staff College

With most officer training schools including military ethics as part of their programme, more than ever there is a need for clarity on ethical decision making. Contemporary military conflict is ever changing and with it military practitioners are confronted by new ethical challenges which often puts additional weight on the professional activity of personnel. At a minimum, military professionals need to have a clear knowledge of the laws that underpin their profession in order to evaluate situations quickly.

The series explores the complexities of acting ethically within the military system. It is not a philosophical debate on military ethics nor is it a general introduction. Instead, this series aims to provide real world guidance for military commanders and leaders. Edited alongside King's College London Centre for Military Ethics and the United States Naval Academy, this series brings a unique and relevant combination of practitioner and academic expertise which profoundly enhances the overall effect of the learning experience from its publications.

Other books in this series

Military Virtues
978 1 912440 00 9

Cyber Warfare Ethics
978 1 912440 26 9

Military Space Ethics

Edited by
NIKKI COLEMAN
UNSW Canberra Space

Howgate Publishing Limited

Copyright © 2022 Nikki Coleman

First published in 2022 by
Howgate Publishing Limited
Station House
50 North Street
Havant
Hampshire
PO9 1QU
Email: info@howgatepublishing.com
Web: www.howgatepublishing.com

All rights reserved.

No part of this publication may be reproduced, stored in a retrieval system, or transmitted in any form or by any means including photocopying, electronic, mechanical, recording or otherwise, without the prior permission of the rights holders, application for which must be made to the publisher.

British Library Cataloguing-in-Publication Data
A catalogue record for this book is available from the British Library

ISBN 978-1-912440-26-0 (pbk)
ISBN 978-1-912440-30-6 (ebk - PDF)
ISBN 978-1-912440-31-3 (ebk - ePUB)

Nikki Coleman has asserted her right under the Copyright, Designs and Patents Act, 1988, to be identified as the editor of this work.

The views expressed in this book are those of the individual authors and do not necessarily reflect official policy or position.

CONTENTS

Notes on Contributors viii
Foreword xiv
List of Abbreviations xvi

Introduction 1
   Nikki Coleman

Introduction to Military Ethics in Space 8
   Stephen Coleman

1   Future War: Will it be Conducted by Robots or Space Marines? 20
     Christopher D. Miller

2   Does Just War Theory Extend to the Space Frontier? 36
     Patrick Lin

3   In Space No One Can Hear the Geneva Convention: *Jus in Bello* and
     Warfare in Space 54
     Pauline M. Shanks Kaurin

4   What is Needed to Prevent Conflict in Space? 68
     Daniel Porras

5   Space Debris: Can We Remove the Landmines of Earth Orbit
     Without Starting a War? 80
     Stephen Coleman and Nikki Coleman

6	Ethical Considerations on the Challenges of the Dual-use Satellite Problem *Amy Hestermann-Crane*	95
7	The Growing Threat of Terrorism in Space *Nikki Coleman and Stephen Coleman*	115
8	The Problems Posed by Non-State Groups and Rogue States Exploiting Space *Kaylee Verrier*	124
9	Bioethics and Military Operations in Space *Sheena M. Eagan*	143
10	Star Laws: The Role of International Law in Regulating Civil and Military Space Activities *Cassandra Steer*	159
11	The Woomera Manual: Legitimising or Limiting Space Warfare? *Cassandra Steer*	178
12	Responding on Earth to Kinetic Attacks in Space *Geordie Jacobs*	201
13	Rocket Cargo: The Vanguard of the U.S. Space Force *Nathan Phillips*	219
14	Where Space Is Not an Option: African Ethics and the Options of Non-contenders in Space Warfare *Ibanga B. Ikpe*	231
15	The United States Space Force and Space as a Military Domain *Nathan J. Phillips*	248
16	From Peaceful Uses to Warfighting: The Dangers of the New Military Era in Space *Jessica West*	269

17 Post-Traumatic Stress and Moral Injury in Extreme Remote
    Warfare   287
    Jayden Park

18 What We Have is What We Bring There: Security in Space as
    Utopian Vision   299
    Evie Kendal

19 Vitoria the Universal Thinker: Some Ethical Dilemmas
    Concerning Space Exploration   318
    Francisco Lobo and David Whetham

Index   *331*

NOTES ON CONTRIBUTORS

Chaplain (SQNLDR) Revd. Dr Nikki Coleman

Nikki Coleman is a military ethicist who specialises in space ethics, obedience in the military, military bioethics, military ethics education, moral injury and PTSD. Dr Coleman has consulted with NASA, the ESA, JAXA, the UN, U.S. Space Command, and a multitude of commercial space companies on space ethics over the past decade. She is a Senior Research Associate at the Case Western Reserve University Inamori International Center for Ethics and Excellence, leads the Space Ethics Research Group at UNSW Canberra Space and is the Senior Ethicist for the Royal Australian Air Force. In her spare time Nikki is a hot air balloon pilot.

Dr Stephen Coleman

Stephen Coleman is Associate Professor of Ethics and Leadership in the School of Humanities and Social Sciences, with the University of NSW Canberra, at the Australian Defence Force Academy. He has published on a diverse range of topics in applied ethics, including military ethics, police ethics, space ethics, medical ethics, and the practical applications of human rights. His latest book is *Military Ethics: An Introduction with Case Studies*.

Assistant Professor Sheena M. Eagen

Sheena M. Eagan is an Assistant Professor with the Department of Bioethics and Interdisciplinary Studies in the Brody School of Medicine at East Carolina University (ECU). Dr Eagan holds a PhD in the medical humanities from the Institute for the Medical Humanities at the University of Texas Medical Branch in Galveston as well as a Master of Public Health from the Uniformed Services University. Her research and teaching have focused on medical ethics and the history of medicine, with a subspecialized focus on military medicine. Before arriving at ECU, she worked as a defense sub-contractor providing ethics education to military service-members and their families. Dr Eagan also held a variety of visiting scholar positions for Yale University, the Brocher Foundation, and the University of Belgrade. Dr Eagan is the creator and president of the American Society of Bioethics and Humanities group for Military, Humanitarian and Disaster Medicine and has worked closely with the International Committee of Military Medicine. She has also worked with the NATO Center for Excellence in Military Medicine and maintains close connections with local military installations. Dr Eagan has published articles in peer-reviewed journals, military-specific journals, and contributed to edited books on a variety of topics in military medicine. She has also given talks and lectures in North America, Asia, Europe, and the Middle East. Her research interests include military medical ethics; military women's health; the history of PTSD; Veteran re-integration; moral injury; research ethics; medicine during the holocaust; and the history of military medicine.

Sergeant Amy Hestermann-Crane
Amy Hestermann-Crane is a Sergeant currently serving as an analyst in the Royal Australian Air Force (RAAF). In 2020, Hestermann-Crane became the first enlisted RAAF space analyst within the Australian Space Operations Centre. She also dedicates her time to the RAAF Women's Integrated Networking Group, where she focuses on promoting and providing opportunities within education and development, mentoring and coaching, and broader RAAF and community engagement. Sergeant Hestermann-Crane has a passion for ethics and STEM, putting both interests to use as a member of the International Space Ethics Collaborative Research Team and the Space Generation Advisory Council. She has completed a Bachelor of Communication and is working towards completing a Bachelor of Arts (Hons.), a Bachelor of Historical Inquiry and Practice and, a Master of Space Operations.

Professor Ibanga B. Ikpe
Ibanga B. Ikpe has taught Contemporary Analytic Philosophy and Critical Thinking for the past 30 years and has variously taught at the University of Botswana, The National University of Lesotho, the University of Uyo and the University of Cross River State. He has also been visiting scholar at the University of the West Indies, Jamaica, and Buffalo State College. He served as a Critical Thinking consultant to the Botswana Defence Command and

Staff College and still teaches Critical Thinking and Military Ethics at the college. His research is mainly in the areas of critical thinking, military Ethics, philosophical analysis and philosophical practice and has recently published in *Theoria: A Journal of Social and Political Theory*, *Human Affairs: Post-disciplinary Humanities & Social Sciences Quarterly*, *HASER: Revista Internacional de Filosofía Aplicada*, the *Journal for Peace and Justice Studies* and the *Journal of Humanities Therapy*. He is certified by the American Philosophical Practitioners Association as a philosophical counsellor and is also a certified conflict mediator.

Flight Lieutenant Geordie Jacobs
Geordie Jacobs joined the Royal Australian Air Force in 2011 and studied at the Australian Defence Force Academy completing a Bachelor of Arts and a Masters in Strategy and Security with a focus on the effectiveness of Air Power in countering insurgencies and non-state actors. Since graduating, Flight Lieutenant Jacobs has worked in the Directorate of International Engagement – Air Force, Head Quarters Joint Operations Command and the Air and Space Operations Centre.

Dr Evie Kendal
Evie Kendal is a bioethicist and public health researcher at the Department of Health Science and Biostatistics, Swinburne University of Technology. Dr Kendal's research interests include ethical dilemmas in emerging biotechnologies, space ethics, and public health ethics.

Professor Patrick Lin
Patrick Lin is the director of the Ethics and Emerging Sciences Group, based at California Polytechnic State University, San Luis Obispo, where he is a philosophy professor. Current affiliations include Stanford Law School, Czech Academy of Sciences, Center for a New American Security, World Economic Forum, and the 100-Year Study on AI. Previous affiliations include Stanford Engineering, U.S. Naval Academy, Dartmouth College, University of Notre Dame, University of Iceland (Fulbright), New America Foundation, CAPPE, and UNIDIR. Dr Lin is well published in technology ethics, including AI, robotics, space development, military and policing weapons, cyberwarfare, human enhancement, nanotechnology, and more. He regularly gives invited briefings to industry, media, and governments worldwide; and he teaches courses in ethics, technology, and law.

Francisco Lobo

Francisco Lobo is a Doctoral Researcher at the Department of War Studies, King's College London. His research focuses on military ethics and human rights. He holds an LL.M. in International Legal Studies from New York University (Fulbright Scholar), a Master of International Law and an LL.B. degree from the University of Chile. He is a lecturer of International Law, Human Rights Law, International Criminal Law, and Legal Theory, in Santiago of Chile. He has worked as an NYU Fellow of International Law and Human Rights at the UN International Law Commission (2018), where he assisted the Special Rapporteur on Peremptory Norms of General International Law (*ius cogens*). He has also worked as a legal adviser at the Ministry of Foreign Affairs of the Republic of Chile (2019-2020).

Lieutenant General Christopher D. Miller

Christopher D. Miller, USAF (Ret.) serves as the Helen and Arthur E. Johnson Chair for the Study of the Profession of Arms at the U.S. Air Force Academy Center for Character and Leadership Development. His active military service included leadership as Air Force deputy chief of staff for strategic plans and programs, U.S. Northern Command and NORAD's director of strategy, plans and policy, and as the senior Air Force operational commander deployed in Afghanistan. He also commanded the Air Force's B-2 wing and B-1 bomber units and held a wide variety of positions in policy analysis, international relations, human resources, aviation, and academia. He was a 1980 distinguished graduate of the Air Force Academy and earned graduate degrees from the U.S. Naval War College and Oxford University.

Flying Officer Jayden Park

Jayden Park is an Electrical Engineer in the Royal Australian Air Force. He graduated from the Australian Defence Force Academy in 2018, and received his Bachelor of Electrical Engineering (Hons) in 2019.

Nathan J. Phillips

Nathan Phillips is a former Navy officer of over fifteen years now working as a Principal Consultant in capability delivery. He has a wide variety of experiences in operations, headquarters support, projects and planning. He holds a Bachelor of Arts in History, a Graduate Certificate in International Relations, a Master of Letters, and a Master of Policing, Intelligence and Counterterrorism. In his spare time, he is also a fiction writer and editor specialising in speculative fiction.

Daniel Porras

Daniel Porras' areas of expertise include international space law and policy, emerging technology threats, international law, and political science. His main focus is the progressive development of sustainable norms of behavior for space activities. He was the resident technical expert for multiple UN bodies working on space security issues, including the Group of Governmental Experts on the Prevention of an Arms Race in Outer Space (PAROS) as well as Subsidiary Body 3 of the Conference on Disarmament (on PAROS). Mr Porras is a Board Member of the Space Court Foundation and a Member of the California Bar.

Professor Pauline Shanks Kaurin

Pauline Shanks Kaurin holds a PhD in Philosophy from Temple University, specializing in military ethics, just war theory, and applied ethics. She serves as the Stockdale Chair and Professor of Professional Military Ethics at the U.S. Naval War College in the College of Leadership and Ethics. She also holds a BA in Philosophy and International Relations from Concordia College, MN and a MA in Philosophy from the University of Manitoba, Winnipeg. Recent publications include: *When Less is not More: Expanding the Combatant/Non-Combatant Distinction; With Fear and Trembling: A Qualified Defense of Non-Lethal Weapons, Achilles Goes Asymmetrical: The Warrior, Military Ethics and Contemporary Warfare* and *On Obedience: Contrasting Philosophies for Military, Citizenry and Community*. She was Featured Contributor for *The Strategy Bridge* and has published in *Clear Defense, The Wavell Room, Newsweek, War on the Rocks, Grounded Curiosity, U.S. Naval Institute Proceedings, Just Security*, as well as a variety of academic journals.

Dr Cassandra Steer

Cassandra Steer is a Mission Specialist with the Australian National University Institute of Space (InSpace), and a Senior Lecturer at the ANU College of Law. She has been a consultant to the Australian, Canadian and U.S. departments of Defence on these issues. Dr Steer was formerly Acting Executive Director at the University of Pennsylvania's Center for Ethics and Rule of Law, Executive Director of Women in International Security – Canada, Executive Director of the McGill Institute of Air and Space Law, and Senior Lecturer at the University of Amsterdam. In 2011 Dr Steer was a Fulbright Scholar, and she has a degree in philosophy (UNSW); a civil law degree, a Master of Law and a PhD in International Criminal Law

(University of Amsterdam). Dr Steer is a member of the Australian Space Agency's technical Advisory group for Space Situational Awareness, and is the Canadian representative member on the International Law Association Space Law Committee, an Associate Expert on the Woomera Manual on the International Law of Military Space Operations, and a member of the International Institute of Space Law

Flying Officer Kaylee Verrier
Kaylee Verrier is an Aeronautical Engineer in the Royal Australian Air Force (RAAF). She graduated from the Australian Defence Force Academy in 2018, receiving her Bachelor in Aeronautical Engineering (Hons) in 2019. She is also completing a masters degree in Space Engineering at UNSW Canberra, due to graduate in 2022.

Dr Jessica West
Jessica West is a Senior Researcher at the Canadian peace research institute Project Ploughshares. Her research and policy work focuses on technology, security, and governance with a particular interest in peace and security in outer space. Jessica interacts regularly with key United Nations bodies tasked with space security and space safety issues. She holds a PhD in global governance and international security from the Balsillie School of International Affairs, Wilfrid Laurier University.

Professor David Whetham
David Whetham is Professor of Ethics and the Military Profession in the Defence Studies Department of King's College London. He is the Director of the King's Centre for Military Ethics located at the UK's Joint Services Command and Staff College. David supports military ethics education in many different countries and has held Visiting Fellowships at the Stockdale Center for Ethical Leadership, U.S. Naval Academy Annapolis, the Centre for Defence Leadership and Ethics at the Australian Defence College in Canberra and at the University of Glasgow. He was a Mid-Career Fellow at the British Academy in 2017-18 and is currently a Visiting Professorial Fellow at the University of New South Wales. He is a member of the UK MoD AI Ethics Advisory Panel, and in 2020 he was appointed as an Assistant Inspector-General to the Australian Defence Force to assist in the final stages of the Afghanistan Inquiry. David is the Vice President of the European Chapter of the International Society for Military Ethics (Euro ISME).

FOREWORD

As we gaze up into the heavens, daring to imagine the future, a glance into the rear-view mirror is fitting.

The year was 1494 and Europe's two "world powers" were eyeing the New World as a source of wealth, power, and status. Both countries had spent great fortunes over 75 years to create, test, and perfect the technologies of craft and navigation to travel great empty distances over the globe. The "Age of Discovery" was at hand.

Competing interests of Spain and Portugal ensured eventual conflict. Political leaders turned to the moral authority of their time to be the arbiter of the disputes over the ownership and rights of the vast new world. The Treaty of Tordesillas codified the papal bull of Pope Alexander IV to neatly divide the new continents between Spain and Portugal. While potential clashes were put to rest for these two countries, the edict was patently ignored by other nations like Britain, France, and the Netherlands who would eventually set out to claim their own share of the spoils.

Like Spain and Portugal's requirement for peaceful and productive commerce on the seas, today's space faring nations need a moral construct to frame the conduct of both commercial and military space operations. Space, like it or not, is "contested, congested and competitive." The past seven decades have been punctuated by a quest to travel to the moon and beyond. Billionaires are normalizing space travel. And a domain that was thought to be a vast and peaceful sphere for the conduct of national business is increasingly subject to warfare. And no small complicator is an ever-increasing population of trackable and potentially deadly space debris, which space warfare will worsen exponentially.

So, what does good and respectful behavior in space look like? Anyone who has ever relied on a weather forecast, safely withdrawn cash from an

ATM, streamed a movie, used GPS to navigate or flown safely in an aircraft ought to be immensely interested in an ethical galactic future. Thankfully, Chaplain Nikki Coleman and her team of authors has provided the world with one of the first works to help us all explore the important questions that must be discussed as we move boldly into space, especially as we use that domain for national wealth, power, and status – the very reasons other countries took to the high seas centuries ago. This book seeks not so much to answer crucial questions as it does to frame the necessary dialogue that will increase the probability that both commercial and military applications of space, worldwide, can occur peacefully.

Abraham Lincoln once said, *"Determine that the thing can and shall be done, and then we shall find the way."* This book for many, will guide the journey.

Steven A. Schaick, Chaplain, Maj. Gen. (Ret)
19th Chief of Chaplains, United States Air Force
1st Chief of Chaplains, United States Space Force
December 2021

LIST OF ABBREVIATIONS

AFRL	Air Force Research Lab
AI	Artificial Intelligence
APC	Armoured Personnel Carrier
ASAT	anti-satellite
CBA	cost-benefit analysis
CD	Conference on Disarmament
COPUOS	Committee on the Peaceful Uses of Outer Space
DA-ASAT	Direct-ascent anti-satellite
DFC	directional fragmentation charge
DIA	Defense Intelligence Agency
EMP	electromagnetic pulse
ET	extra-terrestrial
EW	Electronic Warfare
FCC	Federal Communications Commission
FOB	Forward Operating Base
GEO	geostationary orbit
GNSS	global navigation satellite system
GPS	Global Position System
ICBM	intercontinental ballistic missiles
ICJ	International Court of Justice
IED	Improvised Explosive Device
IHL	international humanitarian law
IOM	Institute of Medicine
ISS	International Space Station
ITU	International Telecommunications Union
JWT	Just War Theory
KKV	Kinetic Kill Vehicle

LOAC	Law of Armed Conflict
LEO	Low Earth Orbits
NSG	Non-State Groups
OST	Outer Space Treaty
PAROS	Prevention of an Arms Race in Outer Space
PID	positive identification
PNT	position-navigating-timing
PPWT	Prevention of the Placement of Weapons in Outer Space Treaty
PSTD	Post-Traumatic Stress Disorder
SDG	Sustainable Development Goals
SKKV	Suicide Kinetic Kill Vehicle
SWF	Secure World Foundation
WMD	weapons of mass destruction
UNOST	United Nations Outer Space Treaty
UNSC	United Nations Security Council
USAF	United States Air Force
USSF	United States Space Force

INTRODUCTION

Nikki Coleman[1]

In one way or another, the military has been involved since the very beginning of human spaceflight. During the 1950s and 60s the Cold War between the U.S.S.R. and the U.S.A. and its nuclear arms race fuelled much of the push for space. The very first artificial satellite to orbit the earth, Sputnik 1, was launched into space by a Russian R-7 Semyorka rocket, a modified form of the world's first intercontinental ballistic missile. The first person in space, Yuri Gargarin, was a pilot and Air Force officer, as were all the early Soviet cosmonauts. The first Americans selected for astronaut training, the Mercury Seven, were all military test pilots.

The early military links to space exist not only amongst those trained to travel to space, but also amongst those who helped to get them there. Because the beginnings of the space race came out of the early cold war missile programs, many of the engineers and scientists were either active-duty military members or veterans. This shaping of the early space programs continued for many decades until the dramatic increase in commercial companies operating space programs. For this reason, military ethics have been at the forefront of space operations, right from the beginning. However, space presents some unique challenges in regards to military ethics, and therefore it is important that we turn our focus to the area of Military Space Ethics.

Sputnik 1 may have been the first artificial satellite in space, but it was by no means the first use of space by humans. The work of ancient astronomers has been vital to us humans in understanding not only our place in the universe but how we relate to the earth we all live on. The study of the sky by Indigenous Australians as long as 40,000 years ago dictated how and when to gather food, and guided them in their stewardship of "country" (the land). In a similar way, military organisations, who do not have commercial pressures, are in a unique position to be leading stewards of space for all people on earth and for future generations.

---

[1] The views expressed are those of the author and do not reflect the official policy or position of the Royal Australian Air Force, the Department of Defence, or the Australian Government.

With the advent of the internet and the rise of space based communications, the reliance on space for all people has created a situation where space, as a global common area, requires protection. Space debris from an all out war in space would be catastrophic for much of the population of the earth. It would impact on communication, navigation, food distribution systems, electrical grids, health care, education and transport systems. For this reason much work has been done by groups such as UNIDIR and the Secure World Foundation, as well as many countries individually, to ensure that space remains safe, secure and sustainable for future generations. This book carries on that work, beginning the discussion of the ethical issues raised by the military uses of space.

Over the past decade I have often spoken about military space ethics, initially at academic conferences, and then later with many other people, including school groups. When I introduced myself to one of these school groups as a military space ethicist, one of the participants asked me, in the way only a 14 year old girl can; "Military space ethicist … is that like, a REAL job?!?!" Despite the incredulity of many people, not merely 14 year old girls, about the existence of such a job, military space ethics does exist, and the questions that military space ethicists address are indeed important ones, as I hope the various chapters in this collection will show.

In the second part of the introduction to this book, Stephen Coleman provides an overview of the main ideas of military ethics, with a focus on Just War Theory, and briefly discusses whether those principles might be expected to apply to the potential new domain of conflict in space. He also suggests that since these principles are actually more general ones which apply in a range of situations outside of armed conflict, we perhaps ought to expect that they would apply to any conflict in space.

The chapters in the first main section of the book, *EARTH TO THE MOON*, discuss ethical issues which arise in military operations conducted in the space nearest to Earth, mainly in Earth's own orbit. Christopher Miller begins the discussion, asking whether any armed conflict which might occur in space in the near future will involve humans directly engaging each other, as has been the case in more traditional wars, or if it will be a new type of conflict, conducted mainly by robots and/or remotely operated devices, and involving few, if any, direct human casualties. Patrick Lin continues the discussion by questioning whether the principles of Just War Theory, which were formulated to limit the destructiveness of war on Earth, actually apply, or apply in the same way, to a conflict in space. In particular he considers the so-called "realist" position which would suggest that ethics has no place in such discussions, and that the only questions which the leaders of a space-faring state should be considering are what they have the power to do in space and who has the power to stop them. While Patrick Lin looks at issues across the entire spectrum of Just War Theory, it is *jus in bello* (justice within war, for example, how wars ought to be fought) which is the exclusive focus of the chapter by Pauline Shanks Kaurin. She considers whether the principles of discrimination and proportionality, which lie at the heart of *jus in bello*, would continue to apply in an armed conflict fought in space. She advances her discussion by using three analogies for space war; drone/remote warfare, maritime warfare, and cyber warfare, and seeing what lessons might be learnt for the ethics of space warfare from each. Rather than questioning what rules or principles might apply in the event of an armed conflict in space, Daniel Porras focuses on what can be done in order to avoid such a conflict developing in the first place. He argues that society's dependence on space capabilities, and on the services that such capabilities make possible, means that it is in everyone's best interests to ensure the future sustainability of space operations. Since conflict in space would inevitably threaten such sustainability, it would appear to be in everyone's best interests to act in ways which reduce the possibility of such conflict occurring.

Having started with an examination of the more general issues posed by the potential for armed conflict in space, the later chapters in this section of the book delve into the ethical questions raised by more specific issues which arise in near-Earth space. In the first of these chapters, Stephen and Nikki Coleman discuss some of the problems posed by the debris in Earth's orbit. Some of this orbiting debris is natural, but the majority of it has been created by past space operations. While it is widely recognised that this debris poses a hazard to future orbital operations, there are serious problems involved in its removal since not only is removal a physically difficult challenge, it is also a highly problematic political one since any technology which could be used for debris removal could also be used as a space weapon. Amy Hesterman-Crane examines some of the ethical issues raised by "dual-use" satellites that have important functionality for both military and civilian use. In the event of an armed conflict in space the military function of such satellites would make them a legitimate (that is, legal) military target, but their importance to the civilian community would also need to be taken into consideration. Nikki and Stephen Coleman's second chapter in this section raises the possibility of space terrorism. As they note, some people just want to watch the world burn, and the cheaper and more widely available access to space is, the more likely it becomes that such actors will become involved in space operations which is why such problems need to be considered before they arise, rather than after. We shouldn't wait for the space equivalent of 9-11 before considering the problems that this sort of space terrorism would cause, so this chapter ends by considering some ways by which currently space-faring states might attempt to mitigate the risks posed by such terrorism. The chapter by Kaylee Verrier continues this line of discussion, looking at some of the problems which could arise out of the exploitation of space by rogue states and/or violent non-state groups, who may have no wish to enjoy the benefits of space technology themselves, but may want to deny the benefits of such technology to others. The final chapter in this section, by Sheena Eagen, examines the bioethical problems likely to arise in the event of an armed conflict in space. As she notes, military medicine already encounters a range of ethical problems which are not faced by those who provide regular medical care, and military operations in space will present even greater difficulties.

The chapters in the second main section of the book, RETURN TO EARTH, examine ethical problems which may arise (or have already arisen) on Earth due to military operations, or potential military operations, in space. This section begins with Cassandra Steer's two chapters discussing the legal problems of space operations and space warfare. The first chapter explains why space is anything but a "lawless frontier" and examines how existing laws, including both the Outer Space Treaty and the international laws regarding armed conflict, would apply in the event of a space-based conflict. The second chapter addresses the Woomera Manual, which attempts to clarify International Law regarding Military Space Activities, with her particular focus here being to examine whether such an attempt does anything more than legitimise further militarisation of space. In his contribution to this volume, Geordie Jacobs discusses when, if ever, it might be ethically appropriate for a state that has been the target of an attack in space to respond with a kinetic attack against an Earth-based target. In particular, he considers the ethical issues which arise if the original attack in space was launched not by a rival state, but rather by a non-state group. The final chapters in this section move away from issues related to the direct use of military force. Nathan Phillips looks at some of the possibilities of space-enabled logistics, a facility currently being investigated by the U.S.A. among others. Ibanga Ikpe asks whether the military personnel of African states that currently have no space capabilities, have anything to contribute to the discussion of military space ethics. He concludes that if such personnel do not contribute their voices to this discussion, then an important perspective will be lost.

The chapters in the last section of the book TO INFINITY AND BEYOND, are focussed on ethical issues which need to be considered as we move out of Earth's immediate celestial environment and deeper into space. Nathan Phillips looks at the U.S. Space Force, which he wittily notes is probably the "first branch of the military to be satirised before it had even set up an operational command."[2] The United States see the establishment of this new branch of the military as vital to their continuing strategic interests in space, but other states around the world see it as an open challenge, and as a further step along the road towards the militarisation of space. In her contribution to the discussion, Jessica West considers the problems posed by the current increase in the militarisation of space. The original space race occurred in the depths of the Cold War with its implicit demands for military uses of this new frontier. Yet despite this fact, both military and peaceful uses of space were rapidly developed, a fact highlighted by the language adopted in the Outer Space Treaty. Civilian and commercial uses of space are now more important than ever, but so are military uses of space. Several states now have military branches specifically delegated with responsibility for the protection of the state's strategic interests in the space domain, and West argues that this swing towards the militarisation of space increases the likelihood of conflict and threatens peaceful uses of space in the future. With a focus on the human side of conflict, Jayden Park considers some of the mental health problems which might be triggered by the "extreme remote war" of space-enabled conflict. In her chapter, Evie Kendall examines some ethical issues related to the establishment of future colonies on other celestial bodies, particularly with regard to law enforcement in such colonies. She uses a wide range of entertaining science fiction examples to illustrate the sorts of problems which must be considered when determining how such matters ought to be dealt with. Finally, Francisco Lobo and David Whetham use insights gained from some historical problems to help illuminate concerns that are currently still in the realm of science fiction; namely what ethical duties we would owe to intelligent aliens, should we ever encounter them. To do this, Lobo and Whetham look at the writings of Francisco de Vitoria, who wrote in some depth about the ethical duties owed to the native peoples of America by the

---

2 Through the Netflix series *Space Force*, created by Steve Carrell and Greg Daniels, released 29 May 2020.

Spanish seafarers who ventured to the New World in the sixteenth century. The thoughts of Vitoria and the analogy of voyages to the New World are used to prompt discussion about the ethical requirements placed on any humans who might encounter equivalent life in the Space Age New World.

Acknowledgements

The Royal Australian Air Force is leading the way in the area of Military Space Ethics, choosing to integrate ethical discussions and concepts from the very beginning of the creation of the Space Domain team. Thank you to the Royal Australian Air Force Space Domain Review Team, especially AIRCDRE Nicholas Hogan and COL Michael Hose, who kindly gave me time to bring this book into being.

Thank you especially to all the authors who have contributed to this book. When I started this journey in Military Space Ethics at UNSW Canberra almost a decade ago, I decided that I would not follow the popular academic advice of creating a large body of work in a small area so as to stamp my impact on the academic area. Instead, very early on I decided that I wanted to grow the emerging discipline of Military Space Ethics so that it would not be only discussed by academics, but rather be something to be hotly debated from military space practitioners through to policy analysts. I have been fortunate that my work in developing the growing field of Military Space Ethics was supported by both Professor Russell Boyce at UNSW Canberra Space and then later by the Royal Australian Air Force. I have had the absolute pleasure to be able to speak with the most interesting and passionate people in military, commercial and civilian space teams. I thank you all for your time and willingness to engage with me on issues that have sometimes seemed out of left field, but which always had a purpose of promoting discussion. This book is the result of some of those discussions, with contributions from highly ranked military members, through to enlisted members, and from civilians who specialise in pacifist writing as well as experienced military ethics academics. I have had the joy of working with all these wonderful people and seeing them shape the future discussions around Military Space Ethics.

I would especially like to thank Kirstin Howgate who has expertly shepherded this book through the COVID crisis. Thank you for your kindness, support and patience.

Finally the greatest thanks must go to my research partner Stephen Coleman, who owing to me being very sick with PTSD as a result of my service, picked up the load that I could no longer carry and did the final editing and organisation to see my eight year project finally come to life.

This book is dedicated to all the practitioners who work in the military space field – this book is for you and is just the start of the journey. May this book assist you to add your own voices to this discussion around Military Space Ethics.

INTRODUCTION TO MILITARY ETHICS IN SPACE

Stephen Coleman[1]

Military ethics is not new, but space is a new domain of potential conflict, so does military ethics need to be re-considered for this new domain? In this chapter I will sketch out a broad outline of what military ethics is and of the major principles which contribute to our modern understanding of it, as well as providing a brief argument for why it would be reasonable to expect that the ideas incorporated into military ethics ought to be applicable to this possible new domain of warfare.

Military ethics focuses on the core ethical ideas which govern the conduct of those engaged in armed conflict. It thus incorporates ideas of military professionalism and honour, which tend to focus on the individual, as well as more general ideas about the ethics of war, particularly the long and influential tradition of ethical thought which has come to be known as Just War Theory that forms the basis of most of the modern international law that now regulates such conflict. Both these streams of thought are literally ancient, having been discussed by the Ancient Greeks and by major philosophical and religious figures in both India and China. In the modern world the main focus of military ethics is usually on Just War Theory, so most of this chapter will be devoted to an examination of the basics of that theory and how it might apply to conflict in space. But it is nonetheless appropriate to briefly discuss notions of individual character and virtue in the military first.

---

1 Parts of this chapter are based on material originally published in Stephen Coleman, *Military Ethics: An Introduction with Case Studies* (New York: Oxford University Press, 2013) and Stephen Coleman, "Even Dirtier Hands in War: Considering Walzer's Supreme Emergency Argument" *Research in Ethical Issues in Organizations* 13(2015): 61-73.

Military Ethics and Individual Virtue

As Shannon French discusses in her excellent work *The Code of the Warrior*,[2] the main point of warrior codes is to remind those who must enter into conflict what is expected of them in order that they can fight, and do and see terrible things, without becoming terrible in the process; as Nietzsche said "He who fights with monsters should be careful lest he thereby become a monster."[3] In modern terms, such codes exist so that those who go to war can still live with what they see in the mirror when they come home. Warrior codes are widespread through history and while some are more famous than others, those who go to war almost invariably have at least some sort of code drilled into them by their peers, even in modern times. Some warrior codes are famously lengthy and intricate, but most can be captured in only a few sentences, or even a single phrase, and these ideas certainly persist into modern times. Some modern codes are so widely known, even if unofficial, that those outside the military often think they are written in stone; a good example of this is probably the idea of "Leave No One Behind" which despite its influence cannot be found written in any U.S. military doctrine.

The main emphasis of both historic warrior codes, be they Greek, Roman, Viking, Chinese, Japanese ... as well as of modern military ones, is on what a good warrior (or in modern terms, soldier or sailor or aviator or guardian etc) will NOT do. As Shannon French notes, it is this code which distinguishes the warrior from the murderer.

> When they are trained for war, warriors are given a mandate by their society to take lives. But they must learn to take only certain lives, in certain ways, at certain times, and for certain reasons. Otherwise, they become indistinguishable from murderers and will find themselves condemned by the very societies they were created to serve.[4]

In modern times, the extensive discourse between formerly widely separated cultures means that the modern warrior's code, at least in its more formal parts, has becoming inextricably intertwined with the principles of Just War Theory. So much so, in fact, that many of the principles of Just War Theory, particularly those regarding the way armed conflict is actually

---

2 Shannon French, *The Code of the Warrior* 2nd Ed. (Lanham MD: Rowman & Littlefield, 2017).
3 Friedrich Nietzsche, *Beyond Good and Evil*, Part IV, Aphorism #146.
4 French, *The Code of the Warrior*, 4.

carried out by those directly involved in its conduct, have been formalised into a series of international treaties which can be collectively referred to as the Law of Armed Conflict (LOAC). The ethical principles which underlie these treaties thus form the heart of modern military ethics, and are the focus of the remainder of this chapter.

Just War Theory

The central claim of Just War Theory is that if certain conditions are met prior to the resort to the use of armed force then a state (and possibly also a certain type of non-state group) is ethically justified in engaging in armed conflict. In any armed conflict the armed forces involved must also abide by other conditions, designed to limit the destructiveness of this use of armed force and prevent unnecessary harm to both combatants and non-combatants alike. Modern Just War Theory has evolved out of a long tradition of thought which stretches back at least to Ancient Greece, but also incorporates ideas from many different thinkers from many different cultures. Probably the most comprehensive modern statement of the ideas of the theory is Michael Walzer's *Just and Unjust Wars*,[5] which is referred to by almost all other writers on this topic.

Just war theory is traditionally taken to have two aspects: *jus ad bellum* (justice of war) which deals with when it is right to resort to war rather than attempting to resolve a dispute by other means; and *jus in bello* (justice in war) which deals with the conduct of those who are actually fighting the war, be they soldiers, sailors, aviators, marines, guardians, or even civilians who have taken up arms. In recent times there has also been considerable discussion of a proposed third aspect of Just War Theory known as *jus post bellum* (justice after war), which deals with peace agreements and ending wars; essentially the business of moving from war back into peace, especially with the aim of producing a just and lasting peace after war. In terms of discussions regarding the possibility of armed conflict in space there has been little interest in the ideas of *jus post bellum*, possibly because there has not yet actually been any open conflict in space, so I will not consider *jus post bellum* any further here.

---

5 Michael Walzer, *Just and Unjust Wars* (New York: Basic Books) original edition 1977. Each of the later editions includes a new preface, but the text as a whole has not changed since the first edition.

There is some argument about the exact number of conditions which must be met in order for a party to be justified in going to war, but most modern writers suggest that *jus ad bellum* consists of six conditions, all of which must be met before it can be considered to be ethically appropriate to engage in an armed conflict:

1. there must be just cause for going to war
2. those deciding to go to war must do so with appropriate intentions
3. the war must be authorised by the appropriate authority and publicly declared to all relevant parties
4. war must only be used as a last resort
5. there must be a reasonable probability of success in the war
6. the overall cost of the war, not merely the financial cost but the harm involved, must be proportional to the benefit which will be obtained by going to war.

Though there is again some dispute about the issue, most writers suggest that *jus in bello* consists of two main principles by which the participants in the war must abide: (1) discrimination; and (2) proportionality. As I mentioned earlier, Just War Theory has been so influential that many aspects of the theory, including parts of both *jus ad bellum* and *jus in bello*, have become incorporated into international law regarding the use of armed force, initially through customary international law, then later through treaties which have codified these customary understandings.

Most just war scholars argue that the two main standards of Just War Theory, of *jus ad bellum* and *jus in bello*, are logically distinct and it is therefore perfectly possible for a war to meet one of these standards without meeting both, though again there is some scholarly dispute about this issue.[6] In international law, however, this distinction is very clear. Thus, unjust or illegal wars, which fail to meet the criteria of *jus ad bellum*, may be conducted in a justifiable manner, that is, in accordance with the principles of *jus in bello*. Similarly, wars which are justified, in that they meet the criteria of *jus ad bellum*, may be conducted in a non-justifiable manner, in that they fail to meet the criteria of *jus in bello*. Entire books have been written on what the various principles of the theory amount

---

6 See for example David Rodin, *War and Self-Defense* (New York: Oxford University, 2003) and Jeff McMahan, *Killing in War* (Oxford: Oxford University Press, 2009).

to, so I won't attempt to explain those principles in detail here.[7] However, some brief discussion of each one of the principles is necessary in order to clarify what that principle actually means in practice, beginning with the principles of *jus ad bellum*.

The most basic aspect of *jus ad bellum* is incorporated into its first principle, which insists that a state may only engage in warfare if it has an appropriate reason, that is, a "just cause" for war. In modern times it is generally accepted, as Michael Walzer argues, that the only just cause for war is resistance to aggression.[8] While the clearest examples of this are the defence of one's own country from aggressive attack and the defence of another country from aggressive attack, in more recent times it has also become commonly accepted that protection of innocent citizens from the aggressive attack of their own government also amounts to a legitimate cause for war, an idea captured through the principles of "Responsibility to Protect", endorsed by the United Nations at the 2005 World Summit.[9] It is worth noting that if resistance to aggression is the only just cause for war, then logic dictates that only one side in a conflict can possibly have just cause, since one side must be the aggressor and the other the victim of aggression. However, it is perfectly possible, and historically all too common, that neither side actually possesses an ethically justified cause for engaging in a conflict.

As a matter of international law, self-defence is recognised as an appropriate ground for war in article 51 of the UN Charter, which states: "Nothing in the present Charter shall impair the inherent right of individual or collective self-defence if an armed attack occurs against a Member of the United Nations, until the Security Council has taken measures necessary to maintain international peace and security."[10] One important issue which arises in the context of self-defence is the question of whether a state is permitted, either legally or ethically, to engage in pre-emptive or preventative

---

7 In addition to Walzer, other books of note include Brian Orend, *The Morality of War* (Toronto: Broadview Press, 2006) and *The Ethics of War: Shared Problems in Different Traditions* edited by Richard Sorabji & David Rodin (Aldershot: Ashgate, 2006).
8 Particularly in Part II of *Just and Unjust Wars*, though this idea is the central theme of the entire book.
9 See United Nations, Office on Genocide Prevention and the Responsibility to Protect, https://www.un.org/en/genocideprevention/about-responsibility-to-protect.shtml.
10 United Nations, *Charter of the United Nations*, 24 October 1945, 1 UNTS XVI, https://www.un.org/en/about-us/un-charter/full-text.

attacks.[11] A pre-emptive attack is launched against an enemy who is fully expected to launch their own attack in a very short period of time, perhaps a few days at the most, while a preventative attack is launched against an enemy who is expected to be a threat at some stage in the future, possibly not for a number of years. The question which then arises is whether a state can strike first as a defence against aggression, or whether the act of striking first entail that one is the aggressor? This is an issue to which I will return shortly, when discussing the *jus ad bellum* principle of last resort.

The second principle of *jus ad bellum* requires that war only be entered into with the "right intention"; that one is engaging in armed conflict only because of the just cause for the war. The idea here is to rule out resort to war due to ulterior motives, such as the desire to gain control of new territory, or for ethically problematic motives such as ethnic hatred. This principle can certainly be difficult to assess in practice since the state that engages in an armed conflict is not an individual and thus does not have intentions of its own and the persons who are involved in making the decision to go to war can themselves have numerous different intentions in doing so. However, in some cases it is obvious that those making the decision to go to war have anything but noble intentions in doing so and this principle helps to make it clear why going to war in such cases is unethical.

The third principle of *jus ad bellum* is that war must be declared by a legitimate authority and publicly declared to all affected by the decision. In modern times the "public declaration" part of this principle is considered far less important than was the case in times past, and modern legal understandings regarding when the laws of armed conflict apply essentially suggest that the declaration of war is irrelevant; that war exists when it exists and thus no state can avoid the legal and related ethical obligations of the law of armed conflict simply by not making a formal declaration of war. The "proper authority" part of the principle on the other hand, is wondrously simple in some modern cases, but horribly complex in others. The simple case is a situation of straight-forward self-defence; if a state has been attacked by another state, then the legal and ethical authority for declaring war will lie with a specified person (or persons) holding a defined position within the apparatus of the attacked state. But in the modern world, in cases other than self-defence the only

---

11 Brian Orend uses the term "anticipatory attack", rather than "pre-emptive attack" to avoid any confusion about the distinction between pre-emption and prevention. See for example *The Morality of War* (Orchard Park, NY: Broadview Press, 2006), 75.

legal authority which can authorise an armed conflict is the United Nations Security Council (UNSC). This situation is ethically problematic since the legal requirement for UNSC authorisation for an armed conflict might well fall foul of a self-interested veto by one of the permanent members of the UNSC.[12] The answer to the question of who might be thought to be an ethically appropriate authority and able to authorise an armed conflict if legal authority for such a conflict cannot be obtained due to a veto in the UNSC, is difficult to provide. Fortunately for current purposes, such situations are more likely to arise in complex earth-bound cases involving multiple warring factions and ideals of "responsibility to protect" and less likely to arise in cases involving potential conflict in space.

The fourth principle of *jus ad bellum* is that war must only be used as a last resort. In modern times this will usually mean that other forms of international pressure, such as diplomatic pressure, economic sanctions etc, will have failed to resolve the situation. Since war is so destructive, and inevitably causes serious, if even unintended, harm to innocents, one should only resort to war when all other less destructive options have failed. Strictly speaking the principle is not so much one of last resort as last reasonable resort, since while there is almost always something else which could be tried before resorting to war, there is no real point in doing so if such measures are extremely unlikely to succeed.

As I mentioned earlier, the principle of last resort also has important implications for pre-emptive and preventative attacks. Such an attack can only be thought to be ethically justified in extreme circumstances; customary international law, for example, suggests that a pre-emptive attack in self-defence can be launched only in cases "in which the necessity of that self-defence is instant, overwhelming, leaving no choice of means, and no moment of deliberation."[13] If a state engages in a pre-emptive attack against another state, even in a case where this is fully justified by the events leading up to that attack and it is plausible to argue that a pre-emptive attack is in fact the last reasonable resort in this situation, launching

---

12 One possible example of such a situation is the 1999 NATO Intervention in Support of Kosovo. See case 5.2 in Stephen Coleman, *Military Ethics: An Introduction with Case Studies* (New York: Oxford University Press, 2013).

13 This quotation from Daniel Webster, U.S. Secretary of State, forms the basis of the idea of anticipatory self-defence in customary international law. He was writing to the British Government regarding a claim for compensation regarding the destruction of the U.S. steamboat *Caroline* which had, while in U.S. waters, been destroyed by British troops who were engaged in putting down a rebellion in Upper Canada (now Ontario). See Michael Byers, *War Law: Understanding International Law and Armed Conflict* (New York: Grove Press, 2005), 53-54.

such a strike rules out the possibility of any options other than war for resolving the dispute. Michael Walzer suggests that Israel's pre-emptive attack against the Egyptian Air Force in the Six Day War of 1967 is an attack which meets these requirements.[14] The states surrounding Israel were all allied with one another and poised for war with Israel, and all of Israel's attempts to resolve the situation by diplomatic means had not only failed but had actually increased Israel's sense of isolation. While the principle of last resort poses difficulties for pre-emptive attacks, those difficulties are relatively minor compared to the ones it poses for preventative attacks since it is virtually impossible to argue that a preventative attack is really a last resort. This is probably the reason why preventative attacks are forbidden under international law, even though pre-emptive attacks are sometimes acceptable. One consequence of this is that some states seem to have made a deliberate attempt to muddy the language in this area. Knowing that preventative attacks are always illegal but pre-emptive ones are not, they have tried to claim that the military actions they have taken against other states have been pre-emptive, despite the fact that the state being attacked was certainly in no position to launch an attack of their own in the immediately foreseeable future.

The fifth *jus ad bellum* principle suggests that one is not justified in going to war if the war will not have a significant chance of improving the situation, or in other words that violent action will not be justified if that action is going to be futile. This criterion can be difficult to apply in practice, since estimating the probability of success is often difficult, and there are many cases in history where a small group has eventually achieved victory in a war despite facing overwhelming odds.

The sixth principle of *jus ad bellum* is the "proportionality" principle which demands that the overall cost of the war, not merely the financial cost but the harm involved, must be proportional to the benefit which will be obtained by going to war. In simple terms this principle is asking the question "is it **really** worth going to war over this?" Applying this principle obviously involves a measurement problem in that war involves so many unforeseeable factors and thus determining the likely costs and benefits of a war will often be difficult. Perhaps the best which can be said is that the principle will be easy to apply in extreme cases, where the costs will be small and the benefits large or where the costs will be large and the benefits small.

---

14 Walzer, *Just and Unjust Wars*, 84.

The General Applicability of *Jus ad Bellum*

War is obviously an extremely destructive activity which is probably why discussions of the ethics of war are usually divorced from discussions of the ethics of other human activities; in essence war is considered unique. However, I would argue that it is a mistake to view war in this way because the conditions of *jus ad bellum* are actually much more general conditions which can and should be applied in all cases where a person has to decide whether it is acceptable to act in a manner which would normally be considered ethically wrong.

Consider, for example, an emergency medical situation. An unconscious and unidentified patient has been brought into a hospital and it is rapidly determined that the patient's current lack of consciousness is caused by a serious underlying medical condition. One of the possible treatments for this condition is surgery. The attending Doctor needs to decide how this condition will be treated. Surgery is obviously a quite extreme form of treatment and is problematic in a case like this where the patient is unconscious and unidentified, since operating on this patient without consent (from either the patient or their next of kin) can be considered a violation of that patient's rights. What sorts of things should the Doctor need to take into consideration before deciding to operate? I would argue that the answer is that the Doctor would need to meet the same conditions as are found in *jus ad bellum*, so before operating the Doctor needs to have: (1) just cause (that is, the patient actually has a condition which can be treated surgically); (2) right intention (that is, it is actually in the interests of the patient for the condition to be treated and the Doctor is not operating for other reasons, such as the Doctor wanting to perform an interesting operation); (3) proper authority (that is, the Doctor needs to be qualified to perform such an operation, both in general and at this particular hospital); (4) last resort (that is, there are no other reasonable courses of treatment which could be tried before surgery); (5) probability of success (that is, the operation has a reasonable chance of improving the patient's condition); and (6) proportionality (that is, the benefits of surgery in this case outweigh the risks of the procedure).

The same conditions also apply in other cases which involve infringing on the rights of a particular individual, such as when a Police Officer has to decide whether they are ethically justified in taking a person into custody, an action which will violate the right to liberty of that person. In order for this to be ethically (as opposed to legally) justified, the Police

Officer will need, once again, to have: (1) just cause (that is, a legitimate reason for taking the person into custody); (2) right intention (that is, the Police Officer will only be taking the person into custody because of the just cause); (3) proper authority (that is, the Police Officer is legally authorized by that just cause to take someone into custody); (4) last resort (that is, there are no other reasonable courses of action which will be equally effective in resolving the situation); (5) probability of success (that is, the Police Officer can take the person into custody without creating a worse situation); and (6) proportionality (that is, the situation is serious enough for infringing on this person's liberty to be justifiable). Given the relationship between law and ethics it is unsurprising that in most jurisdictions these conditions are actually written into the laws concerning arrest.

The same conditions also seem to apply in other more extreme circumstances, such as the so-called "ticking bomb" case, where the circumstances are such that it is thought, at least by some writers, to be ethically reasonable to use torture on a person in custody in order to acquire information which will prevent a catastrophic terrorist attack. One such case is presented by Henry Shue:

> Suppose a fanatic, perfectly willing to die rather than collaborate in the thwarting of his own scheme, has set a hidden nuclear bomb to explode in the heart of Paris. There is no time to evacuate the innocent people or even the movable art treasures – the only hope of preventing tragedy is to torture the perpetrator, find the device, and deactivate it.[15]

As Shue notes, a case such as this is designed in such a way as to make the use of torture seem ethically plausible; I would argue that it does this simply because the use of torture in such a case seems to fulfil the six requirements I have been discussing. If any of the conditions are not met in some actual situation in which the use of torture is suggested (and in real life cases I would suggest that these conditions are probably NEVER met) then the use of torture seems, at best, to be ethically problematic. Shue's case presents perhaps the "ideal" case in favour of the use of torture, but even in that case if any of the six conditions are not met then the use of torture immediately starts to look problematic. Consider the condition of legitimate authority for example; torturing the suspect in the Shue case immediately looks ethically problematic if the decision to resort to torture

---

15 Henry Shue, "Torture" *Philosophy and Public Affairs*, 7(1978): 124-143, 141.

is made by a rank-and-file Police Officer, rather than by, say, the Chief of Police or the French President.

Given the generality of these principles, it seems problematic to suggest both that war is unique and also that just war theory doesn't apply to warfare in space simply because space is a new and different domain of conflict. If war is really unusual in some way I would suggest that this is simply because warfare is an ongoing activity which brings with it more widespread destruction over a longer period of time than these other cases; thus what makes war special is *jus in bello* rather than *jus ad bellum*. Wars obviously can last for years, but a doctor performing a particular operation does not take years (or even days) to do so. It is because of their ongoing nature that wars, unlike medical operations, require principles like those of *jus in bello* to limit the destruction caused by the ongoing conflict.[16]

Jus in Bello

The two fundamental concepts of *jus in bello* are those of discrimination and proportionality. While there are a number of other ideas which are sometimes discussed under the heading of *jus in bello*, such as a ban on the use of prohibited weapons, or the use of methods which are *mala in se* (that is, evil in themselves), these can be seen to be derived, at least in substantial part, from the principles of discrimination and proportionality. At their core these principles represent an attempt to limit the inevitable destruction of war and to try to ensure that those people engaged in what modern military personnel often refer to as "killing people and breaking stuff", actually kill the right people and break the right stuff, at least as far as this is possible. The principle of discrimination asserts that the only appropriate targets are those directly concerned with the enemy's war effort, and the principle of proportionality claims that the damage which is done in prosecuting such targets, particularly any collateral damage, needs to be reasonable given the actual military value of the target itself.

Applying these principles in the real world can of course be difficult at times, particularly in the modern world. It can be difficult, for example, to apply the principle of discrimination in conflicts where you are fighting

---

16 It could be argued that medical procedures such as operations in fact do have to comply with something like the *jus in bello* principle of discrimination. A Doctor performing an operation on a patient's heart, for example, would not be justified in just "poking around" and checking the patient's other internal organs unless there was specific reason to do so.

against an irregular force whose members do not always wear uniforms since it is not a straightforward task to differentiate between combatants, who are legitimate targets for attack, and non-combatants, who are not. In any conflict it is likely that some things that are possible military targets also have a significant civilian use thus making applying the principle of proportionality difficult; a single power station, for example, might provide power to an anti-aircraft missile battery as well as to both the local hospital and the local sewage treatment plant.

Difficulties such as these will also inevitably arise in space in the event of an armed conflict in that domain. But some might question whether such principles even apply in space given that the domain is so different from Earth and that there are only a handful of humans in space at any time to be directly affected by any conflict there. These are questions to be addressed later in this volume.

References

Byers, M. *War Law: Understanding International Law and Armed Conflict*, New York: Grove Press, 2005.
Coleman, S. *Military Ethics: An Introduction with Case Studies*, New York: Oxford University Press, 2013.
Coleman, S. "Even Dirtier Hands in War: Considering Walzer's Supreme Emergency Argument" *Research in Ethical Issues in Organizations* 13(2015): 61-73.
French, S. *The Code of the Warrior* 2nd Ed., Lanham MD: Rowman & Littlefield, 2017.
McMahan, J. *Killing in War*, Oxford: Oxford University Press, 2009.
Nietzsche, F. *Beyond Good and Evil*, Part IV, Aphorism #146.
Orend, B. *The Morality of War*, Toronto: Broadview Press, 2006.
Rodin, D. *War and Self-Defense*, New York: Oxford University, 2003.
Shue, H. "Torture" *Philosophy and Public Affairs* 7(1978): 124-143, 141.
Sorabji, R. and D. Rodin, *The Ethics of War: Shared Problems in Different Traditions*, Aldershot: Ashgate, 2006.
United Nations, Office on Genocide Prevention and the Responsibility to Protect, https://www.un.org/en/genocideprevention/about-responsibility-to-protect.shtml.
United Nations, *Charter of the United Nations*, 24 October 1945, 1 UNTS XVI, https://www.un.org/en/about-us/un-charter/full-text.
Walzer, M. *Just and Unjust Wars*, New York: Basic Books original edition 1977.

1

FUTURE WAR

Will it be Conducted by Robots or Space Marines?

Christopher D. Miller[1]

Leading figures in the nation have advocated military space-satellite programs ... the conclusions seem inevitable to some, ridiculous to others, that war and space travel march side by side. However lightly some persons may speak of space travel, there is this grimmer face of the picture with which we seem fated to reckon.[2]

When these words appeared in 1954, Sputnik was three years in the future and public imagination had already been captured for nearly a decade by thoughts of human spaceflight. Like pilots and aircraft had been in the sky, astronauts were widely envisioned as gallant and inevitable explorers of the realm beyond Earth's atmosphere. Caidin's observation that another less noble human activity – war – might accompany those astronauts was a partial precursor to the ethical challenges mankind faces two decades into the twenty-first century; partial because even the thinkers of the 1950s could not have imagined the numbers, complexity, and importance of unmanned space vehicles that have already been launched into orbit or beyond.

The title of this chapter suggests a degree of uncertainty whether or how war will extend into or come from space, and whether it will be

---

[1] The views expressed are those of the author and do not reflect the official policy or position of the Air Force, the Department of Defense, or the U.S. Government.
[2] Martin Caidin, *Rockets and Missiles: Past and Future* (New York: The McBride Company, 1954), 206.

fought by robots, human combatants,[3] or both. For clarity, we will consider a "robot" to be either a remotely directed or autonomous machine; and a "space marine" to be a human being of any nationality or military service physically operating in the space domain. It is tempting to look far into the future, but the cone of technological uncertainty widens dramatically beyond a couple of decades—so this narrative may strain the link between prediction and speculation but will attempt to avoid breaking it. By considering scenarios over the next two decades involving space combat of any form, and what such conflict might mean for those directing or prosecuting it, and those inhabiting the Earth, some useful insights may emerge.

Conflict involving space is evident even at the most basic level. Even an internationally-agreed definition of "outer space" remains unsettled after decades of discussion because of the competing incentives of national sovereignty over airspace, evolving incentives surrounding national use of space, and the differing physics of air and space operations – the definitional boundaries of which are debatable in a technical sense as well. The infinite expanse beyond Earth's atmosphere associated with the Kármán line[4] (a decades-old convention now under review, 100km above Earth's nominal surface) is an oft-cited demarcation and an altitude that some nations have proposed as a consistent legal boundary. Geophysical phenomena make the exact height where the atmosphere's effect becomes negligible on a spacecraft's orbit both variable and something greater than 100km, depending on whether the orbit is circular or elliptical. Yet another competing demarcation is 50 (80km) miles above Earth, used in U.S. military and civil spaceflight to designate astronaut and spaceflight participant status. In short, the space domain's very boundary with Earth remains contentious, 63 years after Sputnik first orbited the earth, because nations continue to vie for advantage in its definition.[5]

---

3 At the time this chapter was first contemplated, the U.S. Space Force had not yet settled on an appellation for its members, who are now referred to as "Guardians". Since several spacefaring nations possess militaries involved in space operations, and their nomenclature varies, the term "space marines" will be used as an (astro)nautical generic term for future military space combatants.
4 Nadia Drake, "Where, exactly, is the edge of space? It depends on who you ask". *National Geographic*, 20 December 2018, https://www.nationalgeographic.com/science/2018/12/where-is-the-edge-of-space-and-what-is-the-karman-line/.
5 Timothy G. Nelson, "Where does space begin? The decades-long legal mission to find the border between air and space", *Space News*, 26 March 2019, https://spacenews.com/op-ed-where-does-space-begin-the-decades-long-legal-mission-to-find-the-border-between-air-and-space/.

A second fundamental shaping factor for "war in space" is that key elements of space infrastructure are not limited to space alone; in March 2021 there were 3,843 active satellites[6] and one permanently manned space mission on orbit, but the number of ground telemetry nodes; control centers; and technology development, testing, training, manufacturing, and launch locations is too numerous to count. In other words, the global space enterprise is both terrestrial and orbital. Unmanned presence on orbit is continuous and significant; human presence in space is almost exclusively episodic and orders of magnitude smaller; human activity on the ground, supporting operations in space, is ubiquitous.

Military space personnel dynamics are another near-term factor affecting future space combat. Despite the importance of the domain to global economic processes and to the United States in particular, the highest personnel strength estimate for the U.S. Space Force is 30,800 total personnel within its first five years,[7] less than one-sixth of the next smallest U.S. Service. It is a militarily significant reality that space is vitally important to modern life, but space is hard to reach, even harder to traverse beyond earth's orbit, difficult to leave, friendly to Keplerian laws of orbital motion—not to laws of aerodynamics—and unforgivingly lethal for unprotected human life. The speeds (and kinetic energy) required to attain orbit or travel to other celestial bodies make any space operation complex and expensive for any platform, manned or robotic. The thermodynamic requirements to leave orbit and return any object to Earth intact are considerable, and the simple requirement for "return to Earth" associated with human spaceflight will affect how the U.S. Space Force – or any nation's equivalent – is likely to allocate resources. Thus, unlike the world's historical experience of warfare, or the visions painted in popular fiction like Robert Heinlein's *Starship Troopers*[8] or George Lucas'– *Star Wars* film series – it is likely to be a very long time before space combat will involve "mass force on force" anything like traditional land, sea and air and air domains. The role of small numbers of human combatants, however, is more debatable.

Finally, perhaps the most important characteristic of future space war is simply its deep and global relevance to civil and military life: in a mere

---

6   T.S. Kelso, 'NORAD Two-Line Element Sets – Current Data' *CelesTrak*, https://celestrak.com/NORAD/elements/.
7   Congressional Budget Office, *The Personnel Requirements and Costs of New Military Space Organizations*, by Jason Coleman, Adam Talaber, and F. Matthew Woodward, 55178, Washington, D.C.: CBO, 2019, https://www.cbo.gov/system/files/2019-05/55178-SpaceForce.pdf, 6.
8   Robert Heinlein, *Starship Troopers* (New York: Ace Books, 2010).

six and a half decades, humanity has become deeply dependent on the panoply of things that space systems provide. The Australian Government, for example, recently popularized six areas where satellites provide tangible benefits to individuals, including remote sensing for agriculture, weather and climate data, navigation, communications, entertainment, and economic transactions.[9] Add to the list military-specific functions fielded by and heavily relied on by some nations, such as early warning to support nuclear deterrence, intelligence collection, military command and control, secure communications, and enhancement of military systems such as weapons guidance. Further, position-navigation-timing (PNT) systems such as the U.S. GPS, Europe's Galileo, Russia's GLONASS, and Chinese BeiDou provide system-wide support to many sectors of national critical infrastructures. These impact of these space-based services may not be evident to many individuals, but they are essential in varying degrees to effective national systems such as transportation, banking, and life-sustaining utilities like electrical grid management.

As these factors have converged, Martin Caidin's 1954 warning above that "war and space travel march side by side" is echoed by Lt Colonel Brandon Davenport in a June 2020 essay, "conflict will likely follow humanity into orbit".[10] Both are consistent with U.S. Vice Chairman of the Joint Chiefs, General John Hyten's 2018 observation that "there's actually no such thing as war in space. There's just war. Space is a place. Space is a magical, wonderful place and people conduct operations in space, and we do too. But we have adversaries that are building capabilities to deny us the use of space, so we have to make it impossible for them to do that".[11]

If war on Earth remains a distinct possibility among spacefaring nations, if work is underway across the globe to prepare for it, and if logic suggests it might then extend into space, the *means* of prosecuting warlike action and their potential *consequences* are worth considering.

---

9 Australian Space Agency, "World Space Week: Six ways satellites improve our lives", *Australian Government, Department of Industry, Science, Energy and Resources*, October 2, 2020, https://www.industry.gov.au/news/world-space-week-six-ways-satellites-improve-our-lives.
10 Brandon Davenport, "On Implementing a Space War-Fighting Construct", *Air and Space Power Journal*, Spring (2020): 64.
11 John Hyten, "John H. Glenn Lecture on Space History" (Speech, Washington, D.C., June 13, 2018), U.S. Strategic Command, https://www.stratcom.mil/Media/Speeches/Article/1559373/john-h-glenn-lecture-on-space-history/.

Means and Consequences of Space Warfare

Likely means of prosecuting space combat operations *involving satellites in earth orbit* are elegantly outlined in a recent open-source CSIS report,[12] which separates space weapon types into kinetic (physically destructive) and non-kinetic (functionally destructive) categories within three employment arenas: Earth-to-Space, Space-to-Space, and Space-to-Earth.

Earth-to-Space kinetic weapons are primarily direct-ascent antisatellite weapons, designed to be launched from ground or air platforms and rise to directly strike and incapacitate a target space system. Both the 1985 US F-15 launch of an antisatellite missile against a US solar observation satellite[13] and the 2007 Chinese antisatellite test[14] demonstrated the ability to destroy satellites in orbits as high as 530 miles (850km) but at the cost of creating significant debris fields. In the case of the Chinese test almost 3,000 individually trackable pieces of debris were created, and the majority remain on orbit as potential threats to other space vehicles.

Space-to-Space kinetic weapons can be similar in kill mechanism to earth-to-earth space systems but are carried aboard space-based platforms and their specific means of destruction of a target system are both potentially more variable and less well-known. As an example, the Chief of Space Operations for the U.S. Space Force noted that "early in 2020, Russia positioned one of its satellites dangerously close to an American satellite and then instructed it to execute a series of provocative and unsafe maneuvers. This summer, [it then] backed away, released a target, and then conducted a weapons test, firing a projectile at that target. This raw display of space combat power was carefully designed as an act of intimidation, right out of the 1950s Soviet playbook".[15] Space-to-space kinetic weapons, like the Earth-to-space example previously mentioned, carry with them the potential for creating orbital debris with lasting implications for all future space systems.

---

12 Todd Harrison, "International Perspectives on Space Weapons" (Washington, D.C.: Center for Strategic & International Studies, 2020), 6.
13 Peter Suciu, "The History Books Missed This: How an F-15 Killed a Satellite in 1985", *The National Interest*, July 1, 2020, https://nationalinterest.org/blog/reboot/history-books-missed-how-f-15-killed-satellite-1985-163866.
14 Carin Zissis, "China's Anti-Satellite Test", Council on Foreign Relations, February 22, 2007, https://www.cfr.org/backgrounder/chinas-anti-satellite-test.
15 John W. Raymond, "How We're Building a 21st-Century Space Force", The Atlantic, December 20, 2020, https://www.theatlantic.com/ideas/archive/2020/12/building-21st-century-space-force/617434/.

Space-to-Earth kinetic weapons represent a constrained category; under the provisions of the 1967 Outer Space Treaty, placing nuclear weapons in orbit around Earth are prohibited as are nuclear or other weapons based on other celestial bodies like the moon or near-Earth asteroids. Early 1960s efforts to plan for nuclear deterrent forces based on the moon,[16] for example, were rendered non-viable by the treaty. Respecting this prohibition, military planners have envisioned kinetic strikes whose significant destructive power would rely on conventional explosives or a projectile's combination of mass and velocity against selected targets.[17] The effects of such kinetic attack weapons on terrestrial targets would certainly be addressed by the laws of armed conflict, but attributing the action requires a degree of technological detection and discrimination on the part of the attacked party or cooperative allies.

Earth-to-Space non-kinetic weapons include uplink jammers, laser dazzlers, and cyberattacks against space systems on orbit; use of these weapons can result in temporary or permanent damage to the targeted system, and attribution can be more or less difficult depending on the attack modality and the capabilities of the attacked system's operator. They are unlikely to create space debris.

Space-to-Space non-kinetic weapons represent a concerning category to most space-faring nations, as they might be difficult to detect and capable of interfering with or destroying space systems' functioning in ways that may be difficult to anticipate or counter. Examples of these kinds of weapons are cross-link jammers and high-power microwave weapons, which would be deployed from co-orbital platforms and could either degrade, temporarily neutralize, or destroy target systems. They are also unlikely to generate space debris.

Space-to-Earth non-kinetic weapons, such as space-based high-powered lasers or jammers, pose both destructive and non-destructive threats to terrestrial systems. Ranging from jammers directed toward air and maritime platform satellite receivers or satellite ground stations, to high-powered lasers intended to destroy ICBMs or other physical targets, such weapons pose technical challenges but present potential tactical or even strategic advantages to their possessors.

---

16 William E. Burrows, "Beyond the Blue Horizon", In *Harnessing the Heavens: National Defense Through Space*, ed. Paul G. Gillespie and Grant T. Weller (Chicago: Imprint Publications, 2008), 31-32.
17 Harrison, "International Perspectives…".

I began this chapter by asking whether space war would be fought by "robots" or "space marines". Notably, every one of the six categories of hostile act mentioned above can be carried out by unmanned space systems – robots – operating either autonomously (for example, pre-programmed and with machine-based decision-making ability) or under real-time, direct human control; and every category could also be deployed or employed by space marines using suitable vehicles and technological capabilities, albeit at far higher cost in nearly every case and with debatable additional effectiveness.

Regardless of the means or the protagonists, any discussion of space warfare must grapple with the kinds of consequences that must be considered as part of any practical or ethical decision-making regarding such attacks. Four criteria may suggest ways to evaluate the merits and demerits of various hostile acts. They are neither conceptually parallel, mutually exclusive, nor necessarily equally important, but each provides a lens for evaluating the risk-benefit calculus that might be considered by warring parties in choosing their course of action. These criteria are:

1. The **impact** of the action – local or systemic, small or large. An example of a small and local effect would be non-kinetic jamming of the link between a communications satellite and its ground station. Presumably, an adversary would wish to have a significant military tactical effect on the larger conflict as a result of this attack, but the satellite/ground station involved would likely remain intact and function could be restored. An example of a large effect would be kinetic or non-kinetic destruction of all ground stations supporting, for example, the GPS constellation; such an action would have potentially severe collateral consequences on the many civil infrastructure and life-sustaining processes that increasingly depend on satellite-based PNT. Additionally, a kinetic attack on ground stations would likely involve loss of life, a traditionally significant *casus bello*.
2. The **persistence** of the action, whether temporary or long-lasting. Non-kinetic effects can range from minutes to permanent depending on whether the function of a space system is temporarily jammed, or its software or circuitry is irrevocably damaged. In the latter case, the damage is permanent but because it was incurred by non-kinetic means, it affects only the targeted system. Conversely, a kinetic strike on a large space platform can result in orbital debris that is, for all practical purposes, likely to pose an enduring threat

to systems anywhere near the destroyed system's orbital plane. For systems in low earth orbit, debris may de-orbit in days to years, but for higher orbits, kinetically destroyed platform debris lingers essentially forever, as does its adverse effect on an expanding swath of orbital space.

3. The **visibility** of the action – the degree to which it involves politically significant consequences. Attacks that involve kinetic damage on national territory or that of allies, whether loss of life is incurred or not, would be considered visible. An attack that destroys a single orbital space system – even an expensive one – could be considered less visible.[18] An attack that results in the death of a human combatant in space is likely to be politically visible and volatile, as would be an attack that deprives any significant sector of the civil population of some space-provided service that lacks immediate redundancy via terrestrial backup. One of the aspects of space warfare that is explicitly worth considering is its potential to cause real damage to the fabric of modern societies, if information and services provided by space infrastructure become missing, confusing, or deceptive on a large scale.

4. Lastly, the **ethical complexity** of an action – the ability to predict outcomes with sufficient certainty to make value judgments. This is not necessarily a simple question when dealing with space warfare, vis-à-vis more conventional physical domains. Given the dearth of international treaties, customary law, or even real-world experience regarding particular kinds of space operations and their effects at scale, decisions on how and when to use space weapons that create a great deal of long-lasting space debris, or which cripple systems with potentially large or lasting effects on civil infrastructure, will likely be very challenging. In comparison to a soldier's decision to fire a weapon or a pilot's decision to withhold a bomb due to concerns over collateral damage, outcomes resulting from space warfare have the potential to be far less certain.

---

18 The U.S. Chief of Space Operations recently alluded to this "out of sight, out of mind" phenomenon, noting "Satellites don't have a mother. You can't hug it. You can't touch it. You can't hear it, you can't love it. … It's hard for the average person to understand just how reliant their life is on space. It's not tangible". Charles Pope, "Raymond and Space Force enter new, ambitious phase as U.S. Space Command changes leaders", *Space Force News*, August 24, 2020, https://www.spaceforce.mil/News/Article/2322445/raymond-and-space-force-enter-new-ambitious-phase-as-us-space-command-changes-l/.

Weighing the impact, persistence, visibility, and ethical complexity of any particular adversarial space operation occurring in Earth orbit and on the surface of the Earth is difficult enough, but space warfare will not be permanently limited to these arenas. Everett Dolman, writing in 2001, posited four astropolitical regions in which nations would compete or clash; the first two were designated as *Terra (Earth)* and *Terran or Earth space*, which extends from the outer limits of Earth's atmosphere to just outside geostationary orbit, roughly 36,000 km. These are certainly the arenas of principal interest in contemporary space operations. Two additional astropolitical regions are material to our consideration of future space warfare, however: *Lunar or Moon space*, which is the region just beyond geostationary orbit to just outside the moon's orbit, and which contains orbitally-significant space in addition to the moon; and *Solar space*, which consists of everything in our solar system (inside the sun's gravity well) beyond the orbit of the moon, and which promises planetary and asteroid resources containing "the raw materials necessary to ignite a neo-industrial age".[19] In the view of some U.S. space power thinkers, Dolman's four astropolitical regions illuminate important differences between the "traditional mind of space" and the "emerging mind of space". In this paradigm, traditionalists maintain a geocentric view largely concerned with Earth and space systems orbiting Earth, and their integration with traditional military operational forces and concepts. The "emerging mind" contemplates uses of space beyond Earth orbit, including exploration, colonization, resource extraction and the like.[20]

Space War

Conflict can be waged in more or less visible, lasting, and consequential kinetic or non-kinetic forms against space-related assets on the ground, on orbit, or beyond; any such conflict is certain to be technically, ethically, and politically complex. The following four hypothetical scenarios may help to

---

[19] Everett Dolman, *Astropolitik: Classical Geopolitics in the Space Age* (London: Frank Cass, 2002), 62-64.
[20] Matthew L. Lohmeier, *"The Better Mind of Space"*, Wright Flyer Paper No. 79 (Maxwell Air Force Base, AL: Air University Press, 2020), 3-6, https://www.airuniversity.af.edu/portals/10/aupress/papers/wf_0079_lohmeier_the_better_mind_of_space.pdf.

illustrate some of the provocations, harms, and conflict dynamics inherent in space warfare.[21]

Scenario 1: Earth-centric conflict

In a near-term case, rising tensions in a regional grey-zone conflict provoke military posturing between nuclear-armed, spacefaring nations. A third major spacefaring nation provides logistical and rhetorical support to one of the two primary protagonists, exacerbating the conflict, and the locus of provocative air and sea military maneuvers shifts to the Arctic. One of the nations involved determines that an adversary reconnaissance satellite in polar orbit is providing that adversary with undesirably high-fidelity information on air and sea force movements, and it decides to neutralize the reconnaissance satellite. Rather than using a direct-ascent antisatellite weapon or non-kinetic means of jamming or blinding the adversary platform, the attacking nation decides to use a co-orbital platform to cause the reconnaissance satellite to de-orbit over a period of days in an effort to avoid orbital debris while taking a relatively low-visibility action to deter further conflict by imposing a clear cost. The reconnaissance satellite re-enters the atmosphere, but because of its mass it does not completely burn up during re-entry, and the residual debris strikes a major urban area with explosive effect, causing thousands of casualties.

In this case, a conflict on earth initiates an action in space. The combatants are earthbound space experts, not space marines. The conflict point is the interaction of two robots – the satellites – thousands of miles above Earth's surface. The intent was to deter escalation of terrestrial conflict by taking a hostile act in space that would be visible to governments but relatively opaque to publics in the involved nations; in terms above, a kinetic attack intended to avoid the collateral damage of orbital debris. Because of unknowns in the exact shape and mass of the target satellite, unpredictability of its reentry, and the highly visible loss of life that results, the conflict between the two protagonist nations could have escalated significantly and could have impacted an otherwise uninvolved third party. This scenario illustrates the potential significant impact as well as the ethical complexity of assessing the consequences of hostile actions in space.

---

21 These scenarios are entirely imaginary and while they are intended to be plausible, are not meant to imply that the decisions, actions, or consequences presented have any grounding whatsoever in existing nations' strategies, plans, or space operational capabilities.

Scenario 2: Conventional war extends into space

A war between major powers on earth is underway, and one adversary decides it is necessary to neutralize enemy satellite position-navigation-timing (PNT) satellites. The intent of the attack on PNT capability is to deprive enemy air, land and sea combatants in the geographic theater of operations of precision weapon guidance and timing-related communications and networking capability, and to impose cost in the enemy's homeland on the civilian population to increase public pressure to cease hostilities. The PNT attack is multi-faceted, using high-power laser attacks on enemy geosynchronous PNT satellites within line-of-sight of the attacker's territory and a cyberattack against ground stations controlling the entire constellation. Both attacks are successful and the enemy's PNT constellation is effectively neutralized. The attacker's protective measures for their own PNT system remain effective, giving them a military tactical advantage in air, land and sea theater combat. Depriving the enemy population of PNT causes widespread disruption in essential services, from land and air transportation to distribution of petroleum and electrical power, and rapid deterioration of financial transactions as the absence of satellite timing signals gradually de-synchronizes financial networks. While there is no immediate loss of civilian life, civil disorder threatens to break out.

In this scenario, the combatants are again earthbound, and the space systems involved are robotic and invisible but highly relevant to terrestrial combatants and populations alike. Arguably, an attack on an entire PNT constellation would be extraordinarily visible to the governments, military forces, and populations targeted. With the combination of attack means selected, lasers would presumably functionally and permanently destroy a portion of the expensive satellites involved and the cyberattack could be recoverable in a relatively short period of time – the persistence of the attack itself is multifaceted. Significant time would be required to replace PNT satellites rendered permanently unusable. The PNT ground station's operations could likely be restored quickly, but the attack's *effect* on the perceptions of the enemy population, however, is likely to persist since its impact on daily life would be large. In terms of the traditional laws of armed conflict, an attack on satellites being used by fielded military forces seems clearly permissible, but the acceptability of an attack on a system for the purpose of depriving a civilian population of essential – and in some cases life-sustaining – utilities is less clear. It could be considered analogous to a blockade; or could be compared to conventional attacks conducted on electrical power, water, rail, or other target systems. The ethical calculus must consider the scale of the attack and the duration of its effects, both of which arguably more difficult to predict than historical antecedents.

Scenario 3: Terrestrial warfare spills over to lunar colonies

In the mid-term future, multiple nations have established permanent scientific and industrial operations on the moon, despite the fact international agreement on use of extraterrestrial resources for other than exploration remains elusive at present. While not all spacefaring nations have ratified the 1979 UN Moon Agreement,[22] which prohibits military bases, fortifications and maneuvers or testing of weapons on the moon, all nations' existing lunar operations have been overtly peaceful in consonance with the agreement. A period of increased tension between two sponsor countries on Earth transitions to limited but active hostilities, in which both countries attack each other's space infrastructure via electromagnetic jamming, optical dazzling, and cyber means, significantly degrading communications, launch telemetry, and various Earth-orbiting capabilities. For one of the adversaries, a cyberattack resulting in software and physical damage to launch infrastructure makes it impossible to dispatch a nationally controlled spacecraft to resupply its lunar outpost. Tensions on Earth and on the moon escalate as constructive communications between the respective lunar outposts of the adversaries break down, and essential supplies begin to run short. While weapons per se are not present on the moon, the inhabitants of the threatened outpost – partially staffed by military space personnel – contemplate an improvised physical attack on the other outpost's storehouses.

In this scenario, non-kinetic warfare on Earth increases the possibility of hostilities conducted by humans in space. The "tactical situation" on the moon might well be somewhat ambiguous to all parties. The locations of storehouses, the degree to which and means by which they are monitored, existence or ability to improvise weaponry, communication difficulties would compound the political maneuvering between sponsor nations, lunar outposts, and relevant members and organizations of the

---

[22] United Nations Office of Outer Space Affairs, *United Nations Treaties and Principles on Outer Space*, ST/Space/11, New York: United Nations, 2002, https://www.unoosa.org/pdf/publications/STSPACE11E.pdf.

international community. In such a situation, human information gathering and decision-making at the point of conflict may be more successful than any autonomous or time-delayed, earth-controlled actor could achieve. The actual impact of one nation's outpost failing and its inhabitants dying[23] without succor from other nations – even the adversary nation – is hard to assess. Casualties as a result of the conflict (lack of supply or combat on the moon) might well create a public outcry and become a *casus bello* triggering wider war; or it could be submerged in the tidal waves of information that characterize the modern world, intentionally minimized by political decision-makers to avoid escalation.

Scenario 4: Interplanetary war

In a highly speculative work entitled *The First Space War*, co-authors Furman Daniel and T.K. Rogers postulate that colonization of Mars is likely to lead, in some two and a half centuries, to a major war between inhabitants of Earth and of Mars.[24] The great distances involved make bi-directional conversation practically impossible; the time required to travel between planets will diminish earth-to-colony contacts and understanding. After decades of supporting the colonies at great cost, Earth will be confronted with an increasingly resentful, terraforming Mars inhabited by colonists who will be increasingly biologically distinct from inhabitants of Earth's higher gravity, and in the authors' scenario, many of the same dynamics that led to American rebellion from British sovereignty will lead to a Martian rebellion. The mechanics of warfare the authors project between Earth's space marines and Mars' insurgents – from the way space marines would need to commandeer existing space transportation between the two planets, to reinforcing or dismissing some of the space war maneuvering popularized by Hollywood – are fascinating. In terms of our criteria for evaluating actions in space – impact, persistence, visibility, and ethical complexity – this scenario's breadth and speculative nature defies easy application. What the authors do offer, however, is a technological and sociological *tour-de-force* that clearly highlights the extraordinary cost,

---

23 From the earliest days of manned space exploration, international agreements have consistently invoked duties to treat astronauts as envoys of mankind and to render all possible assistance in the event of accident, distress, emergency landing, and the like. This scenario would, as an entering argument, require that commitment to be ignored – which may be unlikely, even in the event of terrestrial hostilities.

24 J. Furman Daniel III and T.K. Rogers, *The First Space War: How the Patterns of History and the Principles of STEM Will Shape Its Form* (Lanham, MD: Lexington Books, 2019), 170.

complexity, and constraints that conduct of war in and through space-by-space marines – even assisted by humanoid and special purpose robots – would entail.

Conclusion

This chapter's focus was on the nature of future space war and whether it would be fought by space marines or by robots. This book in its entirety underlines the reality that space is, as often said, "contested, congested, and competitive". It is also clearly an ethically challenging environment, in that human hostility when it occurs is manifest almost entirely through machines, with little risk borne by today's combatants and the potential injury to societies difficult to assess because of the omnipresence of space-supported aspects of contemporary life.

As terrestrial dependence on space-provided services has grown, and many space-related technologies have advanced both the need for, and ease of deploying, more autonomous and ground-controlled systems has grown and will continue to grow in the future. The pace of manned spaceflight, too, is accelerating as space tourism becomes a possibility, launch costs begin to decrease, and more nations, such as India, develop manned spaceflight capability. It is conceivable that exploration and exploitation of asteroids could result in a celestial human presence in near- to mid-term, but the likelihood of space marines fighting in connection with such operations seems low. In the next several decades, then, absent entirely unforeseen breakthroughs in technology that make reaching and operating in space far less arduous and much more attractive, space professionals – those who understand how to design, build, operate, integrate, and leverage highly capable robot space systems for the good of their nations – will be in immense demand.

In a fittingly prescient 1958 observation, Donald Cox and Michael Stoiko wrote in *Spacepower: What it Means to You* that "Space vehicles will thus carry man (or his controlled electronic instruments at first) into the unknown altitude areas where new political, economic, social and military problems will be accrued to him on Earth as a result of his conquest of space".[25]

---

[25] Donald Cox and Michael Stoiko, *Spacepower: What It Means to You* (Philadelphia: The John C. Winston Company, 1958), 142.

Nations already deal with exactly those terrestrial problems. The space marine recruiting station can probably remain unstaffed for a while longer.

References

Australian Space Agency. "World Space Week: Six ways satellites improve our lives". *Australian Government, Department of Industry, Science, Energy and Resources.* October 2, 2020. https://www.industry.gov.au/news/world-space-week-six-ways-satellites-improve-our-lives.

Burrows, W. "Beyond the Blue Horizon". In *Harnessing the Heavens: National Defense Through Space.* Edited by Paul G. Gillespie and Grant T. Weller. Chicago: Imprint Publications, 2008: 25-34.

Caidin, M. *Rockets and Missiles: Past and Future.* New York: The McBride Company. 1954.

Congressional Budget Office. The Personnel Requirements and Costs of New Military Space Organizations. By Jason Coleman, A. and F. Woodward. 55178. Washington, D.C.: CBO, 2019. https://www.cbo.gov/publication/55178.

Cox, D. and M. Stoiko. *Spacepower: What It Means to You.* Philadelphia: The John C. Winston Company, 1958.

Davenport, B. "On Implementing a Space War-Fighting Construct". *Air and Space Power Journal.* Spring (2020): 63-74.

Daniel, J. and T.K. Rogers, *The First Space War: How the Patterns of History and the Principles of STEM Will Shape Its Form.* Lanham, MD: Lexington Books, 2019.

Dolman, E. *Astropolitik: Classical Geopolitics in the Space Age.* London: Frank Cass, 2002.

Drake, N. "Where, exactly, is the edge of space? It depends on who you ask". *National Geographic*, December 20, 2018. https://www.nationalgeographic.com/science/2018/12/where-is-the-edge-of-space-and-what-is-the-karman-line/.

Harrison, T. *International Perspectives on Space Weapons.* Washington, D.C.: Center for Strategic & International Studies, 2020. https://www.csis.org/analysis/international-perspectives-space-weapons.

Heinlein, R. *Starship Troopers.* New York: Ace Books, 2010.

Hyten, J. "John H. Glenn Lecture on Space History". Speech, Washington, D.C. June 13, 2018. U.S. Strategic Command. https://www.stratcom.mil/Media/Speeches/Article/1559373/john-h-glenn-lecture-on-space-history/.

Kelso, T.S. "NORAD Two-Line Element Sets – Current Data", *CelesTrak.* June 16, 2021. https://celestrak.com/NORAD/elements/.

Lohmeier, M. "The Better Mind of Space". Wright Flyer Paper No. 79. Maxwell Air Force Base, AL: Air University Press, 2020. https://www.airuniversity.af.edu/portals/10/aupress/papers/wf_0079_lohmeier_the_better_mind_of_space.pdf.

Nelson, T. "Where does space begin? The decades-long legal mission to find the border between air and space", *Space News*, March, 26 2019. https://spacenews.com/op-ed-where-does-space-begin-the-decades-long-legal-mission-to-find-the-border-between-air-and-space/.

Pope, C. "Raymond and Space Force enter new, ambitious phase as U.S. Space Command changes leaders". *Space Force News*. August 24, 2020. https://www.spaceforce.mil/News/Article/2322445/raymond-and-space-force-enter-new-ambitious-phase-as-us-space-command-changes-l/.

Raymond, J. "How We're Building a 21st-Century Space Force". *The Atlantic*, December, 20 2020, https://www.theatlantic.com/ideas/archive/2020/12/building-21st-century-space-force/617434/.

Suciu, P. "The History Books Missed This: How an F-15 Killed a Satellite in 1985". *The National Interest*, July 1, 2020. https://nationalinterest.org/blog/reboot/history-books-missed-how-f-15-killed-satellite-1985-163866.

United Nations Office of Outer Space Affairs. *United Nations Treaties and Principles on Outer Space*. ST/Space/11. New York: United Nations, 2002. https://www.unoosa.org/pdf/publications/STSPACE11E.pdf.

Zissis, C. "China's Anti-Satellite Test". *Council on Foreign Relations*. February 22, 2007. https://www.cfr.org/backgrounder/chinas-anti-satellite-test.

2

DOES JUST WAR THEORY EXTEND TO THE SPACE FRONTIER?

Patrick Lin

As new technologies open up the frontier of outer space, it's reasonable to be concerned that rising political and economic competition could slip into war in space. Here on Earth, the laws of armed conflict and international humanitarian law have kept today's skirmishes from spilling (further) into peaceful realms, and a good deal of those legal principles are modeled on just war theory: a philosophical tradition that lays out the moral conditions for war and how to limit its horrors, in order to achieve a lasting peace.

Just war theory (JWT) has always been in tension with reality. Looking around the world, it's forgivable to think that powerful people and states can do as they please, regardless of moral philosophy. As an example, it's still under debate whether the atomic bombs dropped on Japan at the end of World War II were ethically justified; nuclear weapons could be categorically unethical.[1] So, it's fair to ask whether JWT even applies in outer space, especially when we have trouble complying with it on Earth.

This chapter explores the realist's position that JWT does *not* extend into space. For instance, at least in near-term scenarios, there's not much horror to limit in the dead of space, if the fighting is only between satellites or robots, so why would JWT be needed? While this and other reasons may have initial appeal, I will conclude that there's practical value in recognizing JWT as a guiding framework for hostilities in space.

---

1 Michael Walzer, *Just and Unjust Wars: A Moral Argument with Historical Illustrations* (New York: Basic Book, 1977), 263-268; David Luban, "Were the A-Bombs the Last Resort?," USNA Stockdale Center for Ethical Leadership, March 1, 2021, https://www.usna.edu/Ethics/blog/2020/Were_the_A-Bombs_the_Last_Resort.php.

Overview of Just War Theory

Because the basics of JWT and military ethics were covered earlier in this book, I won't fully recite them here but will only highlight some relevant principles. I will also adopt the understanding of war as "actual, intentional, and widespread armed conflict between political communities" as a sensible working definition.[2] Thus, fistfights, gangland-battles, accidental border shootings, and other lesser conflicts are not properly "war", especially if they're not conflicts at scale or part of a broader campaign.

1. Traditionally, JWT is conceived as having two parts: justice in going to war (*jus ad bellum*) and justice in waging war (*jus in bello*). In more recent times, scholars have argued that JWT needs additional parts to be complete: justice in winding down a war and rebuilding peace (*jus post bellum*), and justice in using force that falls short of the threshold for war (*jus ad vim*).
2. *Jus ad bellum* seeks to ensure that a decision to go to war is ethical in the first place.[3] One of its key principles is that war must be the **last resort**; you must try everything else first, since war is such a moral disaster. A **just cause** (or *casus belli*) and **right intentions** must also exist for going to war; individual and collective defense are usually the only accepted reasons. And war must be a **proportionate** response; even if your state or political community suffered an unjust attack first, that attack must be severe enough to justify the risk that your counterattack may escalate aggressions too far.
3. *Jus in bello* regulates how war is conducted, once it's been decided to engage in it.[4] Again, we find the same principle of **proportionality** but with respect to an objective or mission. Attacks must also **discriminate** between combatants and noncombatants who should never be targeted given their moral innocence. Thus, weapons of mass destruction (WMDs), including biochemical weapons, are typically *mala in se*, that is, weapons and tactics that are prohibited because they are "evil in themselves"; other examples include poison, landmines, as well as rape and genocide campaigns.

---

2 Brian Orend, "War", *Stanford Encyclopedia of Philosophy* (2005), ed. Edward N. Zalta, https://plato.stanford.edu/archives/spr2016/entries/war/.
3 Orend "War".
4 Orend "War".

4. *Jus post bellum* is aimed at ensuring that a sustainable peace is possible.⁵ It continues the theme of **proportionality**, this time with respect to a peace settlement, punishment, and reparations. That is, we should ensure the terms of surrender aren't so vindictive or unreasonable that resentment and desperation fester in the defeated until they explode again into war. If the point of war is so that we can live in peace, as Aristotle suggested, it misses the point when the peace cannot be sustained.⁶

5. *Jus ad vim* seeks to prevent low-intensity fighting – conflicts that don't rise to the level of war – from escalating further.⁷ **Proportionality** again is invoked to limit the severity of clashes, ensuring they're not unnecessarily provocative. In fact, other principles are so similar to those in *jus ad bellum* and *jus in bello* that some scholars believe *jus ad vim* is redundant and unnecessary; this includes the *jus ad vim* principles of just cause, right intentions, discrimination, likelihood of success, proper authority, and so on.⁸ I won't take a position on that debate, as our discussion doesn't depend on it.

6. JWT is a very old philosophical tradition, but its influence is still felt today in the laws of armed conflict (LOAC) and international humanitarian law (IHL) – sometimes together known as the laws of war. For instance, the Hague and Geneva Conventions prescribe the same kind of general rules, such as to never target noncombatants.⁹ Other treaties also lean on JWT principles, such as the *mala in se* prohibition in banning biowarfare, chemical attacks, and antipersonnel landmines.¹⁰ And the Charter of the United Nations affirms a state's right to self-defense and collective defense as the primary, if not only, reason a state may justifiably attack another state.¹¹

---

5 Orend "War".
6 Aristotle, *The Nicomachean Ethics*, book 10, 1177b 5–6, trans. W.D. Ross (Oxford: Oxford University Press, 2009).
7 Daniel Brunstetter and Megan Braun, "From *Jus ad Bellum* to *Jus ad Vim*: Recalibrating Our Understanding of the Moral Use of Force", *Ethics & International Affairs*, 27, no. 1 (2013): 87-106.
8 Helen Frowe, "On the Redundancy of *Jus ad Vim*: A Response to Daniel Brunstetter and Megan Braun", *Ethics & International Affairs*, 30, no. 1 (2016): 117-129.
9 "Rule 1. The Principle of Distinction between Civilians and Combatants", Customary IHL, International Committee of the Red Cross, https://ihl-databases.icrc.org/customary-ihl/eng/docs/v1_rul_rule1.
10 Morton Dige, "Explaining the Principle of *Mala in Se*", *Journal of Military Ethics*, 11, no. 4 (2012): 318-332.
11 Charter of the United Nations, article 51, https://www.un.org/en/sections/un-charter/un-charter-full-text/.

7. Thus, even if ethics is idealistic and quaint in a modern world full of evil, a realist should want to abide by LOAC and IHL anyway, which is an indirect endorsement of JWT. That is, even if you're a moral nihilist, it's rational to *not* want your adversaries to target your own noncombatants (such as children), start rape campaigns, torture prisoners of war, and engage in other prohibited conduct against your side. The laws of war, and underlying JWT, are meant to protect everyone concerned.

But the question for this chapter is: do the remoteness and emptiness of outer space mean that human ethics can't reach off-planet? The subject of what military theories apply in space isn't a new issue, but JWT is usually either assumed or ignored: it's either presumed to apply in space even if JWT needs some interpretation,[12] or that ethics is dispensable in favor of realpolitik, such as political theories on power and deterrence.[13] Here, I will take the time to develop and critique the realist's case against JWT in space.

The Realist's Case Against Just War Theory in Space

In making the realist's argument, let's first lay out some possible scenarios, which give rise to different reasons against JWT in space.

Scenario 1: Robot Wars

As of February 2021, only seven people exist in space; they're all on the International Space Station in low Earth orbit, which is barely in outer space.[14] Given this paucity, any armed conflict in the foreseeable future

---

12 Keith A. Abney, "Space War and AI", *Artificial Intelligence and Global Security: Future Trends, Threats and Considerations*, ed. Yvonne Masakowski (Bingley: Emerald Publishing House, 2020), 66-82; Matthew Beard, "Militarising Space: Weapons in Orbit", *Commercial Space Exploration: Ethics, Policy and Governance*, ed. Jai Galliott (New York: Routledge, 2016), 197-210; International Society for Military Ethics, 2019 Conference: Space War and Ethics, Colorado Springs, CO, June 29-30, 2019, https://www.internationalsocietyformilitaryethics.org/2019-conference.html.
13 LaToya Tate, "The Status of the Outer Space Treaty at International Law during War and Those Measures Short of War", *Journal of Space Law,* 32, no. 1 (2006): 177-202; Steven Freeland, "The Laws of War in Outer Space", *Handbook of Space Security: Policies, Applications and Programs*, ed. Kai-Uwe Schrogl et al. (New York: Springer-Verlag, 2015), 81-112; Howard Kleinberg, "On War in Space", *Astropolitics*, 5, no. 1 (2007): 1-27.
14 How Many People Are In Space Right Now?, https://www.howmanypeoplearein spacerightnow.com/.

could very well be conducted by Artificial Intelligence (AI) systems or robots, which are already distributed as far as we can send them. Not only have we launched satellites and other vehicles into deep space, we have also made Mars into the only planet populated entirely with robots.[15]

So, the argument here is relatively straightforward: the things that JWT seeks to protect are simply not there in outer space. At least for the foreseeable future, there are no noncombatants, ecosystems, economies, or other things needed to sustain human life in outer space – the protection of which is the point of proportionality, distinction, and other principles in JWT. War-machines are as lifeless as the moons and planets we know about; so, if JWT is aimed at holding back the suffering and horrors in war, there's nothing out there that can suffer or be horrified.

Scenario 2: Space Terrorists

As more people enter space for research, tourism, economic development, or other reasons, a terrorist organization could possibly manage to send human saboteurs into space. They might be directed to blow up the International Space Station, or an orbiting space hotel, or vital communications satellites, or other strategic targets. Besides the destruction of those valued objects, an attack in orbit would create thousands of new bits of space debris that could threaten spaceflight for everyone.[16]

But for the foreseeable future, there would be relatively few souls who could be targeted in space, including the terrorists, and this puts a natural limit on the scale of any attack. Arguably, these clashes wouldn't rise to the level of "widespread armed conflict between political communities," per our definition of war earlier. That is, these attacks and counterattacks wouldn't be in the context of war *per se*, but they would merely be security actions against space criminals; and a grand ethical theory to govern those (rare) actions is probably not needed.

If any law enforcement or oversight exists in space at all, JWT may be overkill here, since security actions could already be governed by basic law, including due process and human rights protections. Or if there is no such law enforcement in space, then JWT would seem irrelevant anyway – or so a realist could argue.

---

15 Alexander McNamara, "Mars: Oodles of interesting facts, figures and fun questions about the Red Planet", *BBC Science Focus*, September 27, 2020, https://www.sciencefocus.com/space/mars-facts-figures-fun-questions-red-planet/.
16 See Chapter 7.

Scenario 3: Economic Competition

Let's scale up the conflict scenarios in case that matters. There's a growing clamor to open space for asteroid and other mining operations, with water and precious metals as the main prize. Besides policy moves to pave this path, some companies have already been formed (and gone bust) in the business of space mining.[17] It's plausible that some of these operations could have humans on site. And given that space-property rights are still an unclear and contentious issue, misunderstandings and competing economic claims could slip into armed conflict, as they have in Earth's history.

However, even with more development, space would continue to be a frontier because of its sheer vastness and remoteness. As a result, any conflict may still be relatively rare, isolated, and lesser in scale and effects. This would make the concept of "mere frontier incidents" in international law relevant here, as distinguished from "armed attacks" that trigger a right to self-defense.[18] Again, if such conflicts fall short of war or *casus belli*, then JWT would seem to *not* be the right governing framework; and if any law exists in space at all, then basic human rights law and due process could be sufficient.

Scenario 4: Colonial Wars

Let's scale up the scenarios even more to resemble war. Someday, perhaps there will be meaningful numbers of people living in planetary and lunar settlements, not just on space stations. More than economic contests getting out of hand, there could be political communities that might wage war as a "continuation of policy by other means" and at a greater scale than in previous scenarios.[19] In this scenario, there would be many more things to protect from the devastation of war, not just more people but also more essential infrastructure and economic activity that support human life in space, such as mining for water or generating oxygen.

---

17 Atossa Araxia Abrahamian, "How the Asteroid-Mining Bubble Burst", *MIT Technology Review*, June 26, 2019, https://www.technologyreview.com/2019/06/26/134510/asteroid-mining-bubble-burst-history/.
18 International Court of Justice, *Nicaragua v. United States of America*, judgment of June 27, 1986, para. 195, https://www.icj-cij.org/en/case/70/judgments.
19 Carl von Clausewitz, "On War", book 1, chapter 1, section 24, trans. Michael Eliot Howard and Peter Paret (Princeton: Princeton University Press, 1976), 87.

Surely, JWT would be relevant in this scenario, right? Maybe, or maybe not. A realist could invoke something like legal positivism in jurisprudence: laws require an arbiter and enforcer (such as a police force and court system), which is to say that international laws require a broad authority with transnational jurisdiction, but also that an unenforceable law isn't properly "law." Even if a legal system might exist *within* a space colony, there's unlikely to be a legal system *across* colonies, especially if they have different national origins.[20] Hence, there's no space law beyond that which can be locally enforced, according to the legal positivist.

In the vast cosmos without law or the means to broadly enforce law (or ethics), a realist might say that "anything goes", and they would be right insomuch as offenders can evade punishment while in space. At minimum, an Interplanetary United Nations may be needed to serve as this legal bulwark against lawlessness, but that requires goodwill and cooperation that we can barely keep together on Earth; so the prospects of a sufficient interplanetary legal regime – never mind an ethics regime – to govern space seem dim.

Summary of Reasons

From the range of scenarios above, we can extract several reasons a realist might give to deny that JWT applies in outer space. These include:

1. *Robot wars*: Whatever JWT seeks to protect *simply does not exist* in space, such as human suffering, ecological devastation, undue economic harms, and so on. Therefore, JWT is not needed.
2. *Space terrorists*: Even with possible human victims in space, the *limited scale* of terrorist attacks in the foreseeable future means that JWT is still not needed. If any security or law enforcement exists in space at all, they can deal with terrorists as mere criminals, not combatants in war. Alternatively, if there's no legal regime in the first place, then it doesn't make much sense to talk about an ethics regime like JWT.
3. *Economic competition*: Even with more development, space would still be a frontier, given its vastness and remoteness. Any conflicts would still be limited in scale and not rise to the level of

---

20 Patrick Lin, "Look Before Taking Another Leap for Mankind – Ethical and Social Considerations in Rebuilding Society in Space", *Astropolitics,* 4, issue 3 (2006): 281-94.

war. Indeed, international law already recognizes "mere frontier incidents" as a less serious use of force that is short of war; they are *more easily justified and tolerable* than JWT would seem to normally allow. Again, if it's not war, then JWT may be excessive here.
4. *Colonial wars*: Even if "actual, intentional, and widespread armed conflict between political communities" were to occur, JWT still wouldn't make sense pragmatically without a reliable arbiter and enforcer of laws or ethics. Without something like an Interplanetary United Nations, insisting on JWT in space would put the *proverbial cart in front of the horse*.

Other conflicts would fit in the spectrum of cases here, which span from a limited attack with no human victims to full-blown war at the population scale. At least one of the realist's reasons would likely apply to those other scenarios in the middle. Taken together, they add up to a presumptive case against JWT in outer space in a full range of conflicts.

Critique of the Realist's Case

Here, we will examine the reasons offered above, focusing on three main areas of contention.

Critique 1: Space is Not a Fully Distinct Domain

In the realist's case against JWT – specifically in the scenarios about robot wars, space terrorists, and colonial wars – there's an underlying assumption that outer space is a distinct domain from Earth. That may be true in many respects, but space is arguably not so detached from terrestrial life as assumed. If so, then it's an overstatement that "the things that JWT seeks to protect are simply not there" and that terrestrial laws have no jurisdiction in space.

For one thing, we could conceive of our space activities as an *extension* of terrestrial life, not separate from it. There would still be direct connections between our space activities and Earth, as opposed to (say) scenarios of autonomous robot battles on the other side of the galaxy that are truly disconnected from any human control or impact. But a robot battle in low Earth orbit or on Mars would have at least an *indirect* impact on human autonomy and endeavors here on Earth.

That is, even if machines are fighting only each other in space, this is just a *proxy war* among humans. Losing such a war would harm the interests of those affected humans, such as plans to study an area for mining, which could sustain research, settlements, and other missions. Win or lose, *everyone's* interest in exploring space, regardless of their individual motivation, would be hindered by a kinetic battle in space that creates more floating debris. Space debris can rip through a spacecraft at more than 30,000 kilometers per hour (or about 20,000 miles per hour) like a bullet through a tin can.[21]

So, while a space battle might not damage existing ecosystems if there's nothing but barren landscapes, it would create an environmental disaster in littering space with these virtual minefields. The severity of this harm, which could block future efforts in exploring space, would seem to warrant a legal regime to govern the conduct of space conflicts. Further, damaged satellites will affect civilian life on Earth since we heavily rely on GPS and other signals for navigation, financial services, and more. Since there's already a connection between Earth and outer space, the existing regime of LOAC and IHL (as instantiations of JWT) may be the most obvious solution here.

Another connection is that human astronauts and terrorists are still tethered to Earth and need to return to it at some point. Outer space and other celestial bodies cannot yet sustain human life for long. Thus, even if outer space is a lawless frontier for now, its residents are temporary and would eventually come back home, if they don't perish first. If justice could be meted out on Earth for bad conduct in space, that *effectively extends the reach of the law* into space.

This extraterritorial jurisdiction already occurs with domestic laws on Earth, so it's not a great stretch to apply them in space. For instance, if U.S. citizens were to travel abroad, they are generally no longer obligated to obey (say) U.S. traffic and alcohol consumption laws; instead, they should comply with laws in whatever local jurisdiction they travel to. However, some domestic U.S. laws continue to apply even outside the country, and offenders may be prosecuted when they return or are captured; these include laws against sex with children, enlisting in foreign militaries, evading U.S. taxes even if earning a living abroad, counterfeiting or laundering money, bribery of foreign officials, conspiring against the U.S. even if from abroad, and other such crimes.[22]

The Earth's atmosphere may be a natural border, like a river or mountain range, but it doesn't automatically or necessarily limit the reach of

---

[21] "Space Debris", NASA Headquarters Library, https://www.nasa.gov/centers/hq/library/find/bibliographies/space_debris.
[22] Anthony J. Colangelo, "What Is Extraterritorial Jurisdiction", *Cornell Law Review*, 99, no. 6 (2014): 1303-52.

our laws. For instance, there's no breathable atmosphere underwater either, yet our domestic laws still apply to crimes committed by scuba divers. The atmosphere as a physical border prevents humans from easily entering the space, but it's politically porous and would allow space hostilities to flow back to Earth. So, an attack by state A on state B's moon-base isn't only constrained to that lunar battlefield, but it could affect their relationship back on Earth in either new or continuing armed conflicts.

Besides this blowback, space-based weapons could eventually strike targets on Earth, and possibly vice versa; the atmosphere won't always be much of a barrier for military exchanges, just as submarine-launched missiles transcend the domains of sea, air, and land. Along with guarding against an environmental "tragedy of the commons" in space, all this weighs in favor of thinking that JWT continues to apply in space, even if there's no local arbiter or enforcer in outer space.

Critique 2: Frontier Incidents Still Imply JWT

The economic-competition scenario raises a unique issue not addressed in the preceding analysis: the idea of a "mere frontier incident" where some behaviors are tolerated more than they would be in non-frontiers, that is, tolerated more than JWT would seem to allow. Before we look at the origin and intent of the idea, which is a term of art in international law, let's first elaborate more on its plausibility and relevance here.

Consider this analogy: If you were jostling through a crowd, you may get bumped by other people, both intentionally and not.[23] Either way, that unwanted physical contact usually doesn't rise to the level of assault on your bodily integrity. Yes, that contact may hurt, but it's not so serious an offense that you would be justified in punching back and starting a melee in the name of self-defense; that would be a disproportionate response, as opposed to merely pushing back as a natural reaction. Likewise, your incidental bumping into others doesn't make you an assailant, and it's usually not considered criminal to bump back in a crowd.

This is an analogy to a frontier incident because it takes place in an environment to which no one involved has a firm claim, and the conflict is limited in its scale and effects. This unsettled environment contains friction that affects your claim to bodily autonomy or state sovereignty, but

---

23 Patrick Lin, "Ethics of Hacking Back: Six Arguments from Armed Conflict to Zombies", US National Science Foundation-funded report, September 26, 2016, 16.

that friction isn't part of a deliberate and broader campaign to harm or subjugate you. By nature, frontiers (and chaotic crowds) are fraught with tests and misunderstandings, and this weighs in favor of tolerating certain transgressions more than normal. Such incidents are an *assumed risk* in any frontier.

Connecting this to JWT for the realist, if the moral tradition is supposed to govern conduct in wars, then it doesn't seem needed for mere frontier incidents which are lower-intensity and possibly inadvertent conflicts. Likewise, JWT would be too much to govern shoving matches between people; it's not meant to oversee any kind of fight, just the most serious kinds. As with the space-terrorists scenario, if any basic law exists in space at all, that may be enough to handle frontier incidents; and if there is no basic law, then leapfrogging from zero to JWT seems to be an implausible proposal.

Connecting this to international law for the realist, we need to first recognize a couple key legal principles, which also exist in JWT. The United Nations Charter's article 2(4) generally prohibits armed attacks: "All Members shall refrain in their international relations from the threat or use of force against the territorial integrity or political independence of any state, or in any other manner inconsistent with the Purposes of the United Nations."[24]

Nonetheless, it is also within the natural rights of the attacked state to defend its sovereignty and territorial integrity. The UN Charter's article 51 also allows: "Nothing in the present Charter shall impair the inherent right of individual or collective self-defence if an armed attack occurs against a Member of the United Nations, until the Security Council has taken measures necessary to maintain international peace and security."[25]

Against this background, the term of art "mere frontier incident" first appeared in the International Court of Justice's (ICJ) ruling in the 1986 case of *Nicaragua vs. United States of America*, in which the court distinguished it from the more serious "armed attack" or "use of force" that would trigger UN Charter's article 51 to justify a counterattack.[26]

This doesn't mean that the victim cannot take any countermeasures in a frontier incident, only that it cannot invoke its right to self-defense and prolong or escalate the skirmish more than necessary, as if it were part of a broader and deliberate campaign in war. Rather, a frontier incident is more

---

[24] Charter of the United Nations, article 2(4).
[25] Charter of the United Nations, article 51.
[26] International Court of Justice, *Nicaragua v. United States of America*, para. 195.

of an *ad hoc* and even unintentional, regrettable clash when adversaries encounter each other in a disputed territory, similar to our crowd-jostling analogy earlier. If self-defense cannot be invoked, then JWT would seem to be the wrong operating framework, since self-defense is typically the only legitimate reason JWT recognizes for fighting.

Unfortunately, as weighty as the term of art was in the ICJ ruling, it was mentioned only once – almost in passing – and not explicated in this ruling or others, but it has inspired a good deal of controversy and discussion. And if "mere frontier incident" is relevant anywhere, it's relevant to actual frontiers like outer space, which are more frontier-like than the borderlands of Central America.

To critique the realist's case, we don't need to deny that the existence of "mere frontier incidents" or that it may operate under different norms than in a full-blown war, but we can argue that the term of art *already implies JWT and does not obviate it*; they needn't be separate frameworks. That is, even if these incidents fall short of war, they're still part of a sliding scale of conflicts that is governed by JWT, even if far lesser conflicts are not, like shoving matches.

Similarly, some scholars have argued that a "mere frontier incident" isn't much different than other attacks to warrant the legal distinction; they all fall into the same spectrum of armed conflicts, albeit at different levels of intensity, duration, scale, and effects.[27] This tracks the debate mentioned earlier when it's argued that *jus ad vim* is not a distinct part of just war theory but already logically implied by it.

Insofar as the distinction is meant to help ensure frontier incidents don't escalate into war, that implied principle of non-escalation is essentially a principle of proportionality – and proportionality is already a rule in JWT as well as LOAC and IHL.[28] Likewise, it's been argued that *jus ad vim* doesn't need to be added to JWT, because proportionality and other principles in *jus ad vim* already exist in JWT.

In other words, JWT can handle low-intensity conflicts that don't rise to the level of war, without carving out a new theoretical space. But even if *jus ad vim* is needed to complete JWT, that doesn't help the realist's case in this chapter, either; if space conflicts are frontier incidents, then *jus ad vim* is exactly the part of JWT we need in space at minimum, if not the whole theory.

---

27 Christine Gray, "International Law and the Use of Force: 4th edition" (Oxford: Oxford University Press, 2018), 153-7.
28 Gray "International Law".

Therefore, even if we recognize a need to tolerate "mere frontier incidents" to prevent climbing the escalatory ladder, that doesn't support the realist's case. Just the opposite, it seems to imply that JWT *is* needed – or at least a part of it – since a goal of JWT is precisely to prevent this escalation. Whether in a frontier or urban battles, the principle of proportionality in JWT already implies proportionality in countermeasures and some degree of toleration: to not push back harder than you are pushed or is necessary to achieve your military objective. Besides avoiding unnecessary harm, it makes moral and practical sense to not be so unreasonably aggressive that it provokes more intense fighting against you or poisons the possibility for a lasting peace afterward.

Critique 3: Norm-Setting is Enough, Even If Unenforceable

At this point, the realist could concede that JWT is needed in space but also maintain, per the colonial-war scenario, that if there's no broad authority that can enforce those rules – as opposed to local authority to keep the peace in a colony, base, or another limited territory – then it makes little sense to insist on JWT (or its incarnation as interplanetary laws of war) when even basic law can't be enforced.

This is consistent with *legal positivism* in jurisprudence, which recognizes a command as "law" only if it emanates from a person or body that is habitually obeyed and has the power to enforce compliance by way of sanctions or punishments, among other defining features.[29] And if no such authority exists over cosmic battlefields, then those "laws" are toothless and therefore not real laws.

Legal positivism can be understood as the opposite of the natural law tradition, which asserts that some laws – such as "do not murder" and other ethical principles – exist naturally, irrespective of governments, courts, social contracts, and so on.[30] Those natural laws are intrinsically tied to morality, and we are bound by them no matter where we are in the universe. Hence, under the natural law tradition, immoral or bad laws are oxymorons and not real "laws" to be obeyed; they ought to be resisted.

In contrast, legal positivism is grounded in empiricism and is more scientific in character. It doesn't deny that bad laws can exist, but it

---

29 Leslie Green and Thomas Adams, "Legal Positivism", *Stanford Encyclopedia of Philosophy* (2019), ed. Edward N. Zalta, https://plato.stanford.edu/entries/legal-positivism/.
30 Mark C. Murphy, "Natural Law Jurisprudence", *Legal Theory*, 9, issue 4 (2003): 241-67.

decouples morality from law. To a realist, the threat of legal punishment is more compelling than the vague and uncertain social penalties of not abiding by a given moral system; the law has a real bite. Without a reliable arbiter and enforcer of laws (or ethics), the teeth needed to nudge behavior isn't there. So, for a realist, it might not make sense to talk about laws or ethics, including JWT, until those institutions exist.

"First things first" can make sense generally, but as an objection against JWT, it assumes too much. Even unenforced or unenforceable laws can *still have value in nudging behavior by setting norms*, and violations of those norms and expectations could still result in unwanted consequences, even without the hammer of legal justice. Physical and financial penalties can still be levied by the international community by way of peer condemnation, economic sanctions, travel restrictions, and other actions.

For instance, the U.S. isn't a signatory on certain treaties, such as the Ottawa Treaty against anti-personnel landmines, yet we largely *de facto* abide by those treaties anyway, or at least try not to violate them capriciously.[31] This is related to the idea of "customary international law": norms that bind even non-signatory states by virtue of their widespread compliance.[32] More recently, the international campaign against "killer robots" hasn't yet resulted in any treaty or ban on the technology, but it already has a stigmatizing effect that makes nations careful to avoid stepping directly in that controversy and to pledge that humans will always "be in or on the loop" to supervise the lethal machines.[33]

Anyway, if we're imagining a realist who believes that JWT and the laws of war are useful and necessary on Earth but not in space – as opposed to a realist who denies that ethics is *ever* relevant or needed, which is not the target position we're critiquing – that may already be inconsistent with the realist's own objection of "first things first" here.

To wit, the United Nations and other such global bodies arguably don't even meet the requirements of legal positivism, inasmuch as they're not exactly habitually obeyed or have the power to enforce judgments but rely on the voluntary cooperation and goodwill of the states involved, including those punished. In ICJ's 1986 ruling of *Nicaragua vs. United States*

---

31 "Frequently Asked Questions", International Campaign to Ban Landmines, http://www.icbl.org/en-gb/the-treaty/treaty-in-detail/frequently-asked-questions.aspx.
32 "Customary Law", War & Law, International Committee of the Red Cross, https://www.icrc.org/en/war-and-law/treaties-customary-law/customary-law.
33 "Frequently Asked Questions", Campaign to Stop Killer Robots, https://www.stopkillerrobots.org/learn/.

of America discussed above, the U.S. has simply ignored the decision and not paid the $1+ billion fine, citing jurisdictional issues.[34]

Even though prosecuting international crimes is uncertain on Earth, including war crimes and piracy in international waters, it's important that we continue to do it as we can. Requiring an effective authority, then, to exist in space before implementing JWT and associated laws appears to be a double standard that is overly strict and conveniently benefits the realist.

This means that JWT and other laws can still serve as a deterrent and guide for limiting the horrors of war in space. Even if they're unenforceable in space, that may change over time, and future prosecutions may be possible for previous war crimes, for which there's no statute of limitations.[35] And if any states, organizations, or combatants in space have plans to return to Earth or engage in trade with us, then they have an incentive to not run afoul of our norms and laws while in space.

Conclusion

The risk of armed conflict in space is real, whether one-off terrorist events, low-intensity frontier incidents, or open war between political communities. There's initial plausibility in being skeptical about the relevance of military ethics or law in space, especially when the things we value and seek to protect don't exist there, such as human welfare, environmental integrity, life-sustaining economies and institutions.

Even with much more development, our space interests – such as bases, outposts, mining operations, or even colonies – may not be physically close or politically connected enough to each other to "catch fire" if a conflict were to break out in one region; the remoteness of space may be a natural firewall to contain conflicts. If these are "mere frontier incidents", they need to be tolerated more than just war theory would seem to allow. And if conflicts become more widespread, ethics or law seems pointless in

---

[34] Martin Cleaver and Mark Tran, "US Dismisses World Court Ruling on Contras", *The Guardian*, June 28, 1986, https://www.theguardian.com/world/1986/jun/28/usa.marktran; Mariano Castillo, "Nicaragua May Revive $17 Billion Claim Against US", *CNN*, July 22, 2011, http://www.cnn.com/2011/WORLD/americas/07/21/nicaragua.us.claim/index.html.
[35] 1968 Convention on the Non-Applicability of Statutory Limitations to War Crimes and Crimes Against Humanity, United Nations Treaty Collection, https://treaties.un.org/Pages/ViewDetails.aspx?src=TREATY&mtdsg_no=IV-6&chapter=4&clang=_en.

a legal void that is space, where norms and rules can't be easily enforced; to insist on such rules puts the cart ahead of the horse, as it were.

These broad reasons for skepticism are rebutted in the discussion above. No matter where the conflict occurs, the principles of just war theory remain sound: to not only limit the devastation and suffering in war, but to also create the possibility of a lasting peace. If we indeed wage war to live in peace, and if we want peace in space, then just war theory is relevant in space. What's not relevant is whether JWT is fully successful in this aim, just as it's not a good criticism that Earth's laws against murder don't prevent all murders or catch all killers.

The basic response to the realist is that, while the Earth-space divide may be physically great, the reach of our law and norms could be long enough to govern or at least *nudge behavior* in space. Even if they don't, ethics and law can play an *aspirational role* and provide a basis for *future prosecution* of any misconduct and abuse when that capability becomes available. Since frontiers are usually permeable and increasingly so over time, it's a thin line between Earth and space, much thinner than we might think.

References

Abney, K.A. "Space War and AI" in *Artificial Intelligence and Global Security: Future Trends, Threats and Considerations*, edited by Yvonne Masakowski. Chapter 4. Bingley: Emerald Publishing House, 2020.

Abrahamian, A.A. "How the Asteroid-Mining Bubble Burst". *MIT Technology Review*, June 26, 2019. https://www.technologyreview.com/2019/06/26/134510/asteroid-mining-bubble-burst-history/.

Aristotle. *The Nicomachean Ethics*, translated by W.D. Ross. Oxford: Oxford University Press, 2009.

Beard, M. "Militarising Space: Weapons in Orbit" in *Commercial Space Exploration: Ethics, Policy and Governance*, edited by Jai Galliott. Chapter 15. New York: Routledge, 2016.

Brunstetter, D. and Megan Braun. "From *Jus ad Bellum* to *Jus ad Vim*: Recalibrating Our Understanding of the Moral Use of Force", *Ethics & International Affairs* (2013) 87-106.

Campaign to Stop Killer Robots. "Frequently Asked Questions". https://www.stopkillerrobots.org/learn/.

Castillo, M. "Nicaragua May Revive $17 Billion Claim Against US". CNN, July 22, 2011. http://www.cnn.com/2011/WORLD/americas/07/21/nicaragua.us.claim/index.html.

Charter of the United Nations. https://www.un.org/en/sections/un-charter/un-charter-full-text/.

Cleaver, M. and Mark Tran. "US Dismisses World Court Ruling on Contras", *The Guardian*, June 28, 1986. https://www.theguardian.com/world/1986/jun/28/usa.marktran.

Colangelo, A.J. "What Is Extraterritorial Jurisdiction", *Cornell Law Review* (2014): 1303-52.

Dige, M. "Explaining the Principle of *Mala in Se*", *Journal of Military Ethics* (2012): 318-332.

Freeland, S. "The Laws of War in Outer Space", in *Handbook of Space Security: Policies, Applications and Programs*, edited by Kai-Uwe Schrogl et al. Chapter 6. New York: Springer-Verlag, 2015.

Frowe, H. "On the Redundancy of *Jus ad Vim*: A Response to Daniel Brunstetter and Megan Braun", *Ethics & International Affairs* (2016): 117-129.

Gray, C. "International Law and the Use of Force: 4th edition". Oxford: Oxford University Press, 2018.

Green L. and Thomas Adams. "Legal Positivism" in *Stanford Encyclopedia of Philosophy*, edited by Edward N. Zalta, 2019. https://plato.stanford.edu/entries/legal-positivism/.

How Many People Are In Space Right Now? https://www.howmanypeopleareinspacerightnow.com/.

International Campaign to Ban Landmines. "Frequently Asked Questions". http://www.icbl.org/en-gb/the-treaty/treaty-in-detail/frequently-asked-questions.aspx.

International Committee of the Red Cross. "Customary Law", War & Law. https://www.icrc.org/en/war-and-law/treaties-customary-law/customary-law.

International Committee of the Red Cross. "Rule 1. The Principle of Distinction between Civilians and Combatants", Customary IHL. https://ihl-databases.icrc.org/customary-ihl/eng/docs/v1_rul_rule1.

International Court of Justice. *Nicaragua v. United States of America*. Judgment of June 27, 1986.

International Society for Military Ethics. 2019 Conference: Space War and Ethics. Colorado Springs, CO, June 29-30, 2019.

Kleinberg, H. "On War in Space", *Astropolitics* (2007): 1-27.

Lin, P. "Ethics of Hacking Back: Six Arguments from Armed Conflict to Zombies". Report funded by US National Science Foundation. September 26, 2016.

Lin, P. "Look Before Taking Another Leap for Mankind—Ethical and Social Considerations in Rebuilding Society in Space". *Astropolitics* (2006): 281-94.

Luban, D. "Were the A-Bombs the Last Resort?", USNA Stockdale Center for Ethical Leadership, March 1, 2021, https://www.usna.edu/Ethics/blog/2020/Were_the_A-Bombs_the_Last_Resort.php.

McNamara, A. "Mars: Oodles of interesting facts, figures and fun questions about the Red Planet". *BBC Science Focus*. September 27, 2020. https://www.sciencefocus.com/space/mars-facts-figures-fun-questions-red-planet/.

Murphy, M.C. "Natural Law Jurisprudence". *Legal Theory* (2003): 241-67.

NASA. "Space Debris and Human Spacecraft". Last modified on August 7, 2017. https://www.nasa.gov/mission_pages/station/news/orbital_debris.html.

NASA Headquarters Library. "Space Debris". https://www.nasa.gov/centers/hq/library/find/bibliographies/space_debris.

Orend, B. "War" in *Stanford Encyclopedia of Philosophy*, edited by Edward N. Zalta, 2005. https://plato.stanford.edu/archives/spr2016/entries/war/.

Tate, L. "The Status of the Outer Space Treaty at International Law during War and Those Measures Short of War", *Journal of Space Law* (2006): 177-202.

Von Clausewitz, C. "On War", translated by Michael Eliot Howard and Peter Paret. Princeton: Princeton University Press, 1976.

Walzer, M. *Just and Unjust Wars: A Moral Argument with Historical Illustrations.* New York: Basic Book, 1977.

1968 Convention on the Non-Applicability of Statutory Limitations to War Crimes and Crimes Against Humanity. United Nations Treaty Collection. https://treaties.un.org/Pages/ViewDetails.aspx?src=TREATY&mtdsg_no=IV-6&chapter=4&clang=_en.

3

IN SPACE NO ONE CAN HEAR THE GENEVA CONVENTION

Jus in Bello and Warfare in Space

Pauline M. Shanks Kaurin

Introduction

In the film *Star Wars: A New Hope* the planet Alderaan is intentionally targeted and destroyed by the Empire; this is effectively a war crime as the people of the planet were not armed and were targeted as a show of force where a military target was not within range violating both the *jus in bello* principles of discrimination and proportionality of means. This action happens within the context of a powerful political community (the Empire) waging a counterinsurgency against the Rebel Alliance, and it is the powerful committing war crimes against the weaker insurgents, perhaps as an instrument of terror and intimidation. Is this the future of *jus in bello* considerations in space ethics? Or are there other future scenarios that merit consideration which do not fit this arguably conventional war model with humans, artillery, and other weapons operating in space largely as they do in terrestrial battlefields?

This chapter considers whether and how the *jus in bello* considerations of Just War thinking – discrimination, and proportionality of means – apply to space warfare and the role the distance and remoteness of space warfare plays in lowering or raising the threshold for war. First, I consider the character of space warfare primarily in terms of the present, but also what is likely to be the case in the near term in order to see what the context will be like. Second, I will consider three analogies to Earth warfare to see which elements are more salient to space warfare, and what connections there are to *jus in bello* considerations in Just War thinking. Finally, I will consider how these insights gained give insights into how best to approach *jus in bello* in space warfare, opening a larger vista on the question whether, and how *jus in bello* considerations are viable in space warfare and future directions in scholarship for this issue. In particular I want to attend to the question of how central the presence of human belligerents on the battlefield is to *jus in bello* questions, and absent that, whether rethinking the character of these questions is imperative.

Space Warfare

While the militarization of space is not new (and neither is the idea of space warfare), space was also conceptualized early on as a neutral domain not subject to the sovereignty of one nation state or group, but rather as a domain for the Common Good, for exploration and the furtherance of scientific and technical knowledge, and possibly for international cooperation as in the case of the International Space Station. More recently we see space being viewed more for commercial advantage, especially by the wealthier and more powerful nations and their commercial interests who can afford to operate in that domain. What was initially a function of the two Cold War military powers is still largely a domain of an increasing number of wealthier nations (for example China, U.S., Russia) and their attendant commercial interests. This might be analogous to the colonialization by European powers beginning in the fifteenth century, arguably initially spurred by financial interests which then nations come to defend as a matter of preservation of trade as a national interest.

These considerations are important background considerations that do impact the character of space warfare; as in Earth centric warfare, there are a variety of interests and considerations apart from military ones. What comes to mind when we think of space warfare? Is it the scene from the novel *Ghost Fleet* wherein rogue actors take over the Chinese space station?[1] Is it the space battles of the *Star Wars* films with laser special effects and explosions? To begin to answer these questions, consider the character of space warfare. What is involved? Further, what is involved now as opposed to what we might see in the future that requires a moral accounting? Many commentators agree that from a military standpoint space is unlikely to

---

1 August Cole and P.W. Singer, *Ghost Fleet: A Novel of the Next World War* (New York: Mariner Books, 2016).

be the central decisive battlefield; it will be additional to Earth, generating an Earth centric model of warfare. Dominance or control of space is very necessary for other things like communications, air power especially air defenses so its value is largely instrumental, not intrinsic. Australian ethicist Nikki Coleman notes three kinds of military activity and warfare that we might see: 1) Earth-to-Space, 2) Space-to-Space and 3) Space-to-Earth. In addition to direct military attacks, there are both military and commercial satellites in increasing numbers being placed in space thanks in part to the decreasing size and cost (of the satellites themselves and of launching them into space), with attendant shelf life and end of life issues.[23] Related, there is the issue of space debris and its cleanup, which could also be used as a cover for other military attacks and activities – either by conventional or non-conventional military or other forces as we shall see in the discussion below. This may be complicated by concerns about the Kessler syndrome – where space debris eventually makes parts of the space unusable, further impacting space warfare.[4]

Conventional

The most intuitive view of space warfare might be of space as a domain where battles (whether space-to-earth, space-to-space, earth-to-space) and other military activities happen between nation states, non-state actors and other belligerents in a straightforward analogy to warfare and battlefields on earth. Examples from speculative fiction like *Ghost Fleet* and pop culture artifacts like the *Star Trek* and *Star War* franchises come immediately to mind; space warfare would just be this kind of warfare but located in space with advanced technologies and capacities. On this view, we are looking for space analogues for conventional Earth warfare: space architecture like satellites, communication systems, fortifications, colonies (either of combatants or non-combatants, likely mixed with objects as well) and other symbols of military power subject to disruption and attack as part of kinetic hostilities. For example, Russia might have fortifications on the moon to protect a non-combatant colony of scientists and commercial actors harvesting minerals that might need protection by incursions from peer competitors

---

[2] Reverend Dr. Nikki Coleman, "Ethical Issues of Military Uses of Space" July 10, 2019, https://www.youtube.com/watch?v=pReDU9I_Uyc&t=51s.
[3] Coleman, "Ethical Issues".
[4] Louis de Gouyon Matignon, "The Kessler Syndrome" *Space Legal Issues*. March 27, 2019, https://www.spacelegalissues.com/space-law-the-kessler-syndrome/.

and non-state actors wishing to disrupt these activities to put pressure on political negotiations on Earth related matters in Ukraine or some other sphere of interest. Another iteration on this basic concept could be the communal space architecture or infrastructure necessary for commercial enterprises in space to function, either by companies of a particular nation or multinational corporations, the defense of which is maybe judged by the national interest of one or more nations or international organizations. In any case, this mode of warfare is primarily oriented around state actors and their proxies in ways that are still recognizable as conventional warfare.

Unconventional/Asymmetric

In the asymmetric model warfare, it is less about state actors, but more about non-state actors or actors with some level of removal from a state (contractors, privateers.) First, these actors might be engaging in asymmetric conflict like insurgency and guerilla war, perhaps involving terrorism, for either political or commercial ends. The actions of the Rebel Alliance in their insurgency against the Empire from the *Star Wars* films seems to fit this model of asymmetric space warfare. Second, we might consider the operations of pirates and privateers (where the latter have some authority from a state or political entity/community) engaging in activities that target space or operate in space. This could also include smugglers or other organized criminal elements who see space as a lucrative domain – possibly even resulting in something like space gang warfare or battling warlords. (For example, hypothetical space criminal syndicate seeking control of certain orbits or engaged in controlling access to a specific physical resource/minerals?)

Third, there may be less organized efforts to use disruption and sabotage as a weapon of war – rather than just adding friction as conventional militaries might – using sustained disruptive campaigns, analogous to trolling campaigns for efforts on social media, as a primary means of military activity. This could also include the use of non-tradition kinds of deception and misdirection. For example, the need to clean up space debris could be used as front for more conventionally military activities that might look similar, especially from a distance. This might possibly occur and be paired with the aim of loss of dominance and control in a particular area; conventional nation-state powers seem interested in dominance of space, which seems a pretty easy thing to disrupt with even very small activities by small actors for low cost.

Just as with Earth warfare, all of these unconventional activities can achieve outsized effect for comparatively little cost because of the dominance and decisive battle mindset of conventional warfare that many of the larger powers would be committed to, and which their military and political force structures would reflect. The conventional air power dominance model can be seen as one of the closest analogies for the mission of the U.S. Space Force, which is designed to protect U.S. and allies military interests in space through military power, in particular to 1) protect satellites (central to military capabilities of many kinds) and 2) develop a unified warfighting theory including the space domain.[5] What kinds of disruptions or other actions would be viewed by such a power as a threat to their control or dominance in space?

One way to see how this might play out (either in conventional or unconventional terms) is to consider discussions of escalation and deterrence.[6] Zach Cooper and Thomas G. Roberts note five issues to consider: 1) attribution – who is responsible now that there are many more actors in space; 2) reversibility – how seriously are these attacks viewed; 3) resilience – is the attack one that is recoverable in a short period of time without need for extensive repairs or interventions; 4) thresholds – most attacks won't trigger war, some may; and 5) asymmetries – there are many more actors than in the Cold War period and not all of them are state actors. These are all issues that one ought to consider in thinking through what space warfare might look like and what the critical issues are going to be for *jus in bello*.

Due to the issue of asymmetry, attacks that are at least in theory reversible or a minor disruption by one party may be judged as a more serious attack (perhaps one that crosses the threshold to war) by another. There is also the question not just of judging attribution in this environment, but also how one will assess likely damage (to assess proportionality and collateral damage) before the attack and actual damage and impact after the fact. Again, given the distances and lack of human or other assessors on the spot, different parties may judge these differently and they may be difficult to accurately judge at all. In short, there will be fairly serious asymmetries of knowledge involved in this kind of warfare, especially as impacted by temporal and spatial considerations. In addition, the considerations that

---

5 United States Space Force, *Mission*, https://www.spaceforce.mil/About-Us/About-Space-Force/Mission/.
6 Zack Cooper and Thomas G. Roberts, *Deterrence in the Last Sanctuary* https://warontherocks.com/2018/01/deterrence-last-sanctuary/.

Cooper and Roberts lay out could also potentially increase the likelihood of escalation in a way that critics of both remote warfare and non-lethal (less than lethal) weaponry, as well as cyber war, are concerned about with regard to lowering the threshold for the resort to force.[7] The 2017 National Security Strategy notes that attacks on space architecture that directly affect vital U.S. interests are subject to response.[8] How different administrations and actors might interpret this statement could open up the possibility for fairly quick escalation even before the full picture of the impacts of any given action is clear.

Despite these somewhat speculative arguments about the possible character of space warfare, at present space is a domain where states and actors vie for control (especially though technology and technological objects like satellites, vehicles, space debris, the Space Station) from Earth, but we do not have human belligerents fighting in space in an analogous way to battlespaces on Earth. Accordingly, the main characteristics of space warfare salient for this discussion of *jus in bello* considerations thus far seem to be the following: distance, remoteness, increased time issues, increasing number of actors who are not just military and not just state actors, heavily object oriented – especially communications infrastructure for both the military and commercial enterprises, competition for resources, space and dominance as a support an adjunct to Earth battlefields being the central focus.

Thinking by Analogy

If the above arguments are right about the actual and possible (in the near term) character of warfare, how might we think about *jus in bello* questions? One approach is to think by analogy through various Earth warfare contexts to see which ones bear the closest resemblance to space warfare and examine what guidance these analogies could give on how to conceptualize space warfare in terms of *jus in bello*. While none of these analogies will be a perfect fit, there are salient concepts or resemblances that can be instructive about what to focus on for space warfare.

---

7 See Chapter 5 in Pauline M. Kaurin, *The Warrior, Military Ethics and Contemporary Warfare: Achilles Goes Asymmetrical.* (Burlington, VT: Ashgate, 2014).
8 The White House, *National Security Strategy of the United States of America*, December 18, 2017 https://trumpwhitehouse.archives.gov/wp-content/uploads/2017/12/NSS-Final-12-18-2017-0905.pdf.

First, consider the analogy of remote warfare; space warfare seems in many ways like remote warfare, only to a greater degree. In discussing remote warfare, John Williams idea of "drone space" highlights the spatial and ethical distinctiveness of the relationship between drone operators and their targets, where distance and intimacy are important considerations, especially in terms of the impact on reciprocity and autonomy.[9] Williams also notes that remote warfare involves larger physical distances between the operator and their targets than one might see in conventional war, but that unlike other kinds of military activities one (especially in the case of signature strikes) has intimate knowledge of the target as the result of following their activities for days or weeks, which combined with high resolution imagery, means that one can see the immediate impacts of ones attacks. The target, however, has no such knowledge or access to the drone operator's world. Contrast this with the example from the M*A*S*H episode "Dear Sigmund" where the bomber pilot is quite proud of the fact that he does not see the impact and aftermath of his actions.[10] While there is distance involved, that distance also precludes certain knowledge in a way that is quite different from drone operators.

Under the concept of "drone space" autonomy and reciprocity, which Williams takes to be essential to rights-based *jus in bello*, requirements of discrimination and proportionality become problematic to the operator/target relationship along with the issues of risk and vulnerability. In this new spatial and ethical relationship, the vulnerability and risk are all on one side and the autonomy and power are all on the other, rendering it a situation of radical asymmetry. Is this distance a difference of kind or degree compared to manned bombers? Are drones, as some like BJ Strawser have argued, simply a more effective, precise instrument of discrimination and proportionality?[11] In the case of space warfare, we have more distance, but it seems less intimacy and we have to take into consideration the time issue (lag between a decision, action and effects), although that may be less over time as technology improves. Will that change the intimacy issue as we are able to see effects? The distance and reliance on technology seem to be the most salient features of this analogy, although we can note that the intimacy issue seems most pressing when it comes to human targets – which may or

---

9 John Williams, "Distant Intimacy: Space, Drones and Just War" *Ethics and International Affairs*. Vol 29, Issue 1 (Spring 2015), 93-110.
10 M*A*S*H, Season 5, Episode "Dear Sigmund".
11 Bradley J Strawser, "Moral Predators: The Duty to Employ Uninhabited Aerial Vehicles" *Journal of Military Ethics*. Volume 9, 2010 – Issue 4: Ethics and Emerging Military Technologies.

may not be the case with space warfare. This involves the moral status of human targets, empathy and other affective connections and reactions over a longer period of time than might be typical in most combat engagement.

The major salient issues here are risk, avoiding moral dilemmas, seeking destruction without risk. One of the major issues with remote warfare is how it transfers moral burdens and risk to the non-combatants on the ground and away from both the combatants on the ground (reducing their risk) and those who are operating (all but eliminating physical risk, even if there are concerns about moral and psychological risks). We might wonder if the distances involved in space warfare create the same dynamic.

Second, there is an analogy to the high seas and maritime law. With this analogy space is seen as similar to the high seas as a kind of commons, but also with significant commercial and scientific exploration aspects to be considered. Much of the focus on high seas is keeping the common areas open to commercial and other use, while acknowledging the areas that are subject to national sovereignty, so this analogy would involve thinking about whether there are parts of space that ought to be (or are) subject to national sovereignty in that way, and what it means to say the rest of space are part of the commons. Problems like piracy, enforcement of international agreements and what counts as international/common space (as opposed to spaces that are subject to territorial claims) are important questions to consider as a part of space warfare. We would have to consider whatever the analogies to fishing, trade, harvesting of and access to national resources will be in space, the extent to which those provide motivations and flash points for conflict and to what degree those will be protected as non-combatant/civilian related. Finally, as with the high seas there are environmental impacts that go beyond national borders, especially the problem of space debris.

While legal and moral views do not always track in the same way, one legal view sums it up in the following way:

> Similar to the Law of the Sea, objects in space must be provided due regard for freedom of movement. Parties can target each other's space assets during armed conflict. However, belligerent states must give due regard to the rights of neutral states and consider the risk that an attack on a space object would damage the property of neutral states and civilian property.[12]

---

12 Space Law Brief by Jill Goldenziel.

In looking at the near peer competition in Arctic and the South China Sea, these seem to be more analogous to interactions of the major powers (beyond the Cold War adversaries) in space where there is an interest in maintaining dominance on the part of the major powers and in disruption of dominance on the part of lesser powers. The increased commercial presence in space also seems salient here in ways that is not true of the prior drone analogy, which is primarily military in nature, even if operating in mixed environments.

Third, there is an analogy to cyberwarfare where some actions have kinetic effects, but where most actions (kinetic and otherwise) seem to be more oriented towards disruptions. Philosopher George Lucas Jr. makes two points that ought to guide our discussion here. First, he sees cyberwarfare as, "…'unrestricted' warfare carried out by spies and espionage agents who do not think of themselves bound by legal restraints, unlike conventional combatants…"[13] Second, he notes two forms of legitimate cyberwarfare: 1) effects-based cyber conflict, like the Stuxnet attack in which cyber activities produce physical effects comparable to a conventional attack and 2) state sponsored hacktivism which produces political effects in a similar way that conventional attacks in war are designed to impact the will of the adversary to produce political effects.[14]

It is the effects of cyber related actions (rather than the actions themselves) that will impact humans, as opposed to humans being the direct and intentional targets of the military action – either combatant or non-combatant. There are also attribution issues due to the number and nature of actors, both state and non-state, military and commercial as well as individual hackers or criminals. Finally, there is a heavy presence of military and commercial and non-combatant (civilian) systems in what is effectively a mixed environment. It is this mixed environment that may also create motives for criminals and pirate-like individuals since there is money to be made/resources to be attacked for profit or other benefit.

In particular, satellites and other communication infrastructure seem rife for disruption, but not necessarily to the point of terrorism since that seems to require a human audience or impact to induce terror. There is the possibility of space travel and space station supply disruptions, but forms of commercial disruption seem more likely at the present time. (In the future,

---

13 c. George Lucas Jr., *Ethics and Cyberwarfare: The Quest for Responsible Security in the Age of Digital Warfare* (New York: Oxford University Press, 2017) 26.
14 Lucas, *"Ethics and Cyberwarfare"*, 57.

we might consider what happens if travel involving human passengers (especially non-combatants) becomes much more commonplace.) The U.S. Space Force organization at present is much more military oriented and aimed at maintaining U.S. military supremacy and control of space. A further complication, just as in Earth-based warfare, is that the lines between commercial, military and civilian are not as clear as one might like there to be; this is particularly true of services and objects necessary for life support – food, air, water. This might be similar to Michael Walzer's discussion of the farmer whose food supplies both the military and non-combatants, and whether he is a legitimate object of attack as one who contributes to the war effort.[15]

In space there may also be similar questions about the threshold for use of force and what counts as an act of war requiring a kinetic response. Further, how does one 1) identify an attack; 2) identify a responsible party and 3) assess severity in order to calibrate the responses? These difficulties are compounded by the fact of technology and swift advancement in technology that make it difficult for laws and other regulative norms or powers to keep up; our contemporary *jus in bello* principles (and resultant law) emerged in response to immoral actions and were regulative (as most law is) after facts or events that required such regulation emerged and those events were viewed as problematic by a critical mass of a community of practice.[16]

Given these three analogies, what are we to conclude about how to conceptualize what space warfare might be like relative to *jus in bello* concerns? While the issue of distance from the remote warfare analogy seems relevant, the two closest analogies seem to the high seas and cyber ones. There are major powers defending their supremacy, while also trying to claim to guarantee the Commons for all against incursions from emerging peer competitors and non-state actors, who may be commercially, criminally or politically motivated. However, space also shares some important similarities with the cyber realm especially in terms of a mixed environment with increasing commercial use and systems that are subject to disruption – either for profit or for political reasons and it may be hard to delineate the two. Especially as space becomes more crowded with various kinds of actors, there may be more difficulties with attribution as a problem;

---

15 Michael Walzer, *Just and Unjust Wars: A Moral Argument with Historical Illustrations* (New York: Basic Books, 1977), 146.
16 Lucas, *Ethics and* Cyberwarfare", 46.

this attribution problem will not just be one of who is acting, but also in assessing the motives and intents of any actions.

The Shape of *Jus in Bello* in Space

What do the above observations (some rather speculative) and arguments mean for the question of *jus in bello* in space? First, they are quite heavy on objects and space architecture and infrastructure, on largely remote piloted vehicles and less on human beings as present in the battlespace, except in the space to Earth actions and through indirect impacts where they might be collateral damage or indirectly impacted (as in the cyber domain.) This relative lack of humans seems relevant as we think about *jus in bello* considerations which often have human objects and spaces as their object. The character of the *jus in bello* is fundamentally human centric, with objects necessary to human dignity, flourishing or life being secondarily noted for protection after the requirements that directly impact human combatants and non-combatants.

Traditionally, there are two main moral principles in *jus in bello*: 1) discrimination – including non-combatant immunity (one may not intentionally target non-combatants) and moral equality of combatants (they may be targeted as moral equals, bearing no guilt for the *jus ad bellum* decisions of their leaders to go to war) and 2) proportionality of means, requiring no excessive force in means not required to achieve the military end, excluding certain tactics and weaponry (like biological, chemical and nuclear for example.) These two principles yield, for example, prohibitions on direct targeting of non-combatants, a ban on excessive collateral damage, a requirement for humane treatment of POW's (to facilitate the restoration of the peace), prohibitions on targeting of certain kinds of buildings (hospitals, religious or cultural sites) and prohibitions on the use of certain kinds of weaponry that are indiscriminate or produce unnecessary suffering either for combatants or non-combatants.

There is a fair question about whether or to what extent *jus in bello* is even relevant given the limited human presence and interaction in the space domain. Humans are actors in space warfare, but more indirectly through object and technological means where it seems that the intent is more to destroy and disrupt these technological means than to kill/harm humans (which might be a second by product.) How will concerns about collateral damage have to be assessed? Will there ever be any direct targeting of humans or infrastructure necessary to human life and flourishing? It seems that collateral damage, as in cyber, is more likely in the near term than the direct targeting of humans and direct infrastructure – except for possibly the Space Station and commercial space travel.

Further, it is important to consider Chris Coker's point that while the main feature of twentieth century warfare was distance, in the twenty-first century it will be more disconnection; space warfare will make this worse as it will be more object oriented and thus will result in reduced empathy and human connection.[17] There also seems to be an important kind of asymmetry here that may mirror remote warfare more, producing something more akin to piracy or terrorism; these actions are arguably more object oriented with human collateral damage, targeting sources of power and symbols and disruption of "normal" life and the ability of the State power to protect that, thereby producing terror or fear. The asymmetry also may apply to Lucas' first kind of legitimate cyber warfare where it is the disruption (even kinetic) of networks, connectivity, information systems that are necessary for warfare more generally that matters. Given the focus of the U.S. Space Force with control and dominance (military) of space, this seems problematic in roughly the same way that dominance or control of the high seas or air space are in conventional warfare.

If we return to the Alderaan example and take that as indicative of space warfare, then many of the *jus in bello* considerations would apply in roughly the same way. If, as the discussion in this chapter has suggested, space warfare is not to be like that, then it is critical to more carefully consider the object and disruption orientation in space and ask what that means for the two *jus in bello* principles of discrimination and proportionality. For discrimination to be effective, we have to be able to have accurate, reliable and actionable access to information on what humans or targets are combatant versus non-combatant; a significant part of ascertaining that will be sorting out the status of commercial actors who may have contracts with or other connections with the military. If there are time lags (of any significant nature), problems with disruptions, disinformation or inaccuracy of information, then that will impact the assessment of discrimination just as we see in guerilla, insurgency and terror related asymmetric conflicts.[18]

---

17 Christopher Coker, *Warrior Geeks: How 21st Century Technology is Changing the Way We Fight and Think About War* (New York: Oxford University Press, 2013), 45.
18 See my discussion of whether the binary combatant/non-combatant distinction needs to be expanded, in Pauline M. Kaurin, *The Warrior, Military Ethics and Contemporary Warfare: Achilles Goes Asymmetrical* (Burlington, VT: Ashgate, 2014).

Proportionality of means also seems like it will be even more problematic than discrimination due to the difficulty of assessing damage from afar, where reversibility and attribution, as well as intention of the actors involved have to be accurately determined, and where there will be other challenges to predicting impacts and results in a space environment. This would also seem to impact questions of collateral damage, especially where there may be second, third and fourth order impacts that are hard to foresee or predict. Add to this a busy mixed environment with a wide range of actors, objects and space debris producing a myriad of variables that have to be considered in any judgment of proportionality of means, which must include both military necessity and unnecessary suffering calculations.

To conclude the above arguments about the impact on *jus in bello*, considerations ought to be seen in the context of the assessment of the role of humans in space warfare. For now, space is an adjunct to Earth centric warfare in which humans are optional or at least a secondary consideration to objects, vehicles, and space architecture. Should climate change or other events necessitate humans living or working in space as a primary domain for survival and flourishing, that would change these assessments in a way that might make it more analogous to urban warfare.

References

Cole, A. and P.W. Singer. *Ghost Fleet* (New York: Mariner Books, 2016).

Coleman, N. "Ethical Issues of Military Uses of Space" July 10, 2019, https://www.youtube.com/watch?v=pReDU9I_Uyc&t=51s.

Coker, C. *Warrior Geeks: How 21st Century Technology is Changing the Way We Fight and Think About War* (New York: Oxford University Press, 2013).

Cooper, Z. and T.G. Roberts. "Deterrence in the Last Sanctuary" War on the Rocks, https://warontherocks.com/2018/01/deterrence-last-sanctuary/.

Kaurin, P. M. *The Warrior, Military Ethics and Contemporary Warfare: Achilles Goes Asymmetrical* (Burlington, VT: Ashgate, 2014).

Lucas, G., Jr. *Ethics and Cyberwarfare: The Quest for Responsible Security in the Age of Digital Warfare* (New York: Oxford University Press, 2017).

de Gouyon Matignon, L. "The Kessler Syndrome" *Space Legal Issues*. March 27, 2019. https://www.spacelegalissues.com/space-law-the-kessler-syndrome/.

Strawser, B. J., "Moral Predators: The Duty to Employ Uninhabited Aerial Vehicles" *Journal of Military Ethics*. Volume 9, 2010 – Issue 4: Ethics and Emerging Military Technologies.

United States Space Force, Mission. https://www.spaceforce.mil/About-Us/About-Space-Force/Mission/.

Walzer, M. *Just and Unjust Wars: A Moral Argument with Historical Illustrations* (New York: Basic Books, 1977).

The White House, *National Security Strategy of the United States of America*, December 18, 2017, https://trumpwhitehouse.archives.gov/wp-content/uploads/2017/12/NSS-Final-12-18-2017-0905.pdf.

Williams, J. "Distant Intimacy: Space, Drones and Just War" *Ethics and International Affairs*. Vol 29, Issue 1 (Spring 2015).

4

WHAT IS NEEDED TO PREVENT CONFLICT IN SPACE?

Daniel Porras

There is no shortage of scholarly articles that begin by stressing the growing importance of space capabilities to the modern world, and rightfully so. Satellites and space-based services play a key role in the lives of nearly every human being. This is especially true of the military, who utilize satellites increasingly as a force multiplier through applications such as communications, reconnaissance and even targeting. Satellites also play a key role for strategic forces, providing such applications as early warning detection, tracking and control of warhead delivery systems. It is thus no surprise that a number of countries – among them the U.S., Russia, China and India – are seeking the means to disrupt space systems by a number of technological avenues that enable "counterspace capabilities".[1]

Some of these capabilities are already in use, such as electronic jamming and even hacking. More worrying, though, is the trend that more and more countries are now also seeking such arms and are establishing dedicated military units to use them. While no State has ever "attacked" another State's satellite in a way that would be universally understood to constitute an attack, the major players are no longer the only ones preparing for such an eventuality.

Unlike other domains such as land, sea or air, there is as yet little knowledge about the consequences of open conflict in space. The physics of orbit are unlike those on the surface of the Earth, resulting in phenomena – like persistent long-lived debris – that could have grave consequences for all space users. Moreover, the cost of accessing space has resulted in a co-mingling of civilian and military activities, making it all the more

---

1 See Victoria Samson and Brian Weeden, "Global Counterspace Capabilities: An Open Source Assessment", Secure World Foundation, April 2020, https://swfound.org/counterspace/.

difficult to distinguish between legitimate targets and innocent bystanders. Attacking a satellite could foreseeably impact many non-combatants by cutting off critical communications, disrupting global positioning or crippling emergency response. Many experts guess at the possible consequences of conflict in space, but the truth is that no one knows just how extensive the collateral damage could be.

For this reason, over the past four decades, States have put forward a number of initiatives to reinforce geopolitical stability in outer space and, hopefully, prevent conflict from emerging there openly. In particular, efforts focus on the "prevention of an arms race in outer space" (PAROS), an agenda item of the United Nations (UN) and the Conference on Disarmament (CD). Speaking about an arms race in space implies that it is separate and distinct from the arms race taking place on Earth, which is not the case, given today's modern arsenals. As I concluded elsewhere, there are already plenty of indicators that there is an arms race taking place in the world today, and space systems are simply a part of it.[2] It might be more meaningful to discuss the prevention of conflict in space or mitigating the harmful effects of conflict in space. However, stopping the arms race today will require addressing many more fundamental issues than just counterspace capabilities.

Contrary to popular belief, though, an arms race does not have to lead to conflict. The nuclear arms race has not broken out into a hot war, so it could be argued that arms-racing created stability. However, that stability was also bolstered by numerous agreements and treaties between the U.S. and Soviet Union/Russia. Leadership on both sides made numerous efforts intended precisely to prevent open conflict in the nuclear domain, despite the fact that they were developing more powerful weapons all the time.

In this context, while the objective of PAROS may no longer be fit for purpose, there are other ways of mitigating the worst consequences of "counterspace capabilities", namely preventing open conflict from emerging in space. Given the recent wave of arms-control agreements that have fallen by the wayside, an agreement on space security could be one way to slow the degradation of existing global strategic stability.

I propose that three material concepts, and one ethereal one, could lead to a governance architecture in space that, while not delivering on the

---

2 See Benjamin Silverstein, Daniel Porras and John Borrie, "Alternative approaches and indicators for the prevention of an arms race in outer space", UNIDIR Space Dossier 5, May 2020, https://unidir.org/sites/default/files/2020-05/Space%20Dossier%205%20-%20Final%5B3%5D.pdf.

paradise of "peaceful purposes" many hope for, will at least ensure that future generations may continue enjoying the benefits of space as we do today. The three "material" concepts are: increased global dependence on space assets; a proliferation of counterspace capabilities; and the development of international norms of behaviour for certain space activities. These are referred to as "material" because they are traditional policy trends and recommendations that are regularly discussed in strategic-theory and arms-control circles, and we can act on them in demonstrable ways. The "ethereal" concept is a simple one that is often left out of expert discussions, and that is the need for empathy. All too often, stakeholders seek to out-position their rivals in space, but seldom do we try to put ourselves in the shoes of others. As such, competition in space is quickly becoming a zero-sum game, and that can only increase the likelihood of conflict in space. And while empathy is not something that one can track or verify; we will ignore it at our own peril.

CONCEPT 1: Increased Global Dependence on Space

As is noted many times throughout this book, the world is increasingly dependent on space-based services, though this dependence is not yet universal. It is commonly understood that the U.S. is the country that is both the most active and the most dependent on space capabilities, far more than, say, countries in the global South. This dependence can be measured by a variety of metrics, such as annual revenue generated by a particular country[3] or the number of operational satellites in orbit.[4] One strong metric can also be vulnerability. For example, a recent report noted that if GPS were to shut down for one month, the U.S. economy would hemorrhage on average US$1B per day.[5] The result of the U.S. having major

---

3 See Bryce Technologies, "2019 Global Space Economy at a Glance", https://brycetech.com/reports. The report notes that of the USD$366B generated by the global space economy in 2019, the U.S. Government's share of it was $57.9B, with Europe collectively coming in second at USD$12B, and China in third with USD$11B.
4 See Union of Concerned Scientists "Satellite Database", https://www.ucsusa.org/resources/satellite-database. This catalogue notes that, of the 3,372 satellites currently in orbit, the U.S. is responsible for 1,897, while China is responsible for 412, and Russia for 176. The rest of the world combined is responsible for 887 satellites.
5 Economic Benefits of the Global Positioning System (GPS), Final Report, prepared by RTI International, June 2019.

outages in telecommunications or Earth-observation would also lead to major economic losses and security lapses. As a result, the U.S. treats the security of its space systems as a major national security priority. Whatever one thinks of the U.S.' strategies or tactics, there is little doubt that the U.S. has expended considerable resources in seeking safety and security for all its space assets and their relevant systems. No other country has as much "skin in the game" as the U.S. Ironically, most countries that can afford a military space programme are following suit, even geopolitical rivals. For example, Chinese military experts see space as an Achilles' heel for the U.S., yet also see space-supremacy as vital for strategic dominance in the future.[6]

If nothing else, this proliferation of space capabilities means that all actors are invested in the long-term sustainability of space activities. Rivals may seek to gain the upper hand in orbit, but not at the cost of their own critical infrastructure. At the very least, mutual investment in satellites and relevant space systems eliminates at least some behaviours as viable options for disrupting or destroying objects in space. For example, the U.S. discovered during Operation Starfish Prime that detonating nuclear weapons in space would disable all satellites within reach of the electromagnetic blast, including their own.[7] Understanding that their own capabilities would be at risk if nuclear weapons were used in space, the U.S. ultimately agreed to include outer space in the Limited Test Ban Treaty (1963).[8] Today, States might see similar value in limiting the use of kinetic anti-satellite missiles, which create debris that can be harmful to all. And while the major military space powers may not be willing to give up possessing or even using them in a conflict, the time may be ripe for guidelines on testing ASAT weapons at the very least.[9]

---

6 Kevin Pollpeter, "Motives and Implications Behind China's ASAT Test", *China Brief* Volume: 7 Issue: 2, 9 May 2007, https://jamestown.org/program/motives-and-implications-behind-chinas-asat-test-3/.
7 Phil Plait, "The 50th Anniversary of Starfish Prime: The Nuke that Shook the World", *Discover Magazine*, 9 July 2012, https://www.discovermagazine.com/the-sciences/the-50th-anniversary-of-starfish-prime-the-nuke-that-shook-the-world.
8 U.S. Department of State, "Treaty Banning Nuclear Weapon Tests in the Atmosphere, in Outer Space, and Under Water", last visited on 16 February 2021, https://2009-2017.state.gov/t/avc/trty/199116.htm.
9 Daniel Porras, "Towards ASAT Test Guidelines", UNIDIR Space Dossier 2, May 2018, https://www.unidir.org/files/publications/pdfs/-en-703.pdf.

Of course, being a major military space power is not a pre-requisite to having a stake in the sustainability of space activities. Every country in the world today benefits from space-based data, whether it be weather monitoring or disaster relief. But as States continue to invest and depend on space-based services, the greater the lengths States and other stakeholders will be willing to go in order to stabalise and preserve the space environment. As such, emerging States and new space actors will have an increasing influence on the shape of multilateral discussions on space activities. This will happen not only because emerging actors have telecommunications and remote-sensing satellites of their own, but because they will have access to at least some counterspace capabilities themselves.

CONCEPT 2: Proliferation of Counterspace Capabilities

Each year, Secure World Foundation (SWF) publishes "Global Counterspace Capabilities: An Open Source Assessment". One of the troubling trends that SWF tracks in this report is the proliferation of capabilities able to interfere with objects in space. The technology is diversifying and becoming increasingly available to more and more actors. Moreover, following the announcement by the U.S. that it would establish a Space Force, a number of countries (such as France, India and Japan) followed suit, establishing their own dedicated space units. SWF is now adding new chapters to its assessment to account for the new players in the counterspace game. By all accounts, the proliferation of such weapons should increase the likelihood that conflict will emerge in space. However, as noted earlier, arms racing does not necessarily mean that there will be open conflict.

During the Cold War, both the U.S. and the Soviets actively developed and tested counterspace capabilities. Perhaps because they understood the potential sensitivities involved with attacking one another's satellites, neither ever actively used them in space. And while there was widespread support for discussions around PAROS in the early 1980s, the position of Western countries changed in the early 90s and remained that way until the early 2000s. According to several States, but most notably the U.S., there was no need to discuss PAROS because there was no indication that there was a space arms race at all or that any country was even actively seeking counterspace capabilities.[10] While there is plenty of evidence to indicate that, at the very least, the U.S. was actively seeking these capabilities, it was not until 2007 (when China conducted its own ASAT demonstration) that discussions once again picked up momentum. The sense of urgency around PAROS has only increased in recent years as more and more countries are acquiring the necessary expertise to interfere with satellites.

It is not surprising that counterspace capabilities should proliferate, given that much of the requisite technology is both dual-use and multi-use. For example, a kinetic direct-ascent ASAT missile is in essence simply a missile interceptor. In 2008, when the U.S. carried out its ASAT missile demonstration during Operation Burnt Frost, it modified an SM-3 Block IA interceptor to hit a satellite instead of a missile simply with a software patch.[11] The same is essentially true of India's kinetic ASAT capabilities.[12] Likewise, any country pursuing Active Debris Removal or related rendezvous proximity operations has the fundamental basics for more nefarious activities against a target.

Yet all of these capabilities are fairly complex when one considers that electronic interference or even hacking can be extremely effective in terms of disruption and degrading services.[13] This technology is already widely available and was used on several occasions, including the jamming of a Eutelsat satellite to block transmission of foreign news broadcasting[14] and the disruption of global positioning near certain airports.[15] More recently, reports have surfaced of interference with NATO exercises, likely from

---

10 Conference on Disarmament, Report of the Ad Hoc Committee on Prevention of an Arms Race in Outer Space, document CD/1271, 24 August 1994, §10.
11 Victoria Samson and Brian Weeden, "Global Counterspace Capabilities: An Open Source Assessment", Secure World Foundation, April 2020, 3-9, https://swfound.org/counterspace/.
12 Victoria Samson and Brian Weeden, "Global Counterspace Capabilities: An Open Source Assessment", Secure World Foundation, April 2020, 5-2, https://swfound.org/counterspace/.
13 See Rajeswari Pilai Rajagopalan, "Electronic and Cyber Warfare in Outer Space", UNIDIR Space Dossier 3, May 2019, https://www.unidir.org/files/publications/pdfs/electronic-and-cyber-warfare-in-outer-space-en-784.pdf.
14 Stephanie Nebahay, "U.N. tells Iran to end Eutelsat satellite jamming", Reuters, 26 March 2010, https://www.reuters.com/article/us-iran-jamming-itu/u-n-tells-iran-to-end-eutelsat-satellite-jamming-idUSTRE62P21G20100326.
15 "North Korea jamming hits South Korea flights", BBC World Service, 2 May 2012, bbc.com/news/world-asia-17922021.

Russian sources.¹⁶ At this rate, it would not be surprising to see jamming and interference more routinely as a means of harassing but not directly engaging a rival or opponent.

While the proliferation of counterspace capabilities should indicate a growing likelihood that they get used, it might also create pressure for certain key stakeholders to finally come to the table and negotiate. One could argue that part of the reason there hasn't been much progress on arms-control in space is that certain key actors seek to maintain as many options as possible for their own operations. However, if enough other actors acquire the same capability, then possessing them might no longer create an advantage. It might then make more sense for certain key actors to restrain themselves, provided they have a verifiable and enforceable prohibition on the use of a particular technology by others. As noted above, this was the case with the Limited Test Ban Treaty, and might again be the case with guidelines or prohibitions around ASAT testing. Indeed, this very limited issue might be the best current option to be the subject of an explicit norm of behaviour.

CONCEPT 3: Development of Norms of Behaviour

There is no official definition of a "norm of behaviour" in the space community, but a recent survey showed that:

> *participants consistently identified a core set of concepts that operationalize norms, including standards, behaviours, and best practices. They also revealed that norms are social and value-laden: social and moral underpinnings activate a sense of obligation to conform to norms, distinguishing norms from other types of rules. This is reflected in the emphasis on concepts such as shared values, collectivity, common understandings, consensus, and expectations.*¹⁷

In the face of political roadblocks that have held back any progress on the question of a legally binding instrument on space security, some in the

---

16 Gerard O'Dwyer, "Finland, Norway press Russia on suspected GPS jamming during NATO drill", DefenseOne, 18 November 2018, https://www.defensenews.com/global/europe/2018/11/16/finland-norway-press-russia-on-suspected-gps-jamming-during-nato-drill/.
17 Jessica West and Gilles Doucet, "From Safety to Security: Extending Norms in Outer Space – Workshop Report", Project Ploughshares, January 2021, 9.

multilateral community (including many of the space-faring nations) are looking to elaborate norms of behaviour, namely specified forms of conduct (or non-conduct) agreed upon by all to be the standard among space actors. The recent adoption of UN Resolution 75/36, "Reducing space threats through norms, rules and principles of responsible behaviour" is the latest initiative that will likely, though not necessarily, examine areas where non-legally binding norms might be adopted to mitigate the worst threats to space systems. While this proposal does not rule out the possibility of looking at legally binding behaviours, the momentum of multilateral politics would suggest that this Resolution is a first step at developing and adopting norms through politically binding measures. If that is the case, this could be a useful *first* step in establishing a framework that will establish guardrails for conflict in space.

I stress the reference to the development of norms as a "first" step since it is clear from UN voting records that the vast majority of the world (either by number of States or by representative populations) prefer a legally binding treaty to political norms.[18] These States will, in the not too distant future, be space actors and, likely, possessors of counterspace capabilities, even basic ones like jamming and hacking. Any agreement will, therefore, require buy-in from the developing world. As was seen during the consultations around the draft EU International Code of Conduct for Space Activities, the overwhelming preference for a treaty will have to be dealt with. One option could be to refer to a treaty as a long-term yet still undefined goal, while the development of non-legally binding norms is a short-term and focused approach. Again, the adoption of ASAT test guidelines is a good example of a limited but tangible measure that could be explicitly adopted with a view towards benefiting all space actors and beneficiaries.

Yet perhaps the most compelling reason to adopt explicit, positive norms is to prevent the development of permissive, negative norms that permit all the worst consequences of conflict in space. Referring yet again to the development of kinetic ASATs, there was a drastic difference in

---

18 As evidence of this, note that UN General Assembly Resolution 72/250 "Further practical measures for the prevention of an arms race in outer space Statement of financial implications", was adopted by a vote of 108-5-47. The result of that Resolution was to form a Group of Governmental Experts whose mandate was to examine "elements of a legally binding instrument" on PAROS.

the responses to the three major debris generating ASAT demonstrations in the twenty-first century. China, for example, carried out its ASAT demonstration at nearly 1000km in altitude, leaving behind debris that will stay in orbit for centuries. The U.S. demonstration the following year was criticized far less, in part because the U.S. could point to what many see as a pretext (failed NASA satellite with a full tank of dangerous chemicals) but also because the impact took place just above 200km. The debris fully deorbited in less than two years (far more than was originally projected, but considerably better than the previous example). When India conducted its test, just below 300km, there was a noticeable lack of condemnation among the international community. Unfortunately, the message conveyed by the silence is that it is acceptable to the international community for actors to conduct kinetic ASAT tests below 300km. Given that kinetic ASAT tests are among the biggest debris-generating events caused by humans, such a norm could have devastating consequences on Low-Earth Orbit.

Finally, it is important to consider how norms relate to the application of rules that exist today. While we know that, under Article III of the Outer Space Treaty, international law applies in outer space, we don't know how to apply that law. For example, there is no consensus on what is the "threat or use of force" against a satellite, or what is "proportionate" self-defense. This ample room for interpretation means that each State likely has a different threshold not only for when to feel attacked, but even when to feel threatened. As noted above, States have never openly attacked one another with a kinetic ASAT missile, but they are already testing each other with electronic means. It is not beyond the realm of imagination that, eventually, one actor will misjudge another and there could be an escalatory scenario.

Norms can help prevent such scenarios by simply giving actors a sense of what is or is not permissible behaviour in space. Options such as "safety zones" around a satellite – within which another actor may not enter without permission – could create boundaries and guardrails to prevent rising tensions, and a metric by which to measure "hostile intentions". A measure such as this would not prevent all threats (like electronic ones), but it could ease tensions around physical ones. Such measures could go a long way towards stabilizing space until we have perfected the art of divining an adversary's intentions from maneuvers in orbit.

CONCEPT 4: Empathy

The last concept that might prevent the worst-case scenarios of conflict in space is one that is often left out of security discussions as it is seen as being a "soft" tool, and that is empathy. All too often, the rhetoric behind counterspace capabilities and the new "space race" involves placing the blame on someone else for causing the arms race. In the West, it is all too common to blame Russia and China for "weaponising" outer space while not acknowledging the role that Western countries are playing in the arms-race dynamic that exists today. Likewise, it is easy to villainize the U.S. because of the rhetoric used around many of its recent policies, but the accusations are often far from the truth, particularly when it comes to Space Force.

In my experience, what is often lacking is the willingness to accept that a rival has just as much right to defend themselves as we do. We often dismiss the concerns of our rivals as pretext and their measures as provocative, while praising our own actions as defensive and righteous. The result of these attitudes as a premise for negotiations is that no one trusts one another and therefore no one is willing to concede any ground that might then grant their rival an advantage. In essence, everyone is so afraid of being cheated that no one is willing to compromise. In this sense, discussions around space security today resemble those that took place around the Cold War. Without a greater effort for all actors to be empathetic, multilateral efforts to avoid the worst-case scenarios of conflict in space will be stymied by political, cultural and economic stand-offs between Great Powers, between East and West, between North and South, between haves and have-nots. In the end, competition will have to give way to conflict, and that could mean we will no longer enjoy the benefits of human space activities.

Conclusion

It is very likely that conflict already exists in space, as humans took it with them the very first time they reached orbit. Nevertheless, society's dependence on space capabilities and the services they make possible, requires that people ensure the long-term sustainability of their current trajectory. Many of the challenges do not even have anything to do with space, more with politics, but people can at least agree to certain guardrails that will buy time while they continue to resolve their political differences. First, they should encourage and facilitate human space activities, and the reaping of benefits therefrom, to ensure that all actors have a stake in the long-term sustainability of space activities. Secondly, people should recognize that, sooner or later, all States will have some form of counterspace capabilities, and these could prove highly disruptive for all if left unchecked. Thirdly, people should seek to explicitly normalize those behaviours where they can find convergence of interests, such as ASAT test guidelines or "safety zones" for rendezvous proximity operations. Such norms will offer a short-term solution while people seek more robust long-term solutions for stability in outer space. And overall, we must seek to inject our conversations with greater empathy for our rivals. If we are going to share a common space, we must learn to live together or not at all – history has shown us that we do not want the latter.

References

BBC World Service. "North Korea jamming hits South Korea flights", 2 May 2012, bbc.com/news/world-asia-17922021.

Bryce Technologies, "2019 Global Space Economy at a Glance", https://brycetech.com/reports.

Nebahay, S. "U.N. tells Iran to end Eutelsat satellite jamming", Reuters, 26 March 2010, https://www.reuters.com/article/us-iran-jamming-itu/u-n-tells-iran-to-end-eutelsat-satellite-jamming-idUSTRE62P21G20100326.

O'Dwyer, G. "Finland, Norway press Russia on suspected GPS jamming during NATO drill", DefenseOne, 18 November 2018, https://www.defensenews.com/global/europe/2018/11/16/finland-norway-press-russia-on-suspected-gps-jamming-during-nato-drill/.

Plait, P. "The 50th Anniversary of Starfish Prime: The Nuke that Shook the World", *Discover Magazine*, 9 July 2012, https://www.discovermagazine.com/the-sciences/the-50th-anniversary-of-starfish-prime-the-nuke-that-shook-the-world.

Pollpeter, K. "Motives and Implications Behind China's ASAT Test", *China Brief* Volume: 7 Issue: 2, 9 May 2007, https://jamestown.org/program/motives-and-implications-behind-chinas-asat-test-3/.

Porras, D. "Towards ASAT Test Guidelines", UNIDIR Space Dossier 2, May 2018, https://www.unidir.org/files/publications/pdfs/-en-703.pdf.

Rajagopalan, R. "Electronic and Cyber Warfare in Outer Space", UNIDIR Space Dossier 3, May 2019, https://www.unidir.org/files/publications/pdfs/electronic-and-cyber-warfare-in-outer-space-en-784.pdf.

Samson, V. and B. Weeden, "Global Counterspace Capabilities: An Open Source Assessment", Secure World Foundation, April 2020, https://swfound.org/counterspace/.

Silverstein, B., D. Porras and J. Borrie, "Alternative approaches and indicators for the prevention of an arms race in outer space", UNIDIR Space Dossier 5, May 2020, https://unidir.org/sites/default/files/2020-05/Space%20Dossier%205%20-%20Final%5B3%5D.pdf.

Union of Concerned Scientists "Satellite Database", https://www.ucsusa.org/resources/satellite-database.

US Department of State, "Treaty Banning Nuclear Weapon Tests in the Atmosphere, in Outer Space, and Under Water", https://2009-2017.state.gov/t/avc/trty/199116.htm.

West, W. and G. Doucet, "From Safety to Security: Extending Norms in Outer Space – Workshop Report", Project Ploughshares, January 2021, 9.

5

SPACE DEBRIS

Can We Remove the Landmines of Earth Orbit Without Starting a War?

Stephen Coleman and Nikki Coleman

Introduction[1]

The possibility of warfare (or warlike activities) being conducted in space is a very real one. All modern military forces rely on a large number of space based systems which are more or less essential to the conduct of modern war, and these space assets would be a tempting target for opposing military forces.[2] In terrestrial conflict, unexploded ordnance, particularly anti-personnel landmines, can pose a hazard long after the conflict has ended and this has led to a range of efforts to (a) reduce,[3] and even prohibit,[4]

---

1 We wish to acknowledge the support and kind assistance of Professor Russell Boyce (UNSW Canberra Space), Professor Robert McLaughlin (Professor of Military and Security Law, UNSW Canberra), Group Captain Andrew Gilbert (RAAF Air Power Development Centre), and Professor Shannon French (Inamori Center for Ethics and Excellence, Case Western Reserve University). The views expressed are those of the authors and do not reflect the official policy or position of the Royal Australian Air Force, the Department of Defence, or the Australian Government.
2 Phillip Swarts, Space is seen as Increasingly Important to Military Operations, "Air Force Times" 12 September 2015, https://www.airforcetimes.com/news/your-air-force/2015/09/12/space-is-seen-as-increasingly-important-to-military-operations/.
3 United Nations, Convention on Prohibitions or Restrictions on the Use of Certain Conventional Weapons Which May be Deemed to be Excessively Injurious or to Have Indiscriminate Effects (and Protocols) (As Amended on 21 December 2001), 10 October 1980, 1342 UNTS 137, http://www.refworld.org/docid/3ae6b3ac4.html.
4 United Nations, Convention on the Prohibition of the Use, Stockpiling, Production and Transfer of Anti-Personnel Mines and on Their Destruction, 18 September 1997, http://www.refworld.org/docid/3ae6b3ad0.html.

the use of such weapons during future conflicts, and (b) reduce the risk to innocent parties by attempting to remove these "explosive remnants of war" after the conflict has ended. As is the case in terrestrial conflict, any conflict in space would inevitably produce "remnants of war" which would pose a hazard in the future. However, while in terrestrial conflicts it is only the explosive remnants of war which pose a significant hazard in the future, in space warfare every tiny remnant of war is hazardous, and this would pose a massive problem for any attempt to remove these hazards after the conflict has ended. Almost everyone who works in space operations in any way understands that space debris is a significant problem, both because it poses a threat to current and future operations and because it is difficult to remove. The technical issues of debris removal are being addressed in a number of ways, with various methods of removal having been proposed, some of which are currently being developed and tested. However, those who are focused on the technical problems of debris removal often seem to ignore the serious non-technical problems of debris removal, particularly the legal and political ones.

Space Warfare and the Outer Space Treaty

Space based assets are essential to most current terrestrial military activities, with military forces relying on satellite technology for navigation and on other satellite-based systems for communications, information gathering, munition guidance and targeting. Given their importance, the satellites which are used for these military purposes would be considered quite high value targets in the event of a large scale terrestrial conflict, as has been noted in speculative fiction on future warfare.[5] Admittedly there are significant practical problems which need to be overcome if a state wanted to engage in effective military operations directed against enemy military satellites, since these satellites are relatively small, moving very rapidly, and would almost certainly need to be attacked from long range. However, these problems are certainly not insurmountable ones, as was proven by China's test of an anti-satellite missile in 2007.[6]

---

5 P.W. Singer and August Cole, *Ghost Fleet: A Novel of the Next World War*, Boston, Houghton Mifflin, 2015.
6 BBC News, Concern over China's Missile Test, 19 January 2007, http://news.bbc.co.uk/1/hi/world/asia-pacific/6276543.stm.

While some of the international conventions which regulate terrestrial warfare might also apply to military actions in space, the only international treaty of any real significance which specifically mentions the regulation of warfare in space is the Outer Space Treaty (OST).[7] As is noted in the Preamble of the OST, the states who have ratified the treaty have done so out of a desire "to contribute to broad international co-operation in the scientific as well as the legal aspects of the exploration and use of outer space for peaceful purposes" and so the OST does place certain limits on both warfare in, and the militarisation of, outer space. However, these limits are reasonably specific and would do little to prevent destructive warfare in space in the event of a broader terrestrial conflict between states with space-based assets. The OST bans the placing of nuclear weapons or other weapons of mass destruction in Earth orbit, on the moon or other celestial bodies, or elsewhere in outer space.[8] It also limits the use of the moon and other celestial bodies to peaceful purposes, by forbidding the placement of military bases or fortifications on such bodies, and by forbidding military manoeuvres and weapons testing on all non-terrestrial bodies. However, the OST does not restrict the placement, or use, of conventional weapons (including high powered and precision weapons) in orbit or elsewhere in space. Thus, in the event of conflict in space a wide range of attacks would be legally permissible, allowing for ground-to-space, space-to-ground, and space-to-space conflict.

While there may be few legal barriers which prevent a state from engaging in space warfare, the international community has, in various ways, expressed a general desire to avoid such conflicts. For example, in December 2014 the UN General Assembly passed two resolutions on the matter,[9] the first, on prevention of an arms race in outer space, calling "on all States, in particular those with major space capabilities, to contribute actively to the peaceful use of outer space, prevent an arms race there, and refrain from actions contrary to that objective" with the second urging all states to not be the first to place weapons in outer space.

---

7 United Nations Office for Outer Space Affairs, Treaty on Principles Governing the Activities of States in the Exploration and Use of Outer Space, including the Moon and Other Celestial Bodies aka Outer Space Treaty (10 October 1967), 610 UNTS 205/6 ILM 386 (1967)/[1967] ATS 24, https://cil.nus.edu.sg/1967/1967-treaty-on-principles-governing-the-activities-of-states-in-the-exploration-and-use-of-outer-space-including-the-moon-and-other-celestial-bodies/.
8 United Nations Office for Outer Space Affairs.
9 UN press releases, General Assembly Adopts 63 Drafts on First Committee's Recommendation with Nuclear Disarmament at Core of Several Recorded Votes, https://www.un.org/press/en/2014/ga11593.doc.htm.

Space Debris and its Problems

At the current time (2021), all the important military targets located in space are in orbit around the Earth, and this will inevitably be the case for the foreseeable future. Thus, if space-based military assets were to be targeted, these military actions would be occurring in Earth orbit. Military satellites might be attacked with what might loosely be termed "soft kill" weapons, designed to temporarily disrupt, or permanently prevent, the satellite from carrying out its function while still leaving the satellite more or less intact. There are certainly advantages in attacking in this way, since a satellite can be attacked multiple times relatively cheaply, and it may be difficult for the owner of the attacked satellite to be certain with regard to the attribution of the attack. However, in the event of a large scale conflict it is likely that such satellites would be attacked with more conventional weapons, such as missiles, especially if less overt attacks have failed. If a satellite was destroyed this way then the creation of a large amount of space debris would be the inevitable result.

Space debris in orbit around the Earth presents a considerable hazard to both unmanned satellites and manned space missions. Objects in low Earth orbit are travelling at a velocity of 7-8 km/sec, so the force generated by the impact of even a small piece of debris is considerable. Serious damage can thus be inflicted on satellites by relatively small pieces of debris, while depending upon the point of impact, a piece of debris weighing a kilogram is probably capable of destroying a satellite weighing several hundred kilograms. Any weapon used to destroy a military satellite would be of significant size, thus even the initial cloud of debris caused by the destruction of this satellite would be spread over a large area. With the passing of time, variations in orbital velocity and drag of the various pieces of debris would mean that this debris 'cloud' would spread out still further, leaving pieces of debris speeding around the Earth in an enormous range of different orbits and altitudes. This remnant of space war would continue to pose a threat to all types of spacecraft long after the initial conflict has ended.

The only natural way to remove this threat is to rely on the orbital decay of each piece of debris, which will eventually draw the debris into the Earth's atmosphere, where depending on its size, it will either burn up during entry into the lower atmosphere, or if it is large enough, fall to the Earth's surface. Such a process is relatively rapid at lower orbital altitudes, however the greater the altitude at which the debris was created, the longer the potential orbital life it has. Depending upon the initial altitude above the Earth at which the debris was created, pieces of debris might continue to orbit the Earth for years, decades, or even millennia. Shielding spacecraft against the impact of debris is not a straightforward process, and in any case such shielding can only be effective against the threat posed by smaller pieces of debris. Thus, the only way to permanently remove the threat which is posed by larger pieces of debris orbiting the Earth at higher altitudes is to deliberately intervene in some way to remove the debris from orbit.

The debris which might be created by any military conflict in Earth's orbit thus shares several features with one of the explosive remnants of terrestrial war, the anti-personnel landmine. Like space debris, landmines remain a threat long after a conflict has ended, do not discriminate between targets, and are costly, difficult, and dangerous to detect and remove. Given these similarities, it is worth considering whether the problems posed by the use of landmines, and the solutions to those problems, can provide any lessons for the future in dealing with the problem of space debris.

As was noted earlier, one method which has been used to reduce the long-term threat posed by landmines was to restrict the ways in which landmines could be used. Protocol II of the Convention on Certain Conventional Weapons prohibited the use of landmines and similar devices in ways likely to attract civilians, particularly children. It also prohibited the use of mines which were not capable of either self-destruction or self-deactivation, except in areas which were fenced, clearly marked, and monitored. It also required state parties using landmines in a conflict to remove any mines used after the conflict had ended. Unfortunately, these sorts of restrictions would not be of any use in dealing with the problems caused by space debris. Unlike landmines, this debris is not stationary, and so does not need to "attract" a target to cause harm. It is impossible, for obvious reasons, to create self-deactivating or self-destructing debris, and space debris simply cannot be contained within a fenced, marked and monitored area. This leaves only the requirement that parties to a conflict remove any space debris after the conflict ends, a requirement which produces its own special problems, as will be discussed shortly.

The more recent method of preventing landmines from posing a long-term threat has been to prohibit the use of anti-personnel landmines altogether.[10] The only way that this sort of prohibition could be applied

---

10 United Nations, Convention on the Prohibition of the Use, Stockpiling, Production and Transfer of Anti-Personnel Mines and on Their Destruction, 18 September 1997, http://www.refworld.org/docid/3ae6b3ad0.html.

to the case of space debris would be to prohibit space warfare entirely. As was noted earlier, while there is no legal prohibition on the conduct of space warfare per se, it is an aspiration which has been expressed by the international community through things such as the passing of UN General Assembly resolutions on the subject. Interestingly, the most powerful state opponent to such a position, at least so far as UN votes on the issue can indicate, is the USA, one of the two states (along with Israel) to abstain from the vote on the *Prevention of an Arms Race in Outer Space* resolution, and one of four states (along with Israel, Georgia and Ukraine) to vote against the *No First Placement of Weapons in Outer Space* resolution.

The measures which have been put in place to reduce the long-term threat of landmines do not seem to be straight-forwardly applicable to the problem of space debris. But there is more similarity between landmines and space debris when it comes to the problems of removal. The process of removing landmines is dangerous, expensive, and slow. The area in which landmines have been laid needs to be thoroughly searched so that all the mines can be detected, then each mine needs to be disarmed or destroyed. Mines which have been in place for an extended period may have "migrated" from their original position, which may make them harder to detect or safely remove, and the explosive in the mine many have degraded over time, making it more unstable. Therefore, it is relatively easier to clear a recently laid minefield than to clear mines which have been in place for an extended period of time.

Detecting space debris is also difficult, but for entirely different reasons. Landmines are stationary and hidden beneath the ground and need to be detected from close range. Space debris, on the other hand, is moving at high speed and needs to be detected from a distance. Current methods of detection can only be used to detect debris of sufficient size (approximately 10 cm or larger depending upon the type of material of which the piece of debris is composed) but pieces of debris much smaller than this still pose a serious threat to spacecraft. Merely detecting a landmine provides a significant measure of protection against the threat the mine poses; because the mine is stationary, it is easily avoided and once the mine is detected, further risk only arises when an attempt is made to remove and disarm the mine. On the other hand, merely detecting a piece of space debris doesn't really provide any protection against the damage that debris may cause since the problem of ensuring that nothing will be struck by this fast moving object still remains, though detection at least allows the possibility of a satellite engaging in manoeuvres to avoid being struck.

There are some similarities between the problems posed by landmines and the problems posed by space debris, but obviously also some highly significant differences. But perhaps the most important differences between landmines and space debris have not yet been discussed. A landmine which explodes doesn't create new landmines in the process, but when a piece of space debris hits and damages a satellite, new pieces of space debris are likely to be created in the process. Impacts between pieces of debris can also cause the creation of new smaller pieces of debris. If the density of objects in orbit is high enough, then it is possible that collisions between objects could cause a cascade effect, with each collision creating more space debris, thus increasing the likelihood of future collisions, which will in turn produce even more debris. This cascade effect is known as the Kessler Syndrome,[11] and it has been suggested that if the amount of space debris in a particular orbital range exceeds the threshold required to initiate the Kessler Syndrome, this makes collisions in that orbital range so likely that it would become impossible to sustain satellite operations.

Low Earth orbit (LEO) has the highest current density of satellites and is thus most likely to fall victim to the Kessler Syndrome. Destructive military operations launched against satellites in LEO are thus risky since such operations would dramatically increase the amount of debris in LEO and might initiate the Kessler Syndrome. The Chinese anti-satellite missile test of 2007, for example, was the biggest orbital debris generating event on record, creating more than 3000 pieces of debris of trackable size, and an estimated 150000 debris particles.[12] The destruction of only a few satellites in LEO during a military conflict would create a huge amount of debris, potentially initiating the Kessler Syndrome and effectively denying the use of LEO to all of humanity for generations to come.

Legal and Political Problems of Space Debris Removal

Another significant difference between landmines and space debris is the fact that landmines are located on sovereign territory and can be considered abandoned and unwanted by their original owners (for example, those who planted the mines in the first place). The fact that the mines one is intending

---

11 Donald J. Kessler and Burton G. Cour-Palais, Collision Frequency of Artificial Satellites: The Creation of a Debris Belt, *Journal of Geophysical Research*, 83 (1978) 2637–2646.
12 BBC News, ISS crew take to escape capsules in space junk alert, 24 March 2012, http://www.bbc.com/news/science-environment-17497766.

to remove are located on sovereign territory is important, since it makes it very clear who has the authority to allow the mines to be removed. Debris located in space is quite different in this respect, since space, as Article II of the OST makes clear, "is not subject to national appropriation by claim of sovereignty".

The current prevailing legal view is that every object launched into space remains the property of its registered state owner in accordance with Article VIII of the OST. The removal of pieces of space debris which are of natural origin, such as meteorites and space dust, is unproblematic in legal terms. However, any piece of space debris which was originally launched from Earth through human endeavour, even if it was deliberately abandoned in space, is still the legal property of its registered owner. A bottle which floated in the ocean for years after being deliberately thrown into the ocean from a boat would become the property of anyone who retrieved it, but a bottle which floated in Earth's orbit for years after being deliberately thrown from a spacecraft would still remain the property of its original owner.

The effect of this in terms of space debris removal is that anyone attempting to remove debris registered to another state could be accused of theft, or of causing damage to the property of a foreign state. While certain methods of removal are likely to be more problematic than others in this respect, all methods of removal are open to potential legal problems. Legal problems which might arise with regard to technologies which de-orbit debris, causing it to burn up on re-entry into Earth's atmosphere, for example, would most likely be related to questions of liability if this debris, or some part of it, was to cause damage after surviving re-entry. However, if it were to become possible to recover objects more or less intact from space, then since such objects might contain proprietary technology of a state, an attempt to recover an object belonging to another state would rightfully be considered theft and would be likely to cause a significant international incident. Some of these issues are being discussed as part of the attempt to establish customary international law for space, through creation of documents such as the Woomera Manual, but others are likely to remain problematic for some time. However, overall the legal issues may be far easier to solve than the bigger political problems related to the removal of space debris.

One crucial difference between landmines and space debris which has not yet been discussed is the technology used for removal. Techniques and equipment used for the removal of landmines do not have a great deal of generalisability, in that the technology which has been developed for the safe removal of landmines, or which might be developed in the future, can really only be used for that purpose, or for dealing with other similar explosive devices. The technology can only be used defensively, to prevent people being harmed by these explosives. Technology to remove debris from space is quite different, however, since any technology which can be used to remove debris from space can be used on other objects in space as well. The big political problem of space debris removal is simply that any technology which can be used to remove debris in space can also be used as a weapon, in most cases (perhaps all cases) without even requiring any modification. Any technology which can be used to capture a defunct satellite or remove such a satellite from orbit could also be used on an active satellite. Any technology which can be used from a distance to influence the orbit of a satellite or piece of debris can be used in a disruptive manner just as easily as it can be used in a productive manner, so any technology which could be used to decrease the possibility of a collision in orbit could also be used to increase such a possibility. While space debris removal technologies are not the only space technologies which raise these sorts of issues – the technology for on-orbit servicing of satellites could also be used for hostile purposes – those involved in attempts to remove debris are perhaps less likely to appreciate the possible hostile ways in which removal technologies could be used. The removal of space debris would benefit everyone, so those working on debris removal technologies may well see themselves as working for the common good rather than for the benefit of any particular state, an ideal which has the potential to blind developers of space debris removal technologies to the military possibilities of the technologies they develop.

In December 2019 the USA officially created a new branch of the military, Space Force,[13] and President Donald Trump explicitly stated that for the USA "Space is a war-fighting domain, just like the land, air, and sea".[14] Given this viewpoint it is no surprise that the USA in particular, as well as other major powers like Russia and China, tend to view space debris removal technologies as weapons in the first instance, with their secondary use being the removal of space debris, rather than the other way around. This is especially the case when considering the technologies being developed, or already possessed, by their rivals in space.

---

13 The Bill which created Space Force was signed into law on 20th December 2019, with General John "Jay" Richmond being officially sworn in as the first Chief of Space Operations on 14th January 2020.
14 Speech to Service Members at Marine Corps Air Station Miramar, 13 March, 2018.

Humanity faces many "problems of the commons"; problems to which everyone has contributed, which affect everyone on the planet to some extent. Climate change and oceanic pollution are classic examples of such problems. These sorts of things are difficult and/or expensive to deal with, and states don't want to act unilaterally to deal with such problems since they don't want to bear the expense of cleaning up the problem for other states. With most common problems states are happy for others to act unilaterally and accept the expense of dealing with the problem. So, if state X was to develop a method of cleaning up oceanic pollution (such as the Great Pacific Garbage Patch) and was willing to pay to implement that clean up, other states would be happy to see the problem dealt with and to reap the benefits of that clean up, especially if state X was going to bear all the expense.

No one objects to the development of better technology to remove landmines, in the same way that no one would object to a group developing technology to clean up the plastics in the Great Pacific garbage patch. Neither of these technologies poses a threat. But while space debris might be considered garbage in the same way that the Pacific plastics are garbage, any new technology developed to remove the garbage of space debris would be considered quite differently. As a problem of the commons, space debris is perhaps unique, since this is a problem which states DON'T want other states to be able to deal with. This is because the dual use nature of space debris removal technologies would mean that any state able to deal with the problem of space debris would therefore have a military edge in space.

Looking for a Solution

Any technology created to clean up space debris could also be used as a weapon. It is this simple fact, perhaps more than anything else, that ends up making space debris even more difficult to deal with than landmines. However, the problems involved with the removal of landmines do suggest one lesson which we would do well to heed when looking at the problem of space debris. It is possible to remove landmines, but it is slow, expensive and dangerous, and the better solution to the problem is to not allow landmines to be used in the first place. If the landmines can't be removed, the damage they do can be minimised by not allowing people into areas which have been mined. Space debris is even more difficult to deal with than landmines, and space debris cannot be contained or cordoned off. Thus, the best solution is to avoid creating debris in the first place.

The political problems which are raised by space debris removal, and specifically by the dual use nature of the technologies which could be used to remove that debris, are difficult but not impossible to deal with. Given the serious risks to future space operations that space debris poses, some solution does need to be found, and with that in mind here are some proposed first steps towards that solution.

If a state wished to remove from orbit objects that state had placed there, and to which the state held legal title under the OST, there are no legal or political impediments to such removal other than the fact that demonstrating the capability of removing such materials from orbit might increase political and military tensions with rival state powers. But removing any material that is the legal property of another state is a very different matter. In considering this sort of issue, formal international agreements regarding the removal of materials from orbit, while not absolutely essential, would be of great assistance in dealing with this problem. These agreements would probably begin by acknowledging the need for space debris removal, but the most important parts would be agreements on the type of material which could/should be removed from orbit by parties other than the legal owner of the material, as well as agreements on the type of material which MUST not be interfered with. Obviously currently active satellites would fall into the category of orbiting objects which must not be interfered with, but we are sure that the major space powers would insist that many more objects ought to fall into this category as well. Recently launched satellites which are not active due to a failure to properly deploy, for example, are likely to be something which only the state owner ought to be interfering with, since these satellites might well contain important proprietary technology.

Formal agreements on the need to remove debris from Earth orbit might plausibly also incorporate some changes to the legal status of particular space objects. For example, at this time, in accordance with the OST, all material in space remains the property of a state in perpetuity (even if the state itself is defunct), but perhaps it might be reasonable to suggest that material which has been inactive in space for more than a set number of years might have its legal status changed so that it is considered legally abandoned and thus able to be freely removed by anyone with the capability of doing so.

Even in the absence of formal agreements which might help to regulate the use of space debris removal technologies, some informal guidelines for the use of such technologies seem to be required in order to prevent an escalation of tension which might lead to a potentially catastrophic war in space. Having said this, it should be noted that different types of debris removal technologies require somewhat different guidelines for use. All debris removal technologies are able to be weaponised, so debris removal technologies which could be used against a wide range of potential targets, such as technologies which could directly affect large active satellites (for example, those designed to capture or decelerate large pieces of space junk) are more obviously problematic than those debris removal technologies that could only be used in limited ways or against a limited range of targets, such as technologies which can only affect smaller pieces of debris (for example, relatively low powered Earth based lasers). Military forces will, of course, be concerned about any weapons which might be used against them, but they will be more concerned about weapons which can permanently destroy or disable a wide range of vital military assets, than about weapons which only have temporary effects or can only be used against a few targets.

Given that technologies which can (a) be used against large pieces of debris are much more of a military threat than technologies which can (b) only be used against small pieces of debris, it seems clear that while it is always prudent to be cautious in the use of such technologies, more caution is needed when using type (a) technologies than when using type (b) technologies, and thus it is quite plausible that type (a) technologies will also require more guidelines for use than type (b) technologies do. Nevertheless, the basic guidelines for the use of all types of space debris removal technologies seem to be the same, and that is all that will be discussed in this chapter.

Conclusion: Basic Guidelines for the Use of Space Debris Removal Technologies

There are three basic guidelines which need to be applied to the use of debris removal technologies if one wishes to use such technologies without dramatically increasing the political tension between the major space powers and thereby risking a potentially catastrophic war in space. The first of these is transparency of use. Given the potential for their weaponisation, the mere existence of debris removal technologies will almost inevitably increase tension between the major space powers. This is the case even if those technologies are developed and used by other parties, such as corporations or other less powerful space-faring states. But secrecy surrounding the use of these technologies will dramatically increase that tension. If operators of space debris removal technologies keep their operations a close secret, then if an unexpected and/or unexplained incident was to occur in orbit to a satellite or other piece of equipment belonging to one of the major space powers, it is almost inevitable that questions would be asked about whether the operator of the debris removal technology was involved in same way. Thus, if these technologies are to be used for their intended purpose, the removal of space debris for the benefit of all users of space, those using them will need to adopt a policy of absolute transparency, similar to open source software, so that space powers will always be able to verify how and when the technology is being used, and thus assure themselves the technology is not being used for nefarious purposes.

The second major guideline for the use of debris removal technologies exists to ensure that such technologies are not inadvertently used in a manner which causes alarm to active space powers. This is the suggestion that operators of such technologies respect a reasonable "safe zone" around the active satellites of other states, especially military satellites. This is a particular concern for technologies which are actually placed in orbit, such as debris removal satellites which need to be maneuvered close to a piece of debris before capturing it, but would also apply to the use of other technologies as well. Space assets are vital to modern military operations, but these assets are also highly vulnerable, are impossible to replace quickly, and it is very difficult for states to respond rapidly to protect such assets from perceived threats. Given these factors, if debris removal technologies are ever used in the close vicinity of vital military space assets there is a serious risk this will be viewed as a threat to a vital military asset of the state; such a situation might well induce a highly hostile response from the state in question.

The third and final suggested guideline recommends operators of debris removal technologies exercise extreme caution not only in where and how such technologies are used, but when such technologies are used. As is evident throughout history, during times of heightened political tension unwary actions can have massive effects. The conduct of military exercises near a disputed terrestrial border may be relatively inconsequential during times when cordial diplomatic relations are being maintained between the states in question; such actions are likely to be seen as little more than political posturing. But conducting similar exercises at a time of dramatically increased tension between those same states would be a very different matter; such actions would be thought to be extremely provocative, possibly even tantamount to a declaration of war. As was mentioned a moment ago, for modern military operations space assets are vital, yet vulnerable. Since it is difficult for states to respond rapidly to protect these vital assets from perceived threats, the injudicious use of debris removal technologies during times of heightened political tension, even when such use is NOT in the vicinity of important space assets, is likely to be perceived as a military threat, even if only as an implied one. Thus, operators of debris removal technologies need to always be conscious of the international political environment within which such technologies are being operated, lest incautious use trigger a catastrophic response.

The common theme of all of these guidelines is, of course, caution. In this case, caution with regard to how such technologies are used, and where, and when. One of the fundamental precepts of military planning is to hope for the best, but plan for the worst. The military forces of the world's leading powers now rely on space based assets in almost all of their operations, so they are highly wary of any potential threats to those assets. Every type of space debris removal technology can be weaponised so the use of these technologies will inevitably be perceived, at minimum, as a potential military threat, regardless of the intentions of the operator of that technology.

Space debris removal technologies can be developed and used without causing a war, but only if the developers and operators of such technologies are aware of the potential problems and act at all times in a manner which reduces the threat, both real and perceived, that these technologies produce. Lack of caution when using debris removal technologies raises the risk of rapidly escalating retaliatory military responses and of all-out war in Earth orbit, a situation which has potentially disastrous consequences for generations to come.

References

BBC News, Concern over China's Missile Test, 19 January 2007, http://news.bbc.co.uk/1/hi/world/asia-pacific/6276543.stm.
BBC News, ISS crew take to escape capsules in space junk alert, 24 March 2012, http://www.bbc.com/news/science-environment-17497766.
Kessler, D.J. and B.G. Cour-Palais, Collision Frequency of Artificial Satellites: The Creation of a Debris Belt, *Journal of Geophysical Research*. 83 (1978) 2637–2646.

Singer, P.W. and A. Cole. *Ghost Fleet: A Novel of the Next World War*. Boston, Houghton Mifflin, 2015.

Swarts, P. Space is seen as Increasingly Important to Military Operations, "Air Force Times" 12 September 2015, https://www.airforcetimes.com/news/your-air-force/2015/09/12/space-is-seen-as-increasingly-important-to-military-operations/.

United Nations, Convention on Prohibitions or Restrictions on the Use of Certain Conventional Weapons Which May be Deemed to be Excessively Injurious or to Have Indiscriminate Effects (and Protocols) (As Amended on 21 December 2001), 10 October 1980, 1342 UNTS 137, http://www.refworld.org/docid/3ae6b3ac4.html.

United Nations, Convention on the Prohibition of the Use, Stockpiling, Production and Transfer of Anti-Personnel Mines and on Their Destruction, 18 September 1997, http://www.refworld.org/docid/3ae6b3ad0.html.

United Nations Office for Outer Space Affairs, Treaty on Principles Governing the Activities of States in the Exploration and Use of Outer Space, including the Moon and Other Celestial Bodies aka Outer Space Treaty (10 October 1967), 610 UNTS 205/6 ILM 386 (1967)/[1967] ATS 24, https://cil.nus.edu.sg/1967/1967-treaty-on-principles-governing-the-activities-of-states-in-the-exploration-and-use-of-outer-space-including-the-moon-and-other-celestial-bodies/.

UN press releases, General Assembly Adopts 63 Drafts on First Committee's Recommendation with Nuclear Disarmament at Core of Several Recorded Votes, https://www.un.org/press/en/2014/ga11593.doc.htm.

6

ETHICAL CONSIDERATIONS ON THE CHALLENGES OF THE DUAL-USE SATELLITE PROBLEM

Amy Hestermann-Crane[1]

In the twenty-first century much of society relies heavily upon satellites for essential services such as navigation, emergency beacon location, food distribution, health systems, environmental protection, communication, and education. Many of these satellites are considered dual-use,[2] with customers residing in both the military and civilian spheres.[3] The dual-use nature of satellites and the environment in which they operate, combined with their essential function within society prompts the question of whether satellites should have a protected status in war and war-like conflict, similar to hospitals and power stations. Understanding this context of dual-use satellite integration within society is pivotal in grasping the importance of the need for protected status for the continued betterment of the global community. This chapter will examine this dual-use problem in conjunction with ethical considerations on military space activities. This includes an overview of current potential difficulties faced when conducting these operations against dual-use satellites and their possible impacts on continued civilian use of a satellite. Furthermore, a consideration of the ethical responsibilities placed upon a State over their space launch corridors and the space environment will be undertaken. In

---

[1] The views expressed are those of the author and do not reflect the official policy or position of the Royal Australian Air Force, the Department of Defence, or the Australian Government.
[2] Dual-use satellites are those that can be used for civilian/commercial and military purposes.
[3] Jinyuan Su and Zhu Lixin, "The European Union draft Code of Conduct for outer space activities: an appraisal", *Space Policy* 30 (2014): 34; Civilian spheres in this chapter will be used to encompass both non-military government use and commercial companies unless stated otherwise.

addition, the potential for the State and commercial companies to use a virtue ethics cost-benefit analysis framework for future decision making and policy creation will be explored. The transition to a virtue ethics cost-benefit analysis is designed to ensure that the cost of future endeavours upon the global community are considered beyond monetary means.

The history of space is founded in military expansion, even before the launch of the first satellite, Sputnik, in October 1957.[4] Near-earth orbits were first utilitised as a military domain with the U.S. and Russia launching imaging, position, and communication satellites specifically for military use as far back as the late 1950s.[5] However, the concerns surrounding the increased military use of space are rightly justified. The global community relies on many dual-use satellites for benefits in numerous facets of life including emergency beacons, education, health care, environmental protection, and disaster prevention and identification.[6]

It is important to establish that even from the beginning, military and civilian cooperation was necessary to "capitalise" on technical expertise and the avoidance of effort duplication.[7] This encompasses scenarios such as launch vehicle use, telecommunication services providing both dedicated civilian and military frequencies through the same satellite, and services where the same product can be given to both civilian and military customers such as mapping or earth observation data.[8] This integration is not unique to the space domain, with governments such as the United States, relying on civilian/military integration in other areas such as national security.[9]

Developing a code of ethics for the use of space and our continuing exploration is an enormous task. However, it should not seem like a foreign concept to require such a code. Ethics affects every aspect of living, with the majority of professions holding ethical codes of conduct; medicine,

---

4 Elizabeth S. Waldrop, "Integration of military and civilian space assets: legal and national security implications" *Air Force Law Review* 55 (2004), 159; Neil DeGrasse Tyson and Avis Lang. *Accessory To War: The Unspoken Alliance Between Astrophysics and the Military* (New York: W.W. Norton & Company, 2019), 32-33.
5 Waldrop, "Integration of military", 159.
6 Cassandra Steer, "Global commons, cosmic commons: implications of military and security uses of outer space", *Journal of International Affairs* 18 (2017), 9; NASA, "Emergency systems save tens of thousands of lives", *NASA Spinoffs*, 2012, https://spinoff.nasa.gov/Spinoff2012/ps_3.html.
7 Waldrop, "Integration of military", 162-163; DeGrasse Tyson and Lang. *Accessory To War*, 20-21; Joan Johnson-Freese, *Space as a Strategic Asset* (New York, Columbia University Press, 2007) 28.
8 Waldrop, "Integration of military", 162-163.
9 Waldrop, "Integration of military", 164-165.

journalism, militaries, and often commercial businesses.[10] "Space ethics" is an all-encompassing term, where one can compartmentalise segments for consideration such as dual-use satellites with a focus on their impact on society, our actions in space in relation to and towards these satellites, and on the space and terrestrial environments. While the argument has been raised that not all elements of space development demand our "moral approbation",[11] a look at the integrated nature of dual-use satellites and military conflict will show this facet of space development demands ethical standards.

The militarisation of space has already occurred and will continue to occur as the space domain is integrated into every other military domain and offers an unquestionable edge in operations.[12] The question is not *if* the military should use space, but *how* should military operations be limited within the space domain. Historically, governments have focused on the "non-aggressive militarisation of space" compared to the "weaponisation of space".[13] The balance of this concept holds that a commercial, dual-use, or military imagery satellite used for military operations can be considered non-aggressive military use, compared to a military command satellite that is armed with interceptor missiles being "weaponised".[14]

To fully appreciate the need for this class of satellites to gain protected status one must understand the magnitude of dual-use satellite integration into civilian life, military operations, and most importantly, the impacts these services provide towards global betterment.

---

10 Australian Medical Association, 2017, *New code of ethics for doctors*, https://ama.com.au/media/new-code-ethics-doctors; Australian Army, n.d, *Our Values & Contract*, https://www.army.gov.au/our-people/our-values-contract; College of Policing, n.d, *Code of Ethics*, https://www.college.police.uk/ethics/code-of-ethics; Media, Entertainment & Arts Alliance, 2021, *Journalist Code of Ethics*, https://www.meaa.org/meaa-media/code-of-ethics; Microsoft, n.d., *Compliance and Ethics at Microsoft*, https://www.microsoft.com/en-us/legal/compliance.
11 James S. J. Schwartz, "Prioritizing scientific exploration: a comparison of the ethical justification for space development and space science", *Space Policy* 30 (2014), 203.
12 Steer, "Global commons", 11; Todd Harrison, "The Future of Security in Space", in *Air Power in a Disruptive World: Proceedings of the 2018 Air Power Conference*, ed. Department of Defence (Canberra, National Library of Australia), 44.
13 The author acknowledges that difference between these terms is difficult to clearly define and is a slippery concept in and of itself; where does a satellite used for missile strike guidance or relaying of military communications sit on the continuum between non-aggressive use and weaponisation? The scope of this chapter does not touch on this topic.
14 DeGrasse Tyson and Lang. *Accessory To War*, 273-274.

Dual-use Satellite Services Overview

Global Positioning System

One of the first and well-known dual-use services is the global navigation satellite system, Global Position System (GPS). Starting as a military technology with a lot to prove, GPS has quickly integrated into both military and civilian life after reaching operational capability in the mid 1990s.[15] GPS is no longer the only global navigation satellite system (GNSS), with Russia, China, and the European Union completing or near completion of their own GNSS constellations.[16] Despite these constellations, GPS is still the dominant GNSS service either partially or wholly relied upon.[17] Aside from preventing people getting lost on city streets, the project provides numerous important services to the global community and environmental conservation efforts. It is these additional services that truly require protected status during conflict, as their destabilization within society would cause detriment to the human population if removed.

Emergency locator beacons are an essential GPS service. Initiatives such as COSPAS-SARSAT, an international constellation of distress signal trackers, has significantly improved distress beacon location.[18] These distress signal trackers are particularly useful for people who are in need of rescue in remote, difficult to reach, or contested environments and would not be easily found otherwise. Beacons with GPS significantly improve signal location accuracy, where beacons with GPS are accurate to 120m compared to the 5km accuracy of non-GPS enabled beacons.[19] The COSPAS-SARSAT infrastructure aided in saving over 30,000 lives since becoming globally operational in 1985 via decreasing location detection times and increasing accuracy in beacon location.[20]

---

15 Johnson-Freese, *Space*, 39-41; DeGrasse Tyson and Lang. *Accessory To War*, 332-336.
16 Air Power Development Centre, *AFDN 1-19 Air-Space Integration* (Canberra, Royal Australian Air Force, 2019), p. 32; DeGrasse Tyson and Lang, *Accessory To War*, 337-339.
17 DeGrasse Tyson and Lang, *Accessory To War*, 339.
18 NASA, "Emergency systems".
19 Australian Maritime Safety Authority, "Why is GPS best", *Beacons*, n.d., https://beacons.amsa.gov.au/purchasing/GPS-best.asp.
20 NASA, "Emergency systems".

Animal research and conservation is commonly aided via GPS tracking.[21] Tracking of sea turtles has led to the identification of critical habitats, providing data to back the implementation of marine protected areas.[22] Animal conservation on the land has been aided from animals as varied as elephants to wild maned wolves, with GPS data aiding in research of animals that are difficult to observe by other means.[23] This information is used to understand the animals' needs, attempt to clear areas of illegal poachers or smugglers, and locate animals potentially in need of care. These efforts tie into humanity's moral obligation to protect the global environment and reverse the significant damage already inflicted upon it.[24] Acts benefiting the environment will be acts that lead to our individual and collective well-being.

Additionally, the military uses of GNSS have evolved from its navigation origins to precision weaponry. Missiles such as the JASSM (Joint Air to Surface Standoff Missile) use GPS to track over great distances, allowing its operator to fire and forget, leaving dangerous conflict zones sooner while still completing missions.[25]

With GPS and other GNSS being used for precision strike capabilities, force navigation, and situational awareness, why should a State allow use of this force enabler to hostiles? Diminished capabilities would allow most civilian uses to continue. Yet this lowered accuracy could have greater collateral damage to one's own people from imprecise weapon strikes. Commercial situational awareness is also important along major airline routes.[26] On the other hand, locking out users from the hostile state would maintain your own military and civilian uses with no degradation. This

---

21 Emily C. Mills, John R. Poulsen, J. Michael Fay, Peter Morkel, Connie J. Clark, Amelia Meier, Christopher Beirne, and Lee J. T. White, "Forest elephant movement and habitat use in a tropical forest-grassland mosaic in Gabon." *PLoS ONE* 13, no. 7 (2018): 2. doi.org/10.1371/journal.pone.0199387; Robin T. E. Snape, Phil J. Bradshaw, Annette C. Broderick, Wayne J. Fuller, Kimberley L. Stokes, Brendan J. Godley, "Off-the-shelf GPS technology to inform marine protected areas for marine turtles", *Biological Conservation* 227 (2018): 301-302.
22 Snape, Bradshaw, Broderick, Fuller, Stokes and Godley, "Off-the-shelf", 301-302, 306; Carlyle Mendes Coelho, Luiz Fernando Bandeira de Melo, Marco Aurélio Lima Sábato, Dália Nogueira Rizel and Robert John Young, "A note on the use of GPS collars to monitor wild maned wolves Chrysocyon brachyurus (Illiger 1815) (Mammalia, Canidae)", *Applied Animal Behaviour Science* 106, no. 1-3 (2007): 260.
23 Mills, Poulsen, Fay, Morkel, Clark, Meier, Beirne and White, "Forest elephants", 2; Coelho, de Melo, Sábato, Rizel and Young, "A note", 260.
24 Matthew Beard, "Space: the final ethical frontier", *Ethics*, September 24, 2020, https://ethics.org.au/space-the-final-ethical-frontier.
25 Lockheed Martin, "JASSM: Dependable, affordable, long-range strike cruise missile", JASSM, n.d., https://www.lockheedmartin.com/en-us/products/jassm.html.
26 airline navigation.

tactic would also remove the force enabler completely from your hostiles' capabilities, which may prevent long-distance weapon use entirely for certain systems.

Yet what negative effects could GNSS denial have on the global community? In economics alone, the June 2019 RIT International economics benefits of GPS report stated that the global loss of GPS could cause financial losses as high as $1 billion dollars per day.[27] On top of this, reduced agricultural harvests affects international food aid to third world countries, seriously harming vulnerable communities.

Imaging and Earth Observation Satellites

Imaging satellites also tie in with search and rescue efforts, such as in the 2010 Haitian earthquake disaster. The multinational rescue team employed the mapping service OpenStreetMaps to coordinate efforts across the island and fuse mapping and social media data.[28] During the crisis, user input continually updated search and rescue personnel on affected areas and rescue efforts, thus maximizing resources and lives saved.[29] Examples such as this illustrate that the importance of imaging satellites is proportionate to the COSPAS-SARSAT network. Imagery from both dedicated reconnaissance and dual-use imagery satellites are also used for mission planning.

Earth observation satellites are also used globally to monitor environmental factors from coastal and soil erosion to a range of natural disasters within the global community such as severe weather warnings, fires, flooding, tsunamis, and more.[30] Earth observation data is able to be used in all stages of the disaster risk management cycle; alerting communities during prevention to minimise impact, preparing emergency services,

---

[27] Alan C. O'Connor, Michael P. Gallaher, Kyle Clark-Sutton, Daniel Lapidus, Zack T. Oliver, Troy J. Scott, Dallas W. Wood, Manuel A. Gonzalez, Elizabeth G. Brown and Joshua Fletcher, "Economic Benefits of the Global Positioning System (GPS) Final Report", *RIT International*, June 2019.
[28] Patrick Meier, 2012, "How crisis mapping saved lives in Haiti", *National Geographic Society Newsroom*, July 2, 2012, https://blog.nationalgeographic.org/2012/07/02/how-crisis-mapping-saved-lives-in-haiti.
[29] Meier, "How crisis".
[30] G. Le Cozannet, M. Kervyn, S. Russo, C. Ifejika Speranza, P. Ferrier, M. Foumelis, T. Lopez, and H. Modaressi, H. "Space-Based Earth Observations for Disaster Risk Management." *Surveys in Geophysics* 41, no. 6 (2020): 1210.

aiding in timely responses, and assisting with recovery operations.[31] This data enables communities to maximise time to prepare, evacuate and undergo rescue operations, allocating resources most efficiently to save lives.

Satellite Telecommunications and Internet Services

Satellite telecommunications bring significant benefits to the global community. Telecommunication satellites form much of the remote educational infrastructure, capable of providing education to developing nations.[32] India's Kerala Infrastructure and Technology for Education (KITE) initiative provides virtual classrooms offering quality higher education to students in remote villages without access to technical institutes, adult literacy programs or training modules.[33] Satellite education programs can be found across the globe from Columbia and Peru to India, and Australia.[34] Nations like Australia have numerous rural and remote communities, which are capable of suffering from educational marginalization.[35] Satellite internet has aided in the provision of education to these communities, including vocational education to adults[36] and are a vital service in ensuring the widest possible access to education across the global community. It is important to note that increased education is a proven method in combating poverty.[37]

Satellite communication infrastructure is also used to improve healthcare systems in remote locations for multiple nations, including Australia. One such example, InStrat Global Health Solutions brings mobile and

---

[31] Cozannet, Keyyyn, Russo, Ifejika Speranza, Ferrier, Foumelis, Lopez and Modaressi, "Space-Based", 1213-1223.
[32] Internet Business News, 2018, "Avanti Communications Satellite to Deliver Digital Education to African Schools", February 07, 2018, ink.gale.com/apps/doc/A526608606/ITOF?u=dixson&sid=ITOF&xid=e2cbd8e2; Imran A. Zualkernan, Shirin Lutfeali, and Asad Karim, "Using Tablets and Satellite-based Internet to Deliver Numeracy Education to Marginalized Children in a Developing Country", *IEEE Global Humanitarian Technology Conference* (2014): 294.
[33] Kerala Infrastructure and Technology for Education, "About Us", https://kite.kerala.gov.in/KITE/index.php/welcome/about_us.
[34] Zualkernan, Lutfeali, and Karim, "Using Tablets", 294; Kerala Infrastructure and Technology for Education, "About Us"; Kylie Twyford, Stephen Crump, and Alan Anderson, "Satellite lessons: vocational education and training for isolated communities", *Rural Society* 19, no. 2 (2009): 127.
[35] Twyford, Crump and Anderson, "Satellite lessons", 127.
[36] Twyford, Crump and Anderson, "Satellite lessons", 128, 133-134.
[37] Global Education Monitoring Report Team, *Global Education Monitoring Report 2020* (Paris, United Nations Educational, Scientific and Cultural Organisation, 2020).

satellite technologies to health facilities across Nigeria.[38] InStrat Global Health Solutions and similar services have improved record keeping, increased treatment hours, and provides remote health workers with video training to increase health care quality.[39] Satellite communication is also able to play a role in providing medical care in the aftermath of natural disasters.[40] This would involve reserved channels for emergency services, however, nations like Japan are already doing this for other emergency services such as the fire department.[41] Satellite communication health services constitute essential infrastructure involved in increasing quality of life across the global community.

These satellite telecommunications and Internet services also benefit the military. Most obviously through establishing communication infrastructure to remote locations where the military is able to construct extended forward operating bases and protection of forward troops.[42]

Protected Status and Proportionality Assessment

We know that protected status shields physical hospitals, nuclear power plants, religious buildings, and even the environment in times of conflict.[43] Efforts are also to be made to protect civilian objects so long as they hold no military advantage for the adversary.[44] However, the previous examples highlight the direct role dual-use satellites can play in saving lives, raising education standards, increasing quality of life, and protecting the global environmental. The immediate and potentially life-threatening impact that satellite destruction or degradation could cause are some of the reasons why military necessity is unlikely to outweigh civilian suffering caused by this action during conflict proportionality assessments.[45]

---

38 InStrat Global Health Solutions, "About Us", accessed March 13, 2021. https://instratghs.com/about-us/.
39 InStrat Global Health Solutions, "About Us"; Devex, 2018, "Satellites for survival: saving lives in Nigeria", Uploaded in April 07, 2018, Youtube video, 4:46min. https://www.youtube.com/watch?v=4Lq3g1l9Nh0&t=69s.
40 Kiyoko Nagami, Isao Nakajima, Hiroshi Juzoji, Kiyoshi Igarashi and Kenji Tanaka, "Satellite communication for support medical care in the aftermath of disasters", *J Telemed Telecare* 12, no. 6 (2006): 275. doi: 10.1258/135763306778558213.
41 Nagami, Nakajima, Juzoji, Igarashi and Tanaka, "Satellite communication", 275.
42 Air Power Development Centre, *AFDN 1-19 Air-Space Integration*, 28-29.
43 Australian Defence Force, *ADDP 06.4 Law of Armed Conflict* (Canberra, 2019), 5-11 – 5-13.
44 Australian Defence Force, *ADDP 06.4 Law of Armed Conflict*, 5-9-5-11.
45 Australian Defence Force, *ADDP 06.4 Law of Armed Conflict*, 2-2, 2-4.

Proportionality assessments are more difficult when looking at diminishing or locking capabilities. This course of action either temporarily locks out customer use or enables services to be usable at either a degraded capacity or by selected users, such as devices geolocated to a State's own territory. It could be argued that emergency location services could not ethically be trumped by a state-on-state conflict, but what of other GPS uses? Other concerns would be sudden jamming in high traffic areas, such as within the South China Sea, affecting commercial aviation and naval services.

Similar benefits and consequences are raised with communication and internet service satellites. The effects of limiting these satellites are more difficult to justify when contrasted against the global society's health care services and potential loss of life. A government would need to determine how reliant hostile forces were on satellite communications compared with other means such as HF radio, mobile networks, or other land-based infrastructure. Each satellite would need to be assessed for its potential to affect vital health-provisions, maritime and emergency communication, and educational services.

This assessment would also have to be conducted on blue force use overriding civilian channels. In the current environment, the need for military uses to override civilian channels is minimal. However, as positions for new satellites in Earth's orbits become more competitive and availability of channels decreases, commanders and service providers need to be as equally careful about commandeering civilian channels. In both these circumstances, the proportionally ethical solution is most likely narrow, targeted diminishing of services, so long as due diligence was performed on assessment.

Ultimately, various terrestrial civilian infrastructure and objects important to daily function or of cultural significance are protected under international humanitarian laws.[46] Ethically and legally, civilians are to be protected as much as possible from the impacts of conflict. One part of that is ensuring essential services to their quality of life are protected; powers, water, health care, and education. This begs the question; are physical schools more valuable than digital schools? Or do virtual education channels deserve the same protections as physical ones? What of health care services? It is argued here that these virtual services are equally as important to many communities across the globe. As these satellite-based

---

46 Australian Defence Force, *ADDP 06.4 Law of Armed Conflict*, 5-9-5-13.

methods are the *only* means of service to these communities, they should be afforded the same level of protection that their physical counterparts are granted during conflict.

However, caveats would apply. Protected status is only granted while the aforementioned facilities are being used in their designated function, similar to other civilian objects. Once a school or other protected building starts being used as a combatant base of operations, or for military means, the facility becomes a lawful target in conflict.[47] The same clause should apply for the space-based versions of these services. It is an unfortunate reality to accept that militaries and other combatant groups take advantage of these protected buildings to further their own causes.

The issue of proportionality assessment has no simple answer, with policy writers facing the challenge of appropriate wording. The danger arises in focusing on extreme cases, instead of writing moderate policy. Cases such as State vs State, or asymmetric conflicts, need to be written into policy as exceptions to the norm, with careful wording ensuring that misuse of State control privileges are minimized.

Military Operations that Affect the Space Domain

Militaries across the globe have begun to integrate the space domain into operations in increasingly complex ways.[48] Mission planning integrates various services, such as imaging services for situational awareness and Earth observation satellites to establish weather. Satellite communication services are used for information transfer, emergency beacon locators can be utilities, and of course GPS for navigation and advanced weaponry. Yet this reliance on space-based assets is not secret, and with all military advances, countermeasures are developed. It is these counter-space assets that have the potential to cause damage to services and the space environment as a whole.

Counter-space weaponry has not been kept secret amongst the global community. Perhaps this information is shared as a means of deterrence from other actors. Regardless, the capabilities of these weapons, both advertised and demonstrated, is troubling. Further, counter-space capabilities and space defence policy rhetoric exists in numerous nations

---

47 Australian Defence Force, *ADDP 06.4 Law of Armed Conflict*, 5-9.
48 Harrison, "The Future", 44.

such as the United States, Russia, Japan, China and India.[49] These weapons fall broadly into non-kinetic, which are generally non-kinetic Electronic Warfare (EW) weapons, and kinetic anti-satellite weapon systems (ASAT).

The majority of EW is jamming or spoofing weapons, those which degrade or limit a service for a period of time.[50] These attacks are designed for a specific radio frequency range and must be within the field of view of the sensor they wish to target.[51] Spoofing also uses a radio frequency designed to target a specific sensor.[52] However, instead of disrupting the signal, this EW countermeasure aims at tricking the target into locking on to the false signal instead of the legitimate user.[53] These forms of attack are entirely reversible, only causing service issues while in use.[54] The issues for militaries here would be ensuring that proportionality assessments and target development were thoroughly completed to minimise harm to global communities.

There are forms of destructive non-kinetic attacks, primarily through direct energy weapons. High-powered laser systems are a type of direct energy system.[55] Unlike jamming, high-powered lasers can cause either temporary or permanent damage, with temporary "dazzling" considered a countermeasure instead of a weapon system.[56] These weapons can dazzle (temporary) or blind (permanent) optical satellites through damaging critical components within the satellite payload's sensors. China claimed to have blinded a satellite in 2005, and to dazzle American satellites when they pass overhead.[57] While the Chinese Government has made these claims, it must be noted there is no verifiable evidence available in the public domain to confirm this and America has not responded to the claim.[58]

Kinetic attacks are of greatest concern, for these attacks attempt to damage or destroy both orbital and terrestrial space assets. Direct-ascent anti-satellite weapons (DA-ASATs) are ground, air, or sea launched missiles

---

49 Steer, "Global commons", 9; Todd Harrison, "The Future", 44.
50 Air Power Development Centre, *AFDN 1-19 Air-Space Integration*, 68; CSIS, "Counterspace Weapons 101".
51 CSIS, "Counterspace Weapons 101"; Brian Weeden and Victoria Samson, 2021, *Global Counterspace Capabilities: an open source assessment*, Secure World Foundation xvii.
52 CSIS, "Counterspace Weapons 101".
53 CSIS, "Counterspace Weapons 101".
54 CSIS, "Counterspace Weapons 101"; Weeden and Samson, 2021, *Global Counterspace Capabilities*, 2-28.
55 Weeden and Samson, 2021, xxxi.
56 Weeden and Samson, *Global Counterspace Capabilities*, 1-22.
57 CSIS, "Counterspace Weapons 101", *Aerospace Security Project*, October 28, 2019, https://aerospace.csis.org/aerospace101/counterspace-weapons-101.
58 CSIS, "Counterspace Weapons 101".

and cause irreversible damage to satellites through force of impact.[59] Additionally, a DA-ASAT creates large orbital debris clouds which can cause damage or destruction to non-target satellites.[60] This form of attack also risks triggering the Kessler Syndrome, with a cascading effect of satellite damage and unavoidable orbital impacts. This is increasingly true for attacks at higher altitude levels, where the decrease in gravitational pull allows debris clouds to remain in orbit for increasing period of time before re-entry burn.

China demonstrated this risk in 2007, conducting a destructive DA-ASAT test on an aged meteorological satellite.[61] Over 3000 trackable debris pieces were created from the event, with many still threating objects in LEO to this day. India followed suit in March 2019. Although Indian officials claimed the debris field "would not last long", the fact remains that there was at least 50 pieces of trackable debris by end of September 2019. These pieces, and the unknown amount of untraceable debris, remain a constant threat to on-orbit assets. These kinetic tests may be legal, but they are exceptionally unethical. The debris clouds cannot be controlled, and the actions of these tests have the potential to indiscriminately damage or destroy other satellites. The other concern is that the acceptance and established precedence of these tests could encourage other nations to perform kinetic ASAT attacks, further increasing the risk to the global community.

Russia has seemed to find a middle ground on kinetic ASAT testing. In 2020, Russia conducted three DA-ASAT missile tests which did not strike a satellite.[62] However, the international community still condemned these attacks as irresponsible. There is not much that needs to go wrong for a DA-ASAT test to cause a massive debris cloud within LEO, risking the space environment.

There is also concern of on-orbit/co-orbital ASAT satellites. America has claimed that Russia possesses one such satellite, with evidence

---

59 CSIS, "Counterspace Weapons 101"; Weeden and Samson, 2021, *Global Counterspace Capabilities*, xxxi.
60 CSIS, "Counterspace Weapons 101"; Weeden and Samson, 2021, *Global Counterspace Capabilities*, xxxi.
61 CSIS, "Counterspace Weapons 101"; Weeden and Samson, *Global Counterspace Capabilities*, 1-2.
62 Hanneke Weitering, "Russia has launched an anti-satellite missile test, US Space Command says", *Space*, December, 2020. https://www.space.com/russia-launches-anti-satellite-missile-test-2020; *The Barents Observer*, "Russia tests anti-satellite weapon from Plesetsk", *The Moscow Times*, December 17, 2020. https://www.themoscowtimes.com/2020/12/17/russia-tests-anti-satellite-weapon-from-plesetsk-a72398.

suggesting Russia has developed a co-orbital ASAT program called Burevestnik.[63] Russia has performed several close proximity operations in geostationary orbit (GEO) and Low Earth Orbit (LEO).[64] Russia announced the event as a successful test of its inspection satellite, designed to inspect other satellites for damage. However, the United Kingdom and America released statements that the events were tests for Russia's on-orbit ASAT satellites, and that the actions were dangerous and could cause harmful debris. China has also conducted rendezvous operations in LEO and GEO, with the first occurring in 2010.[65] Of note, there is no public evidence to suggest that China has conducted a destructive intercept of a satellite or that China's close approach and rendezvous technology is being developed for counterspace use.[66]

These actions by national militaries have so far been against their own on-orbit satellites, while one can safely assume that in a time of conflict the intent to use ASAT weapons would be against other States' military satellites. Yet such intentions mean little in the space environment, where one's action can cause harm many years after the fact to unintended targets. The risks of these unintended actions occurring only increases as the space environment gets more congested, especially as evasive maneuvers becoming more limited to conduct. Militaries must be aware of their actions against their own or other State satellites in the context of the entire near-Earth space environment.

Cost-Benefit Analysis: Environmental Protection Policies

The protection of the space environment has become an increasingly important topic since the turn of the twenty-first century, with increasing numbers of space professionals considering this realm.[67] Of particular

---

63 Weeden and Samson, *Global Counterspace Capabilities*, 2-2.
64 Weeden and Samson, *Global Counterspace Capabilities*, 2-10; Thomas G. Roberts, "Unusual Behaviour in GEO: Luch (Olymp-K)", *CSIS*, March 31, 2021. https://aerospace.csis.org/data/unusual-behavior-in-geo-olymp-k; Jonathan O'Callaghan, "Russia accused of firing 'anti-satellite weapon' from one of its satellites in space", *Forbes*, July 24, 2020. https://www.forbes.com/sites/jonathanocallaghan/2020/07/24/worrisomerussia-accused-of-firing-a-projectile-in-space-from-one-of-its-satellites/?sh=691d8aa165a5.
65 Weeden and Samson, *Global Counterspace Capabilities*, 1-2; CSIS, "Counterspace Weapons 101".
66 Weeden and Samson, *Global Counterspace Capabilities*, 1-2.
67 Su and Lixin, "The European Union", 38; Beard, "Space: the final ethical frontier".

interest is near-Earth orbits and space debris issues.[68] The concept of space as an environment is important to consider in the development of future space policy and expansion. Yet the space environment is vast and complex. As such, this chapter will provide a limited focus to the most frequent part of the space environment that we are affecting near-Earth orbits. These are facing significant threat from in orbit collisions tripping Kessler syndrome and rendering LEO functionally dead, which in turn could prevent use of the other orbital layers. This topic is not some distant issue for our future generations. The recent saga of space-object 2020 SO illustrates how even far-flung space debris could come back to haunt us.[69] NASA confirmed this object was in fact 54 year old space debris returning to Earth from an orbit around the sun, risking the safety of on-orbit satellites and potentially terrestrial objects or people if it had re-entered.[70]

The space environment is an open system, fragile in its lack of ability to regenerate.[71] Thus a proactive stance is required for the development of space environmental policy, having international expertise at the fore instead of politicians or those with commercial interests, to ensure long-term ethical solutions are in place for the global community's benefit.[72] With space debris being such an imminent and catastrophic thing, it would behoove the international community to produce a pragmatic policy, with agreement on the basic issues and regulations, instead of delaying to design a fine-detailed strategy that runs the risk of coming too late.[73]

To develop meaningful policy, it must first be recognised that space is an environment and that there is value in protecting it.[74] This value could come from numerous avenues; global betterment, economic value, scientific discoveries, or even for its own beauty.[75] After deciding that space is an environment that is worth protecting, it is important to establish the type of decision-making paradigm that is best suited moving forward.

---

68 Mark Williams, "Space ethics and protection of the space environment." *Space Policy* 19 (2003): 47; Beard, "Space: the final ethical frontier".
69 Associated Press, "Fake asteroid? NASA expert identifies mystery object as old rocket", *ABC News*, October 12, 2020, https://www.abc.net.au/news/2020-10-12/fake-asteroid-nasa-expert-ids-mystery-object-as-old-rocket/12752996; Meghan Bartels, "So 'asteroid' 2020 SO was actually 1960s space junk. It may be the first of many to come", *Space*, December 2020, https://www.space.com/2020-so-space-junk-lessons.
70 Bartels, "So 'asteroid' 2020 SO".
71 Williams, "Space ethics", 49-50; Sara Reiman, "Is space an environment?", *Space Policy* 25 (2009): 82, https://doi.org/10.1016/j.spacepol.2009.03.005.
72 Williams, "Space ethics", 50; Schwartz, "Prioritizing scientific", 207-208.
73 Williams, "Space ethics", 50-52.
74 Williams, "Space ethics", 48; Reiman, "Is space an environment?", 81.
75 Reiman, "Is space an environment?", 85.

Baum describes the decision-making paradigm as a procedure for making decisions given both an ethical framework and the requirements of a decision-making scenario.[76] One such framework that encompasses these elements is the cost-benefit analysis (CBA). The CBA is a well-established decision-making paradigm, used both in government and business to justify decisions.[77] Simply, this paradigm has binary outcomes. If a decision's benefits outweigh the costs, the decision passes; however, if they do not, the decision fails.[78]

CBAs are inherently a flexible system, with costs and benefits able to be defined in numerous ways and tailored for each new decision-making scenario. Yet despite this, the common way that CBA is implemented is through monetary cost/benefits paradigms. However, many of the benefits humanity values are outside of market exchanges and the actions that occur in space affect far more than State-based economies.[79] Thus the units for consideration in space policy decision-making should be non-market valuations such as human, environmental, or even philosophical costs and benefits.

These non-market costs should then be considered through a virtue ethics framework, allowing for greater flexibility and adaptability as humanity continues their exploration beyond the near-Earth orbits and into unknown environments.[80] When it comes to ethics, it must be acknowledged that there is a need for proper consideration for *whose* virtue ethics should be used when making decisions in space. The United States stated, in April 2005, that their desire is for space culture to "be Western".[81] Yet, with space affecting the global community, it is not enough for the dominant space nations to impose their ethics on to the rest of the world if we are to progress in a meaningful way.[82] International engagement will continue to create a space environment that is interconnected, with shared

---

76 Seth D. Baum, "Cost-benefit analysis of space exploration: some ethical considerations." *Space Policy* 25 (2009): 75, https://doi.org/10.1016/j.spacepol.2009.02.008; Baum establishes that an ethical framework is "an underlying view of what is fundamentally right and wrong which can be used to evaluate specific decisions".
77 Baum, "Cost-benefit analysis", 75.
78 Baum, "Cost-benefit analysis", 76.
79 Baum, "Cost-benefit analysis", 76.
80 "Near-earth orbits" is an objective term where geosynchronous orbits are 20000km and greater away from the planet. However, they will fall within the terminology of 'near-earth' on the grounds that in the scope of the universe – 20000km is relatively close in the grand scheme of the universe.
81 Linda Billings, "How shall we live in space? Culture, law and ethics in spacefaring society", *Space Policy* 22 (2006): 250.
82 Ibanga Ikpe's chapter in this book is a great starting place to understanding this problem.

benefits and ethical space policy frameworks that resonate with the global community moving forward.[83]

Using virtue ethics in all future space policy development and decision making would allow for a framework that evolves with our expanding exploration of space, instead of having a fixed adherence to where we are now in both space and moral beliefs.[84] Virtue ethics is a moderate approach between the emotional (affective) and cognitive (rational) aspects in decision making, a "rule of thumb" approach.[85] This would add a level of practical flexibility into environmental and expansion policies, with the properties of the agent being a deciding factor, instead of the environment. This provides a more applicable framework to work with as humanity continues into unknown natures of space.[86] Even for the near-Earth orbital environment, this flexibility is necessary. The rate at which technologies are used, and the adaption on how technologies are used and integrate into everyday life, are unpredictable and incredibly fast when compared to the speed in which policies are created and laws are ratified.

The CBA virtue ethics framework method can aid in decision making for policy development and proportionality assessments, imposing ethical constraints on a State's available courses of action beyond monetary considerations.[87] Additionally, removing the standard monetary cost/benefit model is designed to mitigate the injustice that traditional CBA mechanism produce, where a focus on cost dismisses moral content and ethical considerations.[88] Does a State conflict truly out-weight the betterment and improvement of the entire global community? Is a virtual school entitled to the same protections as a physical school? Can the same effect be achieved without degrading the quality of life of citizens reliant on dual-use satellites?

Humanity's drive for space exploration and development must be balanced with the need to preserve the environment, both terrestrial and extraterrestrial.[89] Adoption of a code of ethics, both for military and commercial developments, may also mitigate the "greed and power models so prevalent today".[90] Ensuring that those directly involved in affecting

---

83 Billings, "How shall we live", 253-254; Williams, "Space ethics", 50.
84 Reiman, "Is space an environment?", 85.
85 Reiman, "Is space an environment?", 85-86.
86 Reiman, "Is space an environment?", 85-86.
87 Baum, "Cost-benefit analysis", 78.
88 Baum, "Cost-benefit analysis", 78.
89 Billings, "How shall we live", 252.
90 Billings, "How shall we live", 253.

the space environment are educated in a virtue ethics framework for their CBAs, we limit the potential for harmful choices being made due to out-of-date rules-based systems being adhered to.

Conclusion

With current technology integrating space components and new space-based technology developing, it is apparent that near-Earth orbits are a finite and fragile resource. It is also highly likely that dual-use satellites will increasingly become the norm to maximise customer bases within this limited resource. In turn, this will further increase the importance of States to develop protective policies covering dual-use satellites for the benefit of the entire the global community. Essential services are already protected under other laws and treaties, yet with the betterment of humanity at stake, shouldn't broader considerations be given to services that cannot be provided by other means that affect the quality of life of communities?

Moving forward the continued development of policy is going to remain a complex affair, involving numerous nations with conflicting ideals on the path space doctrine should take. Militaries need to develop stringent rules of engagement policies for actions affecting space-based services or rely on satellites for functionality, particularly dual-use satellites.

Maintaining a focus on virtue ethics, decision makers can assess if actions are desirable for human "wellbeing". A virtue-ethics CBA should also grant a level of flexibility to decision-makers, a necessary aspect to build into any future space policy and decision-making standards. This is particularly true of dual-use satellites, which will only increase their integration into daily life and essential service provisions.

References

Air Power Development Centre, *AFDN 1-19 Air-Space Integration*. Canberra, Royal Australian Air Force, 2019.
Associated Press, "Fake asteroid? NASA expert identifies mystery object as old rocket", *ABC News*, October 12, 2020, https://www.abc.net.au/news/2020-10-12/fake-asteroid-nasa-expert-ids-mystery-object-as-old-rocket/12752996.
Australian Army, "Our Values & Contract", accessed September 22, 2020. https://www.army.gov.au/our-people/our-values-contract.
Australian Defence Force, *ADDP 06.4 Law of Armed Conflict*, Canberra, 2019.

Australian Maritime Safety Authority, "Why is GPS best", *Beacons*, n.d., https://beacons.amsa.gov.au/purchasing/GPS-best.asp.

Australian Medical Association, "New code of ethics for doctors", Last modified March 17, 2017. https://ama.com.au/media/new-code-ethics-doctors.

Beard, M. "Space: the final ethical frontier", *Ethics*, September 24, 2020, https://ethics.org.au/space-the-final-ethical-frontier/.

Bartels, M. "So 'asteroid' 2020 SO was actually 1960s space junk. It may be the first of many to come", *Space*, December 2020, https://www.space.com/2020-so-space-junk-asteroid-lessons.

Baum, S. D. "Cost-benefit analysis of space exploration: some ethical considerations." *Space Policy* 25 (2009): 75-80, https://doi.org/10.1016/j.spacepol.2009.02.008.

Billings, L. "How shall we live in space? Culture, law and ethics in spacefaring society", *Space Policy* 22 (2006): 249-255.

Business Wire, 2018. "Avanti Communications' HYLAS 4 Satellite to Deliver Digital Education to African Schools", February 2, 2018, https://www.businesswire.com/news/home/20180201006437/en/Avanti-Communications'-HYLAS-4-Satellite-to-Deliver-Digital-Education-to-African-Schools.

Coelho, C.M. de Melo, L. F. B. Sa´bato, M. A. L. Rizel, D. N. and Young, R. J. "A note on the use of GPS collars to monitor wild maned wolves Chrysocyon brachyurus (Illiger 1815) (Mammalia, Canidae)", *Applied Animal Behaviour Science* 106, no. 1-3 (2007): 259-264.

College of Policing, "Code of Ethics", https://www.college.police.uk/ethics/code-of-ethics.

CSIS, "Counterspace Weapons 101", *Aerospace Security Project*, October 28, 2019, https://aerospace.csis.org/aerospace101/counterspace-weapons-101/.

DeGrasse Tyson, N. and Lang, A. *Accessory To War: The Unspoken Alliance between Astrophysics and the Military*. New York: W. W. Norton & Company, 2019.

Devex, 2018, "Satellites for survival: saving lives in Nigeria", Uploaded in April 7, 2018, Youtube video, 4:46min. https://www.youtube.com/watch?v=4Lq3g1l9Nh0&t=69s.

Global Education Monitoring Report Team, *Global Education Monitoring Report 2020*, Paris, United Nations Educational, Scientific and Cultural Organisation, 2020.

Harrison, T. "The Future of Security in Space", in *Air Power in a Disruptive World: Proceedings of the 2018 Air Power Conference*, ed. Department of Defence, Canberra, National Library of Australia, 43-48.

InStrat Global Health Solutions, "About Us", March 13, 2021. https://instratghs.com/about-us/.

Johnson-Freese, J. *Space as a Strategic Asset*, New York, Columbia University Press, 2007.

Kerala Infrastructure and Technology for Education, "About Us", accessed March 13, 2021. https://kite.kerala.gov.in/KITE/index.php/welcome/about_us.

Le Cozannet, G. Kervyn, M. Russo, S. Ifejika Speranza, C. Ferrier, P. Foumelis, M. Lopez, T. and Modaressi, H. "Space-Based Earth Observations for Disaster Risk Management." *Surveys in Geophysics* 41, no. 6 (2020): 1209-235.

Lockheed Martin, "JASSM: Dependable, affordable, long-range strike cruise missile", JASSM, n.d., https://www.lockheedmartin.com/en-us/products/jassm.html.

Media, Entertainment & Arts Alliance, "Journalist Code of Ethics", accessed November 15, 2020. https://www.meaa.org/meaa-media/code-of-ethics.

Meier, P. 2012, "How crisis mapping saved lives in Haiti", National Geographic Society Newsroom, July 2, 2012, https://blog.nationalgeographic.org/2012/07/02/how-crisis-mapping-saved-lives-in-haiti.

Microsoft, "Compliance and Ethics at Microsoft", accessed September 22, 2020. https://www.microsoft.com/en-us/legal/compliance.

Mills, E. C. Poulsen, J. R. Fay, J. M. Morkel, P. Clark, C. J. Meier, A. Beirne, C. and White, L. J. T. "Forest elephant movement and habitat use in a tropical forest-grassland mosaic in Gabon." *PLoS ONE* 13, no. 7 (2018): 1-17. doi.org/10.1371/journal.pone.0199387.

Nagami, K. Nakajima, I. Juzoji, H. Igarashi K. and Tanaka, K. "Satellite communication for support medical care in the aftermath of disasters", *J Telemed Telecare* 12, no. 6 (2006): 275. doi: 10.1258/135763306778558213.

NASA, "Emergency systems save tens of thousands of lives", *NASA Spinoffs*, 2012, https://spinoff.nasa.gov/Spinoff2012/ps_3.html.

O'Callaghan, J. "Russia accused of firing 'anti-satellite weapon' from one of its satellites in space", *Forbes*, July 24, 2020. https://www.forbes.com/sites/jonathanocallaghan/2020/07/24/worrisomerussia-accused-of-firing-a-projectile-in-space-from-one-of-its-satellites/?sh=691d8aa165a5.

Reiman, S. "Is space an environment?" *Space Policy* 25 (2009): 81-87. https://doi.org/10.1016/j.spacepol.2009.03.005.

Roberts, T. G. "Unusual Behaviour in FEO: Luch (Olymp-K)", *CSIS*, March 31, 2021. https://aerospace.csis.org/data/unusual-behavior-in-geo-olymp-k.

Schwartz, J. S. J. "Prioritizing scientific exploration: a comparison of the ethical justification for space development and space science", *Space Policy* 30 (2014), 202-208.

Snape, R. T. E. Bradshaw, P. J. Broderick, A. C. Fuller, W. J. Stokes, K. L. Godley, B. J. "Off-the-shelf GPS technology to inform marine protected areas for marine turtles", *Biological Conservation* 227 (2018): 301-309.

Steer, C. "Global commons, cosmic commons: implications of military and security uses of outer space", *Journal of International Affairs* 18 (2017), 9-16.

Su, J. and Lixin, Z. "The European Union draft Code of Conduct for outer space activities: an appraisal', *Space Policy* 30 (2014): 34-39.

The Barents Observer, 2020, "Russia tests anti-satellite weapon from Plesetsk", *The Moscow Times*, December 17, 2020. https://www.themoscowtimes.com/2020/12/17/russia-tests-anti-satellite-weapon-from-plesetsk-a72398.

Twyford, K. Crump, S. and Anderson, A. "Satellite lessons: vocational education and training for isolated communities", *Rural Society* 19, no. 2 (2009): 127-135.

Waldrop, E. S. "Integration of military and civilian space assets: legal and national security implications" *Air Force Law Review* 55 (2004), 157-231.

Weeden B. and Samson, V. 2021, *Global Counterspace Capabilities: an open source assessment*, Secure World Foundation, xxxi. https://swfound.org/media/207162/swf_global_counterspace_capabilities_2021.pdf.

Weitering, H. 2020, "Russia has launched an anti-satellite missile test, US Space Command says", *Space*, December 2020. https://www.space.com/russia-launches-anti-satellite-missile-test-2020.

Williamson, M. "Space ethics and protection of the space environment." *Space Policy* 19 (2003): 47-52.

Zualkernan, I. A. Lutfeali, S. and Karim, A. "Using Tablets and Satellite-based Internet to Deliver Numeracy Education to Marginalized Children in a Developing Country", *IEEE Global Humanitarian Technology Conference* (2014): 294.

7

THE GROWING THREAT OF TERRORISM IN SPACE

Nikki Coleman and Stephen Coleman[1]

Introduction

Technological advances over the past 50 years have made much of society dependent on a safe and secure space environment. Daily communications, GPS systems, economic systems and transactions have all become largely dependent on reliable satellite systems. At the same time the cost of creating and maintaining satellite technologies has decreased to the stage where it is no longer only the "rich" countries of the world who are able to invest in research into these new technologies. This dramatic increase in space capabilities across the globe has enabled new countries to take control of their technological future in relation to space research. The combination of the increasing reliance of individuals and States on satellite technology, combined with the decreasing cost of placing assets into orbit, has dramatically increased the possibility of State or non-state terrorist groups utilising space as an area of operations. This chapter discusses some of the ethical issues raised by potential terrorist attacks on State and commercial assets in space, and some of the measures that may be used to mitigate these risks.

---

1 The views expressed are those of the authors and do not reflect the official policy or position of the Royal Australian Air Force, the Department of Defence, or the Australian Government. We wish to acknowledge the support and kind assistance of Professor Russell Boyce (UNSW Canberra Space), Group Captain Andrew Gilbert (RAAF Air Power Development Centre), and Professor Stephen Latham and Lori Bruce (Yale University Interdisciplinary Center for Bioethics).

Low-Cost Satellites and Space Programs: Benefits and Problems

There is no doubt that the increase in the use of low-cost satellites has the potential to allow more countries to take control of their own space and communication capabilities. In the early days of space flight only very wealthy States could afford to have a dedicated space program, while over 60 States now have a dedicated space agency, and more than 70 have been responsible for satellites in orbit. The capability of these space organisations range from States who merely operate satellites (such as Costa Rica, Azerbaijan and Thailand), through those with the capability to launch satellites into orbit (such as Iran and North Korea), to those capable of manned space flight (such as Russia, China and the USA). Less wealthy States need to rely either on wealthier States or on private providers for launch capabilities, but the decrease in both the cost of constructing satellites and in the ongoing monitoring costs has enabled many more individual States to control the satellites that contribute to their communications and other vital infrastructure.

This international competition and innovation has seen satellite technology improve dramatically in a short space of time, both in terms of the advanced capabilities of the satellites being built, and in terms of size, with satellites which are now being deployed being just as powerful but considerably smaller than had previously been the case. Since the cost of launching a satellite into orbit directly relates to the size of the satellite, particularly to its weight, this miniaturisation of satellites also reduces the cost of deployment, another factor which makes satellites more affordable to less wealthy States. This increased competition and innovation has come about because of the number of new stakeholders working in the field, which in turn has enabled the space industry and those industries that feed into it to flourish and increase in capacity, fostering even more competition and innovation.

The dramatic increase in the number of States having capabilities in space, even if only in terms of the use and control of their own satellites, brings with it a dramatic increase in the number of stakeholders involved in space and in satellite management. This immediately increases the complexity of discussions around issues such as: access to space generally and to specific orbital locations within which satellites can be placed; changes to laws, regulations and protocols regarding the use of space, and of resources located on celestial bodies such as asteroids, and; management of disputes and the potential for military presence or even armed conflict in space. Whilst the situation regarding access to the Earth's oceans is similar, it has the advantage of coming with hundreds of years of precedent and naval law. The space age only began 60 years ago, and the dramatic increase in the number of States involved in space operations has been much, much more recent; a third of all the States who have ever controlled a satellite in orbit have acquired their first satellite within the last 5 years.

The main practical barrier to a State's entry into satellite technology is the difficulty in launching satellites into orbit. While a significant proportion of the world's States control satellites, few States are actually launch-capable, and even fewer are willing to place anything other than their own satellites into orbit. Some States are capable of launching a satellite into Low Earth Orbit (LEO) where the majority of satellites are located but are not capable of launching a satellite into higher orbits, such as geostationary ones. A small number of private companies are also launch-capable, but international law regarding the use of space means that these companies are also subject to regulation by the State in which they operate.

While the difficulty in actually launching equipment into space poses a barrier to States who wish to invest in such technology, it also presents a form of protective barrier to satellite technology, since orbital space is equally difficult to access for those who wish to target satellites with any form of physical attack. However, this does not mean that it is impossible to engage in activities which could damage or destroy space-based technologies.

The Problem of Terrorism

History clearly shows that even quite revolutionary new technologies can become commonplace within a relatively short period of time. As technology develops, costs tend to come down, reliability goes up and what was previously remarkable becomes commonplace, and sometimes is even considered essential to modern life.

Institutions and structures which are seen as symbolic of the power or image of a State can become targets of attack for those who are enemies of that State. Thus, the attacks of September 11 2001 targeted symbols of American military power (the Pentagon) and its financial power (the World Trade Center). The satellites which form one of the foundations of modern communication networks could well be seen as emblematic of the capitalistic system, thus becoming an attractive symbolic target for attack. Modern military forces rely on satellite GPS (Global Positioning System) technology for navigation and on other satellite-based systems for communications, intelligence gathering, munition guidance and targeting. Thus, someone who wished to strike out against the military power of a state such as the USA might well be predisposed to attack the satellites which support that military power, if such an attack were possible. So, it is worth considering whether space-based systems like satellites are vulnerable to a terrorist attack, and if so, how such an attack might be prevented and/or how the effects of an attack might be minimised.

There are many different definitions of terrorism. Some of these would suggest that an act only counts as terrorism if it directly causes death and/or injury to innocent people, while other definitions are much broader. For the purposes of this discussion, a very broad definition of terrorism will be used, thus terrorism will include politically motivated acts that: (a) are intended to directly cause death and/or injury to innocent people (such as bombings); (b) are intended to indirectly cause death and/or injury to innocent people (such as interfering with air traffic control systems), and; (c) deliberately cause indiscriminate and widespread destruction to public and/or private property.

Non-state Terrorism and Space Operations

Satellites are known to be vulnerable to physical attacks, and this has been demonstrated by anti-satellite missile tests conducted by China in 2007 and by the actions of the USA, which shot down a malfunctioning satellite in 2008.[2] However, direct attacks such as these, directed against individual satellites, require an extremely high degree of precision, since a satellite in LEO is likely to be travelling at a velocity of around 7 km/sec. Thus, such attacks are almost certainly beyond the capabilities of anyone but a major State power.

If a non-state group wanted to engage in attacks which would disrupt satellite communication systems, it would be far easier (and cheaper) to carry out such an attack by targeting the communication system's terrestrial

---

[2] BBC News, ISS crew take to escape capsules in space junk alert, 24 March 2012, http://www.bbc.com/news/science-environment-17497766; T. Shanker, Missile Strikes a Spy Satellite Falling From Its Orbit, *New York Times*, 21 February 2008, http://www.nytimes.com/2008/02/21/us/21satellite.html.

based control systems than by attacking the satellites themselves. Even if these control systems are well guarded, the computers which manage the uplinks and downlinks to and from the satellites might be vulnerable to a cyber-attack.

It is feasible at some point in the future a non-state actor might "take up arms" in space, by acquiring a low cost satellite with the intent of using it to launch direct attacks against other satellites in nearby orbits. The satellite would need to be specifically designed to allow such an attack to take place, which would present problems in and of itself, but such a problem would be relatively minor in comparison to the other difficulties such a project would entail. The satellite might be designed to engage in a kamikaze attack on its target, either through deliberately colliding with the target (a very difficult manoeuvre to manage) or by getting close to the target before detonating. Another alternative would be for the satellite to be designed in a way that would allow it to attack its target using an electromagnetic pulse (EMP). The problem with an EMP attack is that a non-nuclear EMP device is relatively low powered and so would probably do little damage unless it was positioned quite close to a vulnerable satellite. A nuclear EMP is more powerful by several orders of magnitude, but even if a non-state group was able to obtain a nuclear weapon which could be used to power an EMP, it is extraordinarily unlikely that they would use a nuclear weapon in this manner given how much more damage they could do by using it on the Earth rather than in orbit.

The major problem facing a non-state group which wanted to launch an attack in this manner, is that even if the group was able to purchase or manufacture the required satellite "weapon", they would still need to procure the services of a launch vehicle in order to place this weapon into space. This would require a long lead time (to prepare for the launch) and would also require the group to pay a substantial amount of money. Most problematically, it would require the operator of the launch vehicle, either a State space organisation or a commercial entity, to be complicit in the attack, since the purpose of the satellite would be obvious from its basic construction. It is impossible to imagine a commercial launch company acting in such a manner, since assisting in an attack like this would be an express route to bankruptcy. Essentially, the only way in which a non-state group could ever manage to get such a satellite into orbit would be with the backing of a rogue launch-capable State. While non-state groups have engaged in sophisticated attacks on high profile targets, the 9-11 attacks being an obvious example, trying to organise a direct satellite-to-satellite attack would seem to be well beyond what any non-state group could manage at this point in time.

State Terrorism and Space Operations

Along with the rapid increase in space innovation comes the risk of a rogue State utilising space for military purposes. North Korea, for example, has recently been testing their ICBMs in order to ascertain if they can use those weapons against Japan and the USA and it is a relatively small step to imagine them engaging in military operations in space, if this was thought to be to their advantage. A launch-capable rogue State such as North Korea could cause widespread devastation through an indiscriminate space attack, and there are at least two ways such an attack could be carried out. North Korea is known to possess nuclear materials, and thus it could produce a nuclear EMP and detonate it in orbit, causing massive and widespread damage to the electronic systems of any satellites within range. However, in the event that North Korea was to develop such a device it seems highly unlikely that it would be utilised in this manner, rather than being loaded into an ICBM. The other way in which a rogue State could carry out an indiscriminate space attack would be to simply introduce a large amount of small but dense material into LEO.

Space debris in orbit around the Earth already presents a considerable hazard to both unmanned satellites and manned space missions. Objects in LEO are travelling at a velocity of 7-8 km/sec, so the force generated by the impact of even a small piece of debris is considerable. Serious damage can thus be inflicted on satellites by relatively small pieces of debris, while larger pieces of debris are capable of rendering inoperable, or even destroying, much larger satellites. Since each impact with a satellite creates new pieces of debris and impacts between pieces of debris can also cause the creation of new smaller pieces of debris, if the density of objects in orbit is high enough, then it is possible that collisions between objects could cause a cascade effect, with each collision creating more space debris, thus increasing the likelihood of future collisions, which will in turn produce even more debris.

This cascade effect is known as the Kessler Syndrome, and it has been suggested that if the amount of space debris in a particular orbital range exceeds the threshold required to initiate the Kessler Syndrome, this makes collisions in that orbital range so likely that it would become impossible to sustain satellite operations.[3] LEO has the highest current density of satellites and is thus most likely to fall victim to the Kessler Syndrome. If a rogue State wanted to instigate such an event, then it would need to do no more than deliberately introduce a large amount of material into LEO. Objects less than 10 cm in diameter are about as small as can reliably be tracked by current methods, so introducing a lot of pieces of debris of about half that size (which would be impossible to track and thus avoid) would be potentially devastating, especially if that material was dense enough to cause major damage on impacting with a satellite.

Some people just want to watch the world burn. Others will resort to extreme actions if threatened, like Saddam Hussein setting fire to the Kuwaiti oil wells during the 1991 Gulf War. It is not beyond the realms of possibility that a State like North Korea might feel they would benefit from pushing the world several decades backward in technological terms, by either initiating the Kessler Syndrome or by taking out satellites through the use of an EMP. It is this sort of State sponsored space terrorism which is, realistically, of most concern at this point in time.

Is it Possible to Mitigate the Risks Without Seriously Impacting on the Benefits?

Until recently the UN non-proliferation of nuclear weapons treaty had been largely successful in stopping the spread of nuclear weapons.[4] This was largely done by regulating not only the use of nuclear weapons, but also the use of nuclear energy and the fuel for nuclear weapons and energy. This regulation of nuclear materials was done through the cooperation of the Nuclear Suppliers Group (NSG). The NSG is a group of nuclear supplier countries which will only sell the materials required for nuclear weapons and energy once they are satisfied that the "transfer would not contribute to the proliferation of nuclear weapons".[5] Unfortunately, the raw materials that go into a low cost satellite are not of the type to make such regulation feasible in regards to stopping rogue States and non-state groups from

---

3 D.J. Kessler & B.G. Cour-Palais, Collision Frequency of Artificial Satellites: The Creation of a Debris Belt, *Journal of Geophysical Research*. 83 (1978) 2637–2646.
4 United Nations Office for Disarmament Affairs – Non-Proliferation of Nuclear Weapons Treaty (NPT), https://www.un.org/disarmament/wmd/nuclear/npt/.
5 Nuclear Supplier Group. About the NSG. https://www.nuclearsuppliersgroup.org/en/about-nsg.

accessing the technology. What is more feasible, however, is the regulation of launch services by States and private space companies with launch capabilities. This is an area in which further regulation and policy work done needs to be done.

Efforts to stop terrestrial terrorism involve multiagency and several approaches through the intelligence and justice systems. These approaches revolve around intelligence gathering to identify and prevent terrorist attacks, as well as limiting access to raw materials which might be used to effect large numbers of people (force multipliers), such as bomb making materials, bioweapon ingredients and nuclear weapon components. Because nuclear material is not freely available, terrorists have had to rely on more crude methods, such as the use of planes, trucks, and home-made bombs. These approaches have been effective in limiting the amount of damage terrorists are able to inflict, and so reduce the effect of terrorist actions. The impact of terrorist actions by rogue States and non-state actors can be limited by similar multi-dimensional approaches, largely through the diplomatic and intelligence community.

Conclusion

It does seem that there is a moral responsibility for the stakeholders in space to come to an agreement regarding the best way to curtail the possible actions of rogue States (and possibly of non-state groups) in relation to the mayhem they could potentially cause in space. This is not only a responsibility to ensure the safety of people currently utilising satellite technology in their day to day lives, but also a responsibility to the future generations who would have to deal with the ramifications of a rogue State and/or non-state group engaging in the sort of space terrorism discussed in this chapter. At the present time the risk of space terrorism is low, but since the risk is only going to increase, discussion of such issues ought to start now. We shouldn't wait for the space equivalent of 9-11 before considering the problems that this sort of space terrorism would cause.

References

BBC News, ISS crew take to escape capsules in space junk alert, 24 March 2012, http://www.bbc.com/news/science-environment-17497766.

Kessler, D.J. and B.G. Cour-Palais, Collision Frequency of Artificial Satellites: The Creation of a Debris Belt, *Journal of Geophysical Research*. 83 (1978) 2637–2646.

Nuclear Supplier Group. About the NSG. https://www.nuclearsuppliersgroup.org/en/about-nsg.

Shanker, T. Missile Strikes a Spy Satellite Falling From Its Orbit, *New York Times*, 21 February 2008, http://www.nytimes.com/2008/02/21/us/21satellite.html.

United Nations Office for Disarmament Affairs – Non-Proliferation of Nuclear Weapons Treaty (NPT), https://www.un.org/disarmament/wmd/nuclear/npt/.

8

THE PROBLEMS POSED BY NON-STATE GROUPS AND ROGUE STATES EXPLOITING SPACE

Kaylee Verrier[1]

Introduction

The military relies on space assets for all war, disaster relief, and peacekeeping activities. One of the most prominent issues of modern warfare is the accessibility and affordability of advanced weapon capabilities, including kinetic, non-kinetic and cyber warfare. As technologies become more widespread, the current position of space dominance is threatened, but more importantly, so is the safety and security of space-related assets. Additionally, even the Earth-based space assets are inherently visible and vulnerable to ground-based attacks, particularly kinetic destruction, electromagnetic interference, or cyber-attack. They can easily be targeted by weapons and technological systems available to any rogue group without the need to gain access to space.

It is feasible for rogue actors to look for new targets and strategic methods to achieve terror-based objectives, including mass losses and irreversible psychological effects by exploiting space assets. Using examples involving attacks in both Space and on the Earth, I will discuss the problems of rogue groups exploiting space for malicious reasons and the potential physical, legal, and psychological consequences.

---

[1] The views expressed are those of the author and do not reflect the official policy or position of the Royal Australian Air Force, the Department of Defence, or the Australian Government.

Space reliance goes beyond military practices and is essential for everyday life. Some of the problems posed by rogue actors exploiting space assets include risks to communications systems, navigation systems, and infrastructure that society is reliant on. As military leaders, to ignore this growing threat to civilian and military assets would be a failure of duty. From an ethical perspective, being aware of the very real threats facing our space environment goes beyond simply protecting assets. It also involves decision making around tenuous political relationships, protecting the space environment for future generations, and avoiding acts that ultimately have the power to (inadvertently or advertently) initiate a major conflict.

Military ethics evaluates military actions based on moral standing. It aims to define the standards of good behaviour for all individual military personnel and develops these standards through training and instructing.² Military ethics asks critical questions of existing laws and regulations in connection with an organised military force. This chapter will focus on military ethics as applied to the laws regarding peaceful uses of outer space. A central concern of military ethics is questioning the use of organised military force. Simply, it deals with the question of "when can a soldier use physical force, or even kill, and how may/must/should they use this force?"³ When considering space systems and potential rogue actors, this question becomes far more complicated. It introduces complications such as the political fight for space dominance, ambiguous and non-legally binding acts, accessibility on a global scale, attribution problems, rapidly developing technology, and extreme remote warfare. This chapter explores why these complications exist, and how dangerous they are for militaries and members of the global rules-based order. It discusses what is already occurring on the potential space battle-front, space's involvement with war-fighting, and the associated physical, political, and ethical risks.

Space as a New Domain of War

For this chapter, the following definitions for non-state groups and rogue states will be used: Non-State Groups (NSG) are defined as violent, non-state actors. They are considered as either individuals or groups who are

---
2 Dieter Baumann, "Military Ethics: A Task for Armies", *Military Medicine*. Vol. 172, 2007, 34-38.
3 C. Stadler, *Military ethics as part of general system of ethics*. In *Civil-Military Aspects of Military Ethics*, by EM Micewski, 2003.

wholly or partly independent of state government and will threaten to use violence to achieve their goals.[4] Similarly, rogue States are defined as a nation or State observed as breaking international law and posing a threat to the security of other nations. They are not bound by either international law or what a reasonable person would deem rational behaviours. Therefore, there is a strong likelihood that rogue States may use weapons of mass destruction against any member of the international community.[5] Both groups can be reasonably assumed to be a risk to space assets and the wider global community through the threat of acts of aggression and/or acts of terrorism. The terms "rogue group" and "rogue actors" is used throughout this chapter to refer to both NSG's and rogue States as a collective.

One of the biggest problems with rogue States and non-state actors becoming involved in space comes from the accessibility and rapid advancements in weapon technology. Additionally, actions in space are not isolated from the other domains but instead pose significant new challenges regarding defensibility, attribution, lawful enforcement, discrimination, and proportionality.[6] Assets on the ground, sea, and in the air all rely heavily on space systems for critical systems such as communication, tracking, and navigation. United States Air Force (USAF) General John Hyten is quoted as highlighting the military's reliance, saying "[t]he loss of U.S. space capabilities would send the U.S. military back to world war two… back to industrial-age warfare."[7] Similarly, Australia's Defence White Paper (2016) specifically highlights that space-based systems are essential "for intelligence collection, communications, navigation, targeting and surveillance", all of which "play a vital role in all ADF and coalition operations".[8] Given this dependency on space capabilities, satellites and space-based assets provide an attractive target for any rogue actors and adversaries.

General Hyten also commented on how "[t]he commercial revolution in space has eliminated the exclusive control of space once enjoyed by national defence, intelligence and government agencies", which points

---

4 Claudia Hofmann and Ulrich Schneckener, "Engaging non-state armed actors in state and peace-building: options and strategies." From *International Review of the Red Cross*. Vol. 93. September 2011.
5 Colonel Scott A. Enold, "Rogue States and Deterrence Strategy." *U.S. Army War College*, 2 April 2009.
6 Daniel R. Coats, *Worldwide Threat Assessment of the U.S. Intelligence Community*. Office of the Director of National Intelligence, 2019.
7 Todd Harrison, Zack Cooper, Kaitlyn Johnson, and Thomas G. Roberts. *Escalation and Deterrence in the Second Space Age*. CSIS, 2017.
8 Australian Department of Defence. *The Defence White Paper*, 2016.

to the growing access to space across the globe.⁹ Following this train of thought, General Robert Kehler, retired USAF general and ex Commander of USAF Space Command is quoted about the essential nature of space assets within the military, saying "[t]he space capabilities we provide today are embedded in all of our combat operations."¹⁰ Due to the technologies involved with space-related activities becoming widespread, a monopoly on space power, or the ability to hold a position of serious space dominance, is neither credible nor sustainable.¹¹

Space is inherently a global domain where the actions of one nation or group can affect all others. Objects consistently pass over the territory of other nations. Any space debris created by one action may affect other objects in orbit, as seen by acts such as China's 2007 Anti-satellite (ASAT) test, which as of October 2016 had produced a total of 3,438 pieces of detectable debris, with and 2,867 still in orbit nine years after the incident. In January 2007, there were 2,864 satellites (active or inactive) in Earth orbit with known positional data.¹² Of these, 1,899 pass through the orbital regime affected by the debris – two-thirds of all payloads in Earth orbit.¹³

Compared with traditional warfighting, the international laws, treaties, and norms of behaviour when acting in space are less mature. There is a pattern to the thoughts and beliefs of military commanders across the coalition forces that confirms the deeply embedded use of space as a new domain of warfare that militaries cannot provide effective capabilities without.

However, deterrence exists in the mind of one's adversary. The way an adversary views attacks and provocations in other domains may be a useful indicator for how they will regard attacks and provocations in space.¹⁴ Space presents several asymmetries that differ from the other domains, hence how people deter actions in space likely needs to differ from the other domains. For example, the threat of equal or greater retaliation may not bother an adversary who has no space assets to lose, such as an NSG. The threat of initiating a war will not deter a rogue actor acting with

---
9 Report to the Commission to Assess United States National Security Space Management and Organisation, 2001.
10 Harrison, et al., *Escalation and Deterrence in the Second Space Age*, 2017.
11 Bruce W. MacDonald, "Crisis Stability in Space: The China Challenge and Implications for U.S. Security." Centre on Contemporary Conflict, the Naval Postgraduate School, 2013.
12 Kelso, TS, "Analysis of the 2007 Chinese ASAT Test and the Impact of Its Debris on the Space Environment", 2007 AMOS Conference, Maui, Hawaii.
13 Brian Weeden, "2007 Chinese Anti-Satellite Test Fact Sheet", Updated November 23, 2010.
14 Harrison, et al., *Escalation and Deterrence in the Second Space Age*, 2017

terrorist ideals, nor will it bother any members who can hide behind the uncertainty of attribution.

Deterrence and defence go hand in hand, although the capabilities that work best for deterrence may not necessarily be the best capabilities for defence. A strong defence reduces the enemy's capability to damage or deprive adversaries,[15] however, space assets are inherently vulnerable. The traditional method of deterrence is a strong ability to retaliate and the possession of a strong offensive capability, hence the existence of militaries. Yet it is ambiguous as to whether weaponisation of space, as a method of deterrence or defence, will ever be ethically acceptable. This is where the primary issue of rogue actors, particularly NSGs, getting involved in space arises – if we cannot deter them, how do we then defend against them?

The Committee on the Peaceful Uses of Outer Space currently have five international treaties and five sets of principles on space-related activities, each with varying numbers of ratifications and signatures. The treaties cover issues such as the non-appropriation of outer space by any one country, arms control, the freedom of exploration, the prevention of harmful interference with space activities and the environment, and the settlement of disputes. Each treaty stresses the notion that outer space and the activities that occur in outer space should be devoted to enhancing the well-being of all countries and humankind, with an emphasis on promoting international cooperation.[16] However, there are no laws or legally-binding restrictions specifically inhibiting States from using space for military purposes, military advantage, political moves and/or weaponising space assets, aside from restrictions against the placement of weapons of mass destruction.[17]

These current "Space Laws" label space as an area free for exploration and use by all States, which has effectively declared the region as available to any party capable and willing to use it, regardless of their intentions. That space activities must be in the interest of all countries is a guideline with very little capacity to be enforced or to prove culpability. While most publications accept that space is the next "high ground", they usually discuss this with the idea that large States are the main actors. They neglect the possibilities of rogue actors and even terrorists getting involved in space.[18]

---

15 LTCOL Peter F. Witteried, "A Strategy of Flexible Response." *U.S. Army War College*, 1972.
16 UNOOSA, Space Law Treaties and Principles.
17 UNOOSA, *The Outer Space Treaty*, 1967.
18 Nina-Louisa Remuss, "The Need to Counter Space Terrorism – A European Perspective." ESPI Perspectives.

Militaries should work with allies and partners to develop tailored deterrence options that are better suited to the increasingly diverse, disruptive, disordered, and dangerous environment of outer space.[19] All space involved organisations require better capabilities for attribution, particularly for non-kinetic attacks against space systems, which are discussed in the next section.

Philosophy Professor Gregory Kavka explores these dilemmas in his book *Space War Ethics* by asking "can we assume it is morally permissible for each superpower to practice some form of nuclear deterrence?"[20] Kavka is suggesting here that given States are allowed to practice such an extreme method of offence, would it then be morally desirable to enhance and practice space-based weaponisation? Would it then be morally beneficial for some space-based weapon systems to replace, or be added to, the existing land, air, or sea capabilities? The answer to such questions should be an absolute no, however, inhibiting any nation or State from delving into the realm of space-related or space-based weapon systems would be an impossible ideology.

We cannot honestly step back and declare that weaponising space will not occur, there is potential that it already has. What governments can do is set up a politically correct method to control and monitor the weaponisation of space, to ensure any rogue groups or adversaries with malicious intent can be accounted for, tracked, and ideally stopped before any malicious interaction with space assets can occur.

An interesting component of space discussions is the fact that policymakers have traditionally and repetitively considered warfare in space to be linked to nuclear warfare.[21] Attacks on critical assets and infrastructure in space are commonly viewed in the gravest terms, regardless of whether they were precursors to attacks on nuclear forces.[22] This mindset is not particularly valid or relevant with modern weapon technology being more than capable of destruction without any nuclear or chemical association. Modern and emerging weapon technology includes laser-based weapons, ballistic and hypersonic missiles, and a rapidly growing use of cyber and electronic warfare based capabilities. These threats are discussed further in the next section.

---

19 Harrison, et al., *Escalation and Deterrence in the Second Space Age*, 2017.
20 Gregory S. Kavka, "Space War Ethics." *Ethics, Special Issue: Symposium on Ethics and Nuclear Deterrence*, 1985, 673-691.
21 Michael Krepon, "Space and Nuclear Deterrence." *The Space Review*, 16 September 2013.
22 Krepon, "Space and Nuclear Deterrence", 2013.

Threats to Space Systems

According to the chapter *Threats to Space Systems*, from the report *Escalation and Deterrence: In the Second Space Age*, modern threats can be divided into four general categories based on the mechanism of attack: kinetic physical, non-kinetic physical, electromagnetic and cyber.[23]

Kinetic physical relates to the generally obvious and visible attacks such as from missiles, bombs, gun systems and other explosive ordinance based systems. As previously mentioned, ground stations are more vulnerable and accessible to rogue groups acting with malicious intent. Space systems can easily be disrupted by targeting their associated Earth-based sites, including support sites such as electrical power grids, water lines, communication lines and so on.

Additionally, existing missile defence systems can be adapted to serve as ASAT weapons, which was demonstrated by the U.S. in 2008 by launching a Standard Missile 3 (SM-3) missile to intercept and destroy a disabled U.S. military satellite.[24] Russia is also developing and testing a new generation of direct ascent ASAT weapons, including an air-launched ASAT missile.[25] These forms of weapons and attack (particularly missile capability) are likely obtainable by both NSGs and rogue states, given 31 countries already own ballistic missiles and eight possess intercontinental ballistic missiles (ICBMs).[26] Therefore, the threat exists globally and must be given the caution and attention it demands. A significant factor of consideration is the growing use of satellites existing in Low Earth Orbit (LEO). The lower altitude orbits are easier for countries with limited missile capabilities to reach, increasing their vulnerability. A key issue associated with kinetic attacks in orbit is the creation of space debris which will likely exist for long periods, effectively denying access to that area of orbit. This can be used to initiate the effect known as "Kessler syndrome", and potentially wipe out entire orbits of operation. The Kessler syndrome is a theory proposed by NASA scientist Donald J. Kessler in 1978.[27] It describes a self-sustaining, exponential rate of collision of space debris in Low Earth Orbit (LEO). This concept identifies the risk of an exponential increase in the amount of space debris due to the effect of mutual collisions. One major implication is that

---

23 Harrison, et al., *Escalation and Deterrence in the Second Space Age*, 2017.
24 Laura Grego, "A History of Anti-Satellite Programs", UCS USA, January 2012.
25 Coats, *Worldwide Threat Assessment of the U.S. Intelligence Community*.
26 Arms Control Association, "Worldwide Ballistic Missile Inventories", 2017.
27 Michelle La Vone, "Kessler Syndrome", *The Space Safety Magazine*, n.d.

the distribution of debris in orbit could render space activities and the use of satellites in the specific orbital ranges impossible for many generations.[28] As such, the effects of ASAT attacks are generally irreversible and have a very high risk of collateral damage.

According to space strategist John Klein, "some strategists question whether non-kinetic and reversible actions are hostile acts or armed attacks that warrant a military response."[29] Non-kinetic physical attacks include systems such as lasers and microwave systems that can act rapidly and create effects that may not be immediately evident.[30] Whilst these forms of technologically advanced attacks are less likely to come from NSGs, they are still a potential threat from any technically advanced state. The technology is expensive and not yet widely available; however, some adversaries may have access to such forms of attack. For example, in 2006, China "illuminated" U.S. satellites with ground-based lasers.[31] United States Defence News reported that China "fired high-power lasers at U.S. spy satellites flying over its territory in what experts see as a test of Chinese ability to blind the spacecraft." Similarly, intelligence indicates Russia is already developing airborne laser platforms,[32] as is the United States and potentially other technologically advanced states.

A nuclear detonation at high altitude could create an electromagnetic pulse to damage and potentially destroy satellites. This would be a completely indiscriminate, challenging, yet highly effective attack. A high radiation environment could damage all the satellites within an orbital regime. Given this would disrupt all other assets in the area, this targeting method would likely only come from a rogue group acting with malicious intent with no personal aversion to destroying their own (if any) space systems. The likelihood of such a group obtaining and actually using these forms of nuclear capabilities is (at this time) highly unlikely, although there are rogue States such as North Korea that are already expected to be in possession of nuclear capabilities. Such a method could be used as a denial of access attack by a peer adversary to potentially "even out" a battlefield, or simply leave a nation or State without access to their essential space systems.

---

28 Louis de Gouyon Matignon, "The Kessler Syndrome", *Space Legal Issues*, 2019.
29 War on The Rocks, "Space Warfare: Deterrence, Dissuasion and The Law of Armed Conflict." *War on The Rocks*, 30 August 2016.
30 Caroline Amenabar, "Counterspace Weapons 101", *Aerospace 101*, 23 July 2020.
31 Glenn Kessler, "Bachmann's claim that China 'blinded' U.S. satellites." *The Washington Post*, 4 October 2011.
32 Leonard David, "China, Russia Advancing Anti-Satellite Technology, U.S. Intelligence Chief Says." *Space Insider*, 18 May 2017.

Electronic attacks are a relatively more easily accessible form of disruption against satellites and space systems. Electronic attacks include jamming and spoofing of signals. Jamming is a denial of capability which is typically not permanent, however, it also results in the jammed section of the electromagnetic spectrum being denied to all users. The Jammer's position is normally given away by jamming activity, whereas electromagnetic spoofing can be utilised to degrade an enemy's capabilities and falsify data. There are several case studies, discussed later, already showing how simple yet effective an electronic or electromagnetic attack can be.

The world's largest battleground, cyber, poses a significant threat to space capability. Already it is difficult to know what exactly causes a State to lose control of a satellite, and when you include the cyber specific characteristics of culpability and traceability, there is a significant level of threat that has proven very difficult to respond to, as discussed below.

According to the NETSCOUT's Threat Intelligence Report, there were 8.4 million denial of service (DoS) attacks observed in 2019. Additionally, there were more than 15 billion records exposed in data breaches in the same year.[33] Attackers, from a cyber-perspective, can conceal their identity or even imitate the identity of another. One of the earliest displays of this for the space world was a cyber-attack on NASA's Terra EOS satellite in 2008. The satellite was disrupted for two minutes in June and nine minutes in October of the same year through insurgents allegedly hacking into a Norwegian ground station. This disruption was similarly experienced by NASA's Landsat-7 in October 2007 and July 2008.[34]

Previous attacks on aircraft systems, such as drones and modern military aircraft, can provide a realistic example of what satellites may experience from an adversary, given the similar technology basis used across most advanced air and space-based platforms. Insurgents in Iraq were able to hack into a U.S. aircraft to view sensor feeds in real-time. They gained access to live video feeds from a U.S. Unmanned Aerial Vehicle through an unprotected communications link. According to media reports on the event, "[t]he breach of the Pentagon surveillance system's security in Iraq is said to have come to light when footage shot by a Predator drone was found on the laptop of an apprehended insurgent".[35] A secondary

---

33 NETSCOUT, *NETSCOUT Threat Intelligence Report*, 2019.
34 Matt Liebowitz, "Hackers Interfered With 2 U.S. Government Satellites." 2011.
35 BBC News "Iraq insurgents 'hack into video feeds from U.S. drones'." *BBC News*, 17 December 2009.

implication of this occurrence is that the U.S. military was not aware that their system had been compromised.

In 2014, the network of the National Oceanic and Atmospheric Administration (NOAA) was hacked, allegedly by China.[36] This event forced the NOAA cybersecurity teams to shut down all access to data, which may have been vital for disaster planning, aviation, shipping, and other crucial applications.[37] While multiple parties claimed that Chinese hackers were behind the event, identifying a State actor is extremely complex in the context of cyber-attacks.

This fascination with developing methods for non-kinetic use of force, particularly from the cyber realm, is an issue posed by historical adversaries, neutral members, and current allies, which should all be considered and treated with caution. The U.S. Air Force formed a partnership with researchers at the annual Defcon hacking conference in 2019.[38] This interesting method of finding satellite vulnerabilities was keyed by the question "after all, who wouldn't want to hack a satellite?"[39] This demonstrates a desire to discover all potential methods of disabling or damaging a satellite, leading to the ability to exploit such vulnerabilities. It also suggests that when considering space threats, it is not isolated to simply responding to rogue actors but must be considered from all other actors in the space environment.

The main potential ethical issue developing from such research is the increasing trend towards space-based aggression. All nations who have the capacity to invest in space asset defences seem to be simultaneously developing offensive capabilities as well, perhaps relying on the idea that the best defence is a strong offence. If one State (rogue or not) decides to put such research efforts to a functional test, such as with a cyber-attack or a deliberate ASAT "test", there could be aggressive responses globally.

Space systems are not designed (and realistically cannot be designed) to be resilient against the full spectrum of threats. The problem of rogue groups comes down to the advancement of technology and the associated ease of access. As militaries advance their technology, potential adversaries continue to do so as well. Cyber threats can be subtle – a breach of critical

---

36 Nayef Al-Rodhan, "Cyber security and space security." *The Space Review*, 26 May 2020.
37 Mary Pat Flaherty, Jason Samenow, and Lisa Rein, "Chinese hack U.S. weather systems, satellite network." *The Washington Post*, 12 November 2014.
38 Al-Rodhan, "Cyber security and space security", 2020.
39 Brian Barrett, "The Air Force Will Let Hackers Try to Hijack an Orbiting Satellite." *Wired*, 17 September 2019.

information, a delayed or misaligned navigation signal on a spoofed guided missile, a small observation of our communications and planning, allowing for evasion or traps to be set. Such impacts on space assets could have devastating consequences.

Violent Non-state Groups and Rogue States

Terrorist groups are both motivated and capable of conducting a terrorist attack in space.[40] As discussed previously, the technology has already proven itself to be dangerous, yet obtainable and readily available. Additionally, a fundamental issue of space warfare is the fact that the threat of "Space Terrorism" is currently, to a large degree, neglected by decision-makers and the current military training, discussions, and operational preparation.

Given the nature of their goals, terrorist groups may use any method to cause disruption, as long as that method is relatively easy for them to acquire and use to achieve their assumed objectives of mass casualties and/or long-lasting psychological effects.[41] It can be reasonably assumed that a rogue State, in comparison, would have more discretion in their methods. This also comes with the general assumption that a rogue nation has greater resources, organisation, and training, making them (currently) a more significant threat that must be observed very closely. Many States are gaining various forms of space capability. According to the National Air and Space Intelligence Centre (NASIC), as of December 2018 nine countries, as well as the ESA, possess independent spacecraft launch capability: China, India, Iran, Israel, Japan, the U.S., North Korea, South Korea, and Russia.[42] Seven of these also have their own launch vehicles.[43] Many more can access those existing capabilities regardless. Four countries have conducted ASAT demonstrations, 31 have ballistic missiles, five have Intercontinental Ballistic Missiles capable of reaching low Earth orbits,[44] and as of 31 March 2020, there are 2,666 active artificial satellite orbiting the Earth.[45] It may sound repetitive to point out the global scale of space

---

40 Remuss, "The Need to Counter Space Terrorism – A European Perspective", 2017.
41 Martha Crenshaw, *Terrorism in Context*, Pennsylvania State University Press 1995.
42 National Air & Space Centre, *Competing in Space*, Published December 2018.
43 OECD, "Space launch activities worldwide." In *The Space Economy at a Glance*. Paris: OECD Publishing, 2011.
44 Sergio Pecanha and Keith Collins, "Only 5 Nations Can Hit Any Place on Earth with a Missile. For Now." *The New York Times*, 7 February 2018.
45 E. Mazareanu, *Number of satellites in orbit by major country as of March 31, 2020*, 2020.

and space warfare, however, this is what makes the domain so much more challenging to protect, control, and operate within.

Space policy arguably still fall behind general security studies by not yet properly accounting for new threats, such as space warfare and space terrorism, and desperately needs to adapt. Similarly, the discussion of the ethical use of space often does not account for the fact space is already militarised. As discussed in the previous sections, space is already a battleground. It is used directly for events such as observations and communications which lead directly to attacks, denials of service, and interruptions, even if no traditional weapons are currently stored in space.

In general warfare and terrorism, targets are usually chosen with regard to a symbolic purpose, or as a method of conveying a message, as opposed to simply "being destructive".[46] Hence, there is generally a very open-ended category of attractive targets, with space assets beginning to stand out amongst them. Historically, specialists have agreed that terrorist operations and tactics reveal a remarkably low degree of innovation in contrast to a very high degree of imitation.[47] Therefore, as space militarisation and weaponisation grow globally, rogue actors will begin to imitate the technology and pose a higher level of threat.

Militaries are putting a significant amount of resources into the protection and defence of their space assets, from internal specialists and significant amounts of technology research to building on offensive deterrence capabilities. The problem lies with the general military operators lacking awareness of how reliant the existing capabilities are on space systems, and just how devastating the loss of them (or loss of access to them) could be.

There is significant evidence of known rogue groups investing in technology and education to ensure they remain a considerable threat.[48] According to the article, *The Role of Technology in Modern Terrorism*, footage obtained by Sky News showed members of the Islamic State already possess sophisticated surface-to-air missiles and are reportedly developing versions that be used in attacks against airliners and military jets flying over controlled territories.[49] Terrorism is assuming an even more global

---

46 Rex A. Hudson, *The Sociology and Psychology of Terrorism: Who Becomes a Terrorist and Why?*, 1999.
47 Remuss, "The Need to Counter Space Terrorism – A European Perspective." 2017.
48 Pierluigi Paganini, "The Role of Technology in Modern Terrorism." *InfoSec Institute*, 2018.
49 Pierluigi. "The Role of Technology in Modern Terrorism", 2018.

connotation due to new technologies and represents one of the main threats to modern society as stated by the World Economic Forum 2016.[50]

According to a paper by the European Space Policy Institute (ESPI), there are three categories of Space Terrorism: attacks against satellites, attacks on launch facilities/ground stations, and attacks on the user/service equipment.[51]

When it comes to considering NSGs or rogue States exploiting space, the threats are not restricted to the stereotypical terror attack which uses kinetic destruction as a motivator. As discussed previously, there are many ways in which rogue groups can disrupt, damage, or destroy our space systems. As well as destruction, simply having space assets and capabilities poses a series of problems that must be addressed. A series of examples for these non-standard forms of disruption are discussed below.

Factual Examples

In the past, a series of jamming and piracy events have occurred in the commercial satellite sector. An example is when the mobile satellite communication signal provided by Thuraya Satellite Telecommunications was jammed from three isolated locations inside Libya, compromising the L-band signal for more than six months in 2006 throughout the area.[52] Similarly, the Liberation Tigers of Tamil Eelam (LTTE) hijacked the Intelsat-12 satellite in geosynchronous orbit to broadcast their propaganda across India. While Intelsat continuously tried to interrupt LTTE's pirating, LTTE was able to continue its satellite piracy for two years.[53] Two similar events have occurred in China, one example being when followers of the Falun Gong spiritual movement overrode the broadcast signals of nine China Central Television stations and ten provincial stations in June 2002 and replaced the programming with their content. This occurred again in 2004 where they disrupted AsiaSat signals for four hours.[54]

More recently, in 2011 Iran "captured" a U.S. RQ-170 Sentinel unmanned aircraft, using a combination of communication jamming and

---

50 World Economic Forum Annual Meeting. *World Economic Forum*, 20-23 January 2016.
51 Remuss, "The Need to Counter Space Terrorism – A European Perspective", 2017.
52 Charles Q. Choi, "Libya Pinpointed as Source of Months-Long Satellite Jamming in 2006", 2007.
53 John C K. Daly, "LTTE: Technologically innovative rebels." *Intellibriefs*, 5 June 2007.
54 Daly, "LTTE: Technologically innovative rebels", 2007.

GPS spoofing. Allegedly, Iranian electronic warfare specialists were able to cut off communications links to the drone and were able to then reconfigure the drone's GPS coordinates to make it land in Iran.[55] This act could also have likely been meaconing. Meaconing occurs by receiving radio signals and rebroadcasting them on the same frequency to confuse navigation. Meaconing stations cause inaccurate bearings to be obtained by aircraft or ground stations.[56]

Incidents were reported throughout 2018 and 2019 of GPS signals for transponders in the main port of Shanghai being inaccurate. These signals are used to broadcast the location and direction of a ship, which is required by international maritime law. One U.S container ship, the Manukai, reported another ship being broadcast as travelling down the channel towards the Manukai at significant speed, whereas the captain could visually see it was docked and stationary. Another ship was being reported almost five kilometres away from its actual location. Todd Austin, leading authority on GPS jamming and spoofing at the University of Texas declared that "to be able to spoof multiple ships simultaneously into a circle is extraordinary technology", especially given each one was spoofed to a different location.[57]

The technology needed to jam many satellite signals is commercially available and relatively inexpensive.[58] Attribution of these non-kinetic attacks is also significantly challenging. In 2015, General John Hyten, commander of USAF Space Command at the time, noted that the U.S. military was jamming their own communications satellites 23 times a month on average.[59]

The ease of satellite tracking was proven during the 1950s, by the "Moon-watch Program" as well as the Kettering Group, who proved that tracking only required minimal technology. Amateur satellite observers used stopwatches, sky maps, personal computers and sometimes binoculars to determine satellites' orbital elements.[60] Hence, even simple tracking can be achieved using common, inexpensive electronics with minimal training. This is in line with U.S. Undersecretary of State Robert Joseph's concern about non-governmental satellite observers tracking satellites and posting

---

55 Nancy Owano, "RQ-170 drone's ambush facts spilled by Iranian engineer", 2011.
56 The Free Dictionary – Meaconing, https://www.thefreedictionary.com/meaconing.
57 CSIS, *Space Threat Assessment 2020*, 15-16.
58 Pavel Velkovsky, Janani Mohan, and Maxwell Simon. *Satellite Jamming*, CSIS, 2019.
59 Sydney J. Freedberg Jr., "US Jammed Own Satellites 261 Times; What If Enemy Did?" *Breaking Defence*, 2 December 2015.
60 Remuss, "The Need to Counter Space Terrorism – A European Perspective." 2017.

their orbits on the internet, which could be used by terrorist organisations.[61] Basic internet research can show that the amateur community maintains orbital elements for most classified U.S. vehicles in Low-Earth Orbit (LEO). Their identified missions can be derived from press stories and supplemented by an analysis of orbits and visual appearance, making secrecy an almost impossible component of satellites.

Why This Is a Problem

Whilst the previous sections speak for themselves, there is a significant relationship between military ethics and the problems posed by NSGs and rogue States exploiting space that must be discussed. Ethics is the process of questioning, discovering, and defending one's values and purpose.[62] In the military, this is pertinent as they are legally authorised to operate with weapons, technology, and skills capable of destruction, up to and including taking life, all of which are actions that may directly involve space assets. Whilst militaries have ethical training and moral restrictions which deliberately tailors actions to be for the benefit of all persons involved in a conflict, this is not the case for rogue actors, and particularly NSGs. Knowing the risks and threats is important because rogue actors are not acting on the same moral values or ethical restrictions which shapes our decision making. Such groups will see a capability weakness, or an opportunity for disruption, and potentially not face the same ethical dilemmas when choosing to act on it that we would. Rogue States will have their own politically driven motives for space dominance, possessing a credible space capability, or simply undermining those of adversary nations.

Aside from the risk of losing capability, the impairment or loss of a significant satellite could very rapidly escalate to a direct conflict or generate other unpredictable and dangerous consequences.[63] The act of destroying a satellite can damage political perspectives, national trust, and international partnerships. It can be the tipping point of tensions and lead to retaliation or perceived retaliation which endangers the entire State.

There are States with either credible technological abilities, significant dedicated research towards space technology, or both, that may not

---

61 Robert Joseph, "The U.S. National Space Policy." The George C. Marshall Institute.
62 The Ethics Centre, *What is Ethics?*, https://ethics.org.au/why-were-here/what-is-ethics.
63 Laura Grego, "A History of Anti-Satellite Programs." UCS USA, January, 2012.

necessarily maintain the understanding of peaceful use that currently exists. Numerous strategists suggest we are heading towards, or arguably already in, another Cold War between major powers. The U.S. gained a lot of attention when it established its Space Force on 20 December 2019. However, Russia has had a dedicated Military Space Force since 1992, now falling under Russian Aerospace Forces. China has had its Space Systems Department since 2015, and France initiated a Joint Space Command in 2010. During 2020 Japan established its Space Operations Squadron, Iran had its Space Command acknowledged, and India established a Defence Space Agency in 2018. Space in the military, and space in warfare, is not a new concept to be faced. However, even though this great-power competition will include rivalry in space, space also involves a great deal of cooperation, such as with the International Space Station.[64] As a result, the most likely scenarios involving attacks against significant coalition interests in space may not come from other States. Instead, they involve non-state actors seeking to challenge the existing international order, overturning the status quo in their own countries, or profiting from the lack of attention paid to them by State governments across the world.[65]

As mentioned, one of the greatest policy implications of this "second space age" is that the availability of advanced technology and space capabilities on the commercial market which could potentially "bring the advantages of space within the reach of rogue nations and non-state actors".[66] Whilst military assets are generally well-defended, dual-use systems (satellites that operate for both civilian and military purposes), and commercial space assets are far less so. Access or damage to one system can have disastrous flow-on effects, and any threat that goes unchallenged is simply opening the door wider for rogue actors to play their hand in space.

References

Al-Rodhan, Nayef. "Cyber security and space security." *The Space Review*, May 26, 2020.
Amenabar, Caroline. "Counterspace Weapons 101." *Aerospace 101*, July 23, 2020.

---

[64] Simon Saradzhyan and William Tobey "US-Russian space cooperation: a model for nuclear security." *Bulletin of the Atomic Scientists*, 7 March 2017.
[65] Gregory D. Miller, "Space Pirates, Geosynchronous Guerrillas, and Non-terrestrial Terrorists." *Air and Space Power Journal* 33(3), 2019.
[66] Harrison, et al., *Escalation and Deterrence in the Second Space Age*, 2017.

Arms Control Association. "Worldwide Ballistic Missile Inventories." December 2017.
Australian Department of Defence. "The Defence White Paper." 2016.
Barrett, Brian. "The Air Force Will Let Hackers Try to Hijack an Orbiting Satellite." *Wired*, September 17, 2019.
Baumann, Dieter. "Military Ethics: A Task for Armies." *Military Medicine*. Vol. 172. no. suppl_2. December 1, 2007. 34-38.
BBC News. "Iraq insurgents 'hack into video feeds from US drones'." *BBC News*, December 17, 2009.
Choi, Charles Q. "Libya Pinpointed as Source of Months-Long Satellite Jamming in 2006." *Space.com*, April 09, 2007.
Coats, Daniel R. *Worldwide Threat Assessment of the US Intelligence Community*. Office of The Director of National Intelligence, 2019.
Commission to Assess United States National Security Space Management and Organisation. "Report to the Commission to Assess United States National Security Space Management and Organisation." Washington, DC, 2001.
Crenshaw, Martha. *Terrorism in Context*. Pennsylvania: Pennsylvania State University Press, 1995.
CSIS. *Space Threat Assessment 2020*. Center for Strategic and International Studies, 2020, 15-16.
Daly, John C K. "LTTE: Technologicallyinnovative rebels." *Intellibriefs*, June 05, 2007.
David, Leonard. "China, Russia Advancing Anti-Satellite Technology, US Intelligence Chief Says." *Space Insider*, May 18, 2017.
Enold, Colonel Scott A. "Rogue States and Deterrence Startegy." Carlisle, PA: US Army War College, April 02, 2009.
Farlex. "The Free Dictionary – Meacloning." n.d.
Flaherty, Mary Pat, Jason Samenow, and Lisa Rein. "Chinese hack U.S. weather systems, satellite network." *The Washington Post*, November 12, 2014.
Freedberg Jr., Sydney J. "US Jammed Own Satellites 261 Times; What If Enemy Did?" *Breaking Defense*, December 02, 2015.
Grego, Laura. "A History of Anti-Satellite Programs." *UCS USA*. UCS. January 2012. https://www.ucsusa.org/sites/default/files/2019-09/a-history-of-ASAT-programs_lo-res.pdf (accessed August 2020).
Grego, Laura. *The Anti-Satellite Capability of the Phased Adaptive Approach Missile Defense System*. Federation of American Scientists, 2011.
Harrison, Todd, Zack Cooper, Kaitlyn Johnson, and Thomas G. Roberts. *Escalation and Deterrence in the Second Space Age*. CSIS, 2017.
Hofmann, Claudia, and Ulrich Schneckener. "Engaging non-state armed actors in stat and peace-building: options and strategies." *International Review of the Red Cross*. Vol. 93. September 2011.
Hudson, Rex A. *The Sociology and Psychology of Terrorism: Who Becomes a Terrorist and Why?* The Library of Congress, 1999.
Joseph, Robert. "The U.S. National Space Policy." Washington, DC: The George C. Marshall Institute, 2006.

Kavka, Gregory S. "Space War Ethics." *Ethics, Special Issue: Symposium on Ethics and Nuclear Deterrence* (The University of Chicago Press) 95, no. 3 (1985): 673-691.

Kelso, TS. "Analysis of the 2007 Chinese ASAT Test and the Impact of Its Debris on the Space Environment." *AMOS Conference*. Maui, Hawaii, 2007.

Kessler, Glenn. "Bachmann's claim that China 'blinded' U.S. satellites." *The Washington Post*, October 4, 2011.

Klein, John J. "Space Warfare: Deterrence, Dissuasion and the Law of Armed Conflict." *War on The Rocks*, August 30, 2016.

Krepon, Michael. "Space and Nuclear Deterrence." *The Space Review*, September 16, 2013.

Liebowitz, Matt. "Hackers Interfered With 2 US Government Satellites." *Space.com*, October 27, 2011.

MacDonald, Bruce W. "Crisis Stability in Space: The China Challenge and Implications for U.S. Security." Monterey, California: Center on Contemporary Conflict, The Naval Postgraduate School, 2013.

Matignon, Louis de Gouyon. "The Kessler Syndrome." *Space Legal Issues*, March 27, 2019.

Mazareanu, E. *Number of satellites in orbit by major country as of March 31, 2020*. Statista, 2020.

Miller, Gregory D. "Space Pirates, Geosynchronous Guerrillas, and Nonterrestrial Terrorists." *Air and Space Power Journal* 33, no. 3 (2019).

National Air & Space Intelligence Center. "Competing in Space." December 2018.

NETSCOUT. *NETSCOUT Threat Intelligence Report*. NETSCOUT, 2019.

OECD. "Space launch activities worldwide." In *The Space Economy at a Glance*. Paris: OECD Publishing, 2011.

Owano, Nancy. "RQ-170 drone's ambush facts spilled by Iranian engineer." *Phys. org (Science X Neworx)*, December 17, 2011.

Paganini, Pierluigi. "The Role of Technology in Modern Terrorism." InfoSec Institute. February 3, 2018. https://resources.infosecinstitute.com/the-role-of-technology-in-modern-terrorism/#gref (accessed June 10, 2019).

Pecanha, Sergio, and Keith Collins. "Only 5 Nations Can Hit Any Place on Earth With a Missile. For Now." *The New York Times*, February 7, 2018.

Remuss, Nina-Louisa. "The Need to Counter Space Terrorism – A European Perspectiv." *ESPI Perspectives* 17. European Space Policy Institute, 2017.

Saradzhyan, Simon, and William Tobey. "US-Russian space cooperation: a model for nuclear security." *Bulletin of the Atomic Scientists*, March 7, 2017.

Stadler, C. "Military ethics as part of general system of ethics." In *Civil-Military Aspects of Military Ethics*, by EM Micewski. Vienna, Austria, 2003.

The Ethics Centre. *What is Ethics?* https://ethics.org.au/why-were-here/what-is-ethics.

UNOOSA. *Space Law Treaties and Principles*. n.d. https://www.unoosa.org/oosa/en/ourwork/spacelaw/treaties.html.

—. "The Outer Space Treaty." *restrictions against the placement of weapons of mass destruction*. 1967.

Velkovsky, Pavel, Janani Mohan, and Maxwell Simon. *Satellite Jamming*. CSIS, CSIS, 2019.

Weeden, Brian. "2007 Chinese Anti-Satellite Test Fact Sheet." Secure World Foundation, November 23, 2010.

Witteried, LTCOL Peter F. "A Strategy of Flexible Response." US Army War College, 1972.

World Economic Forum. *World Economic Forum Annual Meeting*. World Economic Forum. January 20-23, 2016. https://www.weforum.org/events/world-economic-forum-annual-meeting-2016.

9

BIOETHICS AND MILITARY OPERATIONS IN SPACE

Sheena M. Eagan

According to the World Medical Association, medical ethics in times of armed conflict are identical to times of peace.[1] Others have challenged this and argued that the austerity and intensity of combat medicine are unparalleled, warranting unique moral considerations.[2] Despite this disagreement, there remains broad agreement that the intensity and austerity involved in military operations create a morally complicated space – replete with ethical dilemmas that are more extreme and have higher stakes than most. Although international conflict's changing landscape has continuously redefined the battlefield within which these military operations take place, recurring ethical dilemmas have persisted. As evidence, a 2015 report from the U.S. Defense Health Board's Medical Ethics Subcommittee identified the following recurring issues in military medical ethics:[3] (a) Disclosures of private health information to command; (b) Treatment priorities and triage; (c) Humanitarian Assistance, Disaster Response, and Medical Support Missions; (d) Detainee Treatment; (e) Deployments and Professional Support; and (f) Post-Deployment Issues. Although not exhaustive, this list was developed based on a rigorous systematic review of the field. Other recurring themes include

---

1 World Medical Association, *Ethical Principles of Health Care in Times of Armed Conflict and Other Emergencies* (2017), https://www.wma.net/wp-content/uploads/2017/02/4245_002_Ethical_principles_web.pdf.
2 John Moskop, "A Moral Analysis of Military Medicine," *Military Medicine* 163 no. 2 (1998): 76-79; William Madden and Brian Carter, "Physician-Soldier: A Moral Profession." In *Military Medical Ethics*, eds. Thomas E. Beam et al. (Washington, DC: The Borden Institute, 2003), 269-291.
3 Defense Health Board. (2015). *Ethical Guidelines and Practices for U.S. Military Medical Professionals*. United States Department of Defense (2015).

discussions of service-member autonomy; physician complicity in torture and abuse; research ethics; and the growing intersection with the broader field of military ethics more generally. While many of these issues are likely to persist in space, new and novel issues are inevitable.

Throughout the history of modern warfare, we have seen shifts from state-based warfare to counter insurgency and asymmetrical warfare. Still, military operations have only occurred "at home", or on the planet Earth. We know what to expect and how to live on Earth. Space promises a level of uncertainty never before seen in military operations that will complicate risk management. As will be shown in this chapter, military operations in space present unique and significant complexities of relevance to bioethics. This chapter will explore bioethical considerations in this setting, focusing on how space challenges traditional bioethics due to uncertainty and unknown risk that complicate understandings of harm to human health. In light of this significant level of unknown risk, a framework developed by the Institute of Medicine (IOM) is offered as ethically appropriate due to its twin grounding in clinical and research ethics.[4] In the final section, issues specific to military actors in space will be introduced.

The United States Space Force (USSF)

In 2018, the U.S. Pentagon announced plans for a sixth military service – the Space Force. Once limited to the world of Hollywood as science fiction, space combat has leapt from the cinematic screen to become reality. Founded in December 2019, the establishment of the U.S. Space Force (USSF) as a new branch of the U.S. Department of Defence provides scholars and ethicists the unique opportunity to engage in preventive ethics – discussing ethical issues before they become dilemmas. Bioethics represents an important part of this discussion as priorities, values, and understandings of risk may be altered by this new combat environment.

This new branch of military service aims to develop war-fighters, technologies, and tactics for space warfare. According to its official website, the mission of the USSF is "to protect U.S. and allied interests in space and

---

4 Institute of Medicine, *Safe Passage: Astronaut Care for Exploration Missions.* Committee on Creating a Vision for Space Medicine During Travel Beyond Earth Orbit, Board on Health Sciences Policy, Institute of Medicine, Washington, DC: The National Academies Press (2001).

provide space capabilities to the joint force."[5] Its responsibilities include "developing military space professionals, acquiring military space systems, maturing the military doctrine for space power, and organizing space forces to present to our Combatant Commands."[6] Officially part of the U.S. Air Force, the USSF is still in its nascent stages and there is much that we do not know about its priorities, goals and timelines. However, one thing can be sure: this new setting for military operations will usher in unique ethical dilemmas and considerations for those deployed in its service, as well as those who care for them.

Moving Towards Long-Duration Missions and Habitation

Much like the future of space exploration itself,[7] military operations in space will likely be characterized by long-term habitation and long-distance missions. NASA is developing the capabilities needed to send humans to an asteroid by 2025, and Mars in the 2030s.[8] These goals are in line with those outlined in the bipartisan NASA Authorization Act and U.S. National Space Policy.[9] Plans to put service-members in space are already being discussed, with top generals stating that it is "… in the future members of the United States Space Force will go physically, directly, and personally into space…"[10] As further evidence, The Institute for Applied Space Policy and Strategy (IASPS), whose membership is composed of newly minted USSF officers, has specifically examined the possibility of a lunar military base and space soldiers.[11] One of only two IASPS research teams is titled "Military on the Moon" and aims "to evaluate the possibility and necessity of a sustained United States military presence on the lunar surface."[12] The planning has started and while discussions of this type are still speculative,

---
5 USSF, *United States Space Force Mission*, 2020 at https://www.spaceforce.mil/About-Us/About-Space-Force/Mission/.
6 USSF, 2020.
7 Institute of Medicine, 2001.
8 Claudia Jones, email to author, 2019.
9 Claudia Jones, email to author, 2019.
10 M. Weisgerber, *Boots on The Moon Are Going to Have to Wait*. October 1, 2020. https://www.defenseone.com/technology/2020/10/boots-moon-are-going-have-wait-space-force-general-says/168939/.
11 USAFA Institute for Applied Space Policy and Strategy. 2019. *Military on the Moon*, 2020. https://usafaiasps.wixsite.com/space/astropolitics-copy.
12 USAFA, 2020.

they are key to ensuring military operations in space are conducted in the most ethical way possible.

The shift towards long-duration space missions will be the focus of this chapter. Short-term and repeat missions present their own ethical dilemmas, but extended space missions, and habitation serve to both magnify those issues while also ushering in new and novel concerns. After decades of space exploration, we have learned a great deal about its effects on human health. NASA's Human Research Program has identified thirty space-related health risks for humans that range from radiation exposure to behavioural health and psychological impacts.[13] The 2001 IOM report identifies the three most significant risks for long-duration missions as follows: radiation, loss of bone mineral density, and behavioural adaptation.[14] But the most morally complicated feature of these missions is not what we know – it's what we don't know. The long-term nature of these missions introduces significant ethical considerations related to uncertain but probable and considerable risk to human health. The possibility of harm is significant to bioethics and the commonly cited principles of autonomy, beneficence, non-maleficence, and justice.[15] Physicians are supposed to do what will benefit their patients and are under a professional moral obligation to avoid causing them harm. As part of its effort to prevent or mitigate health risks, NASA has relied on established health standards for astronauts.[16] However, long-duration missions are unlikely to meet the existing health standards, and new health standards will need to be developed and bioethics must be a part of the discussion. As mentioned before, long-duration space missions will not only exaggerate existing ethical dilemmas but also introduce new and novel concerns.

Medical Care in Space

While the level of risk to human health associated with long-term space missions is still mostly unknown, we know that those serving in this environment will need medical care. NASA estimates a 90 percent probability that a crew member will suffer severe injury or illness during long-term

---

13 Institute of Medicine, 2001.
14 Institute of Medicine, 2001.
15 Tom Beauchamp and James Childress, *Principles of Biomedical Ethics* (New York, NY: Oxford University Press, 2013).
16 Institute of Medicine, 2001.

space missions.[17] Importantly, this figure focuses on injuries unrelated to the space environment, making the risks to human health even higher. Due to this operational planning's speculative nature, it is difficult to assess and understand risk adequately. Current ethical standards were developed for short-duration and repeat space missions where evacuation or return to Earth was possible within days. However, long-duration/distance missions and habitation will mean crews are isolated and unable to access the same options and resources possible during short-term missions. This resource scarcity will significantly impact medical decision-making and, thereby, ethical analysis. In light of this, the future of military operations in space presents unique conditions for which new ethical standards are needed.[18] The following sections will briefly explore issues unique to long-duration operations in space.

Proximity and Distance

The isolation and distance from Earth will significantly limit the types of care and support that service-members can access. This reality will be an ethically significant departure from current military health systems, which are robust. Even while deployed, service-members have access to advanced care and are quickly repatriated to modern western hospitals. In contemporary conflicts, most care is not provided on the battlefield or even within the combat zone. Modern evacuation has been an integral part of medical care in military operations since Vietnam.[19] This is particularly true of the past two decades of conflicts in Iraq and Afghanistan, where air evacuation was a mainstay of combat casualty care.[20] Many ethical dilemmas (related to triage and other recurring dilemmas in military medicine) have been largely resolved through the advent of air evacuation and the development of fast evacuation protocols. Unlike earlier wars where the limitations of forward operating positions restricted access to care, triage in contemporary conflicts involve stabilizing patients, followed by immediate transport out of the active conflict zone to receive definitive

---

17 C. S. Allen, R. Burnett, J. Charles, F. Cucinotta, R. Fullerton, J. Goodman, A. Griffith, J. Kosmo, M. Perchonok, J. Railsback, S. Rajulu, D. Stillwell, G. Thomas, and T. Tri, *Guidelines and Capabilities for Designing Human Missions*, NASA (2003).
18 Institute of Medicine, 2001.
19 M.J. Novosel. *Dustoff: The Memoir of Army Aviator* (Presidio Press: 2003).
20 Atul Gawande, "Casualties of War – Military Care for the Wounded from Iraq and Afghanistan," *New England Journal of Medicine* 351: 2471-2475 (2004).

care.[21] The average time from injury on the battlefield to arrival in the United States is less than four days – compared to 45 days in Vietnam.[22]

Reliance on evacuation will not be possible in space, especially given the distances and duration expected of future missions. While the long-term goal of space habitation will likely include establishing and equipping extra-terrestrial medical treatment facilities, this will not be possible for some time and only apply to habitation/colonization locations. Therefore, it is likely that all medical care will need to be provided in place. Providing emergency medical care in space will be significantly limited by several constraints related to resource scarcity and the austere environment in which the care will be provided. The modern hospital that generally marks the end of successful evacuation will be difficult, if not impossible, to recreate in space. Put bluntly, we will not be able to bring everything that we need. Beyond that, changes in gravity and the possible physiological changes associated with space will complicate standards of care and best practices (to be discussed in the next section).

Providers themselves will also be a scarce resource. The number of crew members will be limited by necessity, further restricting the types of care and interventions that can be offered in this setting. From an ethical standpoint, this complicates the practitioner's ability to deliver an appropriate standard of care. Even if a physician, or other medical provider, is part of every crew, it will be impossible to anticipate all events and similarly impossible for physicians to be adequately cross-trained across multiple specialties. This is especially true for interventions that require specialized training, including the treatment of rare pathologies or surgical intervention. In light of this, medical care in space will likely continue to rely heavily on a combination of new technology and remote monitoring along with the consultation of specialist medical providers back on Earth.

Surgery in Space

The most challenging type of medical intervention to perform in space will be surgery. However, emergency surgery will inevitably be needed at some point in future space operations. In recognition of this, NASA has been

---

21 This is true for service-members and allied forces. Evacuation and access to definitive care is currently not guaranteed for enemy combatants and civilians.
22 Gawande, 2004.

exploring telesurgery to provide surgical intervention without requiring an onboard doctor. NASA Extreme Environment Mission Operations (NEEMO) researchers have used the Aquarius underwater laboratory to simulate space isolation and confinement.[23]

The preliminary research into telesurgery reveals essential issues of relevance to assessing risk. Of course, this technology is novel and presents similar risks to all new technological interventions. The research conducted by NASA NEEMO has highlighted challenges that may make this option ethically untenable. Specifically, aside from concerns related to the feasibility of bringing these machines into space due to their size and weight, there are also concerns related to data latency, or lag time, in light of the vast distances involved in space travel. In past experiments, a surgeon successfully performed simulated surgery in the Aquarius habitat from thousands of miles away in Canada.[24] While overall successful, this experiment underscored concerns related to latency, as it resulted in a single surgical knot taking 10 minutes to complete.[25] Research on this topic continues and the technology will continue to improve to address these issues. However, the current state of knowledge related to latency and a user's ability to compensate is inconsistent. Notably, no researcher or practitioner has argued that an operator would be capable of compensating for the extended latency required for long-distance manned space missions. Estimates place the Mars latency to be between 6.5 and 44 minutes depending on its orbital positioning.[26] Not only do these issues present new challenges to a surgeon's technical skill, but it also limits the surgeons' ability to react to unexpected complications during surgery. There are additional contextual features that challenge the standard of care, which may result in harm to the patient. The effects of zero gravity and physiological changes resulting from long-term space missions will change how interventions happen and the bodies being intervened upon. Ultimately, we can speculate that the risk to those involved would be substantial.

The expectation of surgery in space highlights the level of complex medical care that will need to take place in support of future space missions.

---

23 R. Thirsck, D. Williams and M. Anvari, "NEEMO 7 undersea mission," *Acta astronautica* 60 no. 2: 512-517 (2007).
24 C.R. Doarn, M. Anvari, T. Low, and T.B. Broderick, "Evaluation of teleoperated surgical robots in an enclosed undersea environment." *Telemedicine journal and e-health* 15: 325-335 (2009).
25 Doarn et al, 2009.
26 T. Haidegger, J. Sándor, and Z. Benyó. 2011. "Surgery in space: the future of robotic telesurgery." *Surg Endosc.* 25 no.3: 681-90 (2011).

But this is only part of the moral landscape of concern to bioethics. While medical care will inevitably be limited in space, this is not the only care required to ensure the health and wellbeing of those going to, living in, and returning from space. The risks to human health are uncertain and long term (Institute of Medicine 2001).[27] A specific and comprehensive health system will be necessary for this population, including preventive medicine, responsive care, and long-term monitoring.

Do Traditional Bioethics Apply In This Setting?

The unique factors explored above completely change what medical care looks like, thereby complicating traditional ethical principles' applicability. According to the 2014 report from the IOM on this topic, new ethical standards are needed for long-duration space missions and habitation precisely because of these issues.[28]

The following section will provide a brief overview of the ethical framework outlined by the IOM in 2014.[29] This report from the National Academy of Science was requested by NASA and is meant to guide the agency's decision making for future long-duration space missions. The IOM committee's report outlines an ethical framework composed of a series of principles.[30] The report also presents recommendations for ethically responsible long-duration space missions.[31]

An Ethical Framework: From the Institute of Medicine to Space

In 2014, the IOM published its report titled *Health Standards for Long Duration and Exploration Spaceflight: Ethics Principles, Responsibilities, and Decision Framework*.[32] The principles offered in this report are based on the Principles of Biomedical Ethics[33] but informed by considerations in both research ethics and clinical ethics. While these principles and recommendations were

---

27 Institute of Medicine, 2001.
28 Institute of Medicine, 2001.
29 Institute of Medicine, *Health Standards for Long Duration and Exploration Space Flight* (Washington, DC: The National Academies Press, 2014).
30 Institute of Medicine, 2014.
31 Institute of Medicine, 2014.
32 Institute of Medicine, 2014.
33 Tom Beauchamp and James Childress, 2013.

developed for NASA, they are relevant to discussing military operations in space and offer a valuable paradigm shift towards a framework informed by both clinical medicine and research.

The IOM report makes a compelling argument that space bioethics shares morally relevant parallels with the issues raised in human subject research.[34] The positioning of space bioethics at the intersection of clinical and research addresses the high levels of uncertainty.

> **Principle of Non-Maleficence/Avoiding Harm**
> **Recommendation:** NASA should exhaust all feasible means to minimize risk to astronauts.

Non-maleficence is a central principle in moral philosophy that has been applied across bioethics. This principle includes duties to avoid causing harm to others, remove existing harms, and prevent likely harms from occurring. Harms are not always physical and should include psychological as well as social harms. In this context, harm is also not limited to the individual, as broader harms to the crew, mission, and society will need to be considered.

According to this principle, ethically sound decisions must exhaust all feasible means to minimize risk to humans in space.[35] Decision-making must be guided by this principle and thereby avoid, prevent, or eliminate harm. This extends beyond decisions specifically related to medical care and could include vehicle design, safety processes, protective technologies, and other risk mitigation strategies.[36] Standing at the intersection of clinical medicine and research, risks also include risks related to health data privacy and confidentiality. As an example, the small sample size of participants in any of these studies could render commonly used strategies for protecting individual privacy (for example, de-identifying data) ineffective. In focusing on harms associated with research, it is essential to note that service-members are recognized as a vulnerable (or captive/special) population due to the history of human subjects' abuse in military research. In accordance with this history and the protections of this group, particular attention must be paid to issues of autonomy and coercion (discussed later).

---

34 Institute of Medicine, 2014.
35 Institute of Medicine, 2014.
36 Institute of Medicine, 2014.

> **Principle of Beneficence/Providing Benefit**
> **Recommendation:** NASA should weigh a mission's potential benefits, including scientific and technical importance, as well as benefits to current and future astronauts and the public.[37]

Beneficence in traditional biomedical ethics refers to acting in ways that benefit patient health. In the context of space operations, benefits are not limited, or even focused on the individual. Long-duration space missions and space habitation will expose service-members to both certain and uncertain risks of harm for broader societal benefits instead of individual ones. In clinical medicine, individuals participate in medical treatments that may bring risks along with desired benefits. Therefore, bioethics in space must include considerations relevant to research where those involved take on uncertain risks for the benefit of broader society. While individuals may gain benefit from participation, the benefits of human spaceflight are societal and include advancements in knowledge, as well as international collaboration. It is important to note that long-duration space missions may also benefit future crews and help avoid harm in future missions. Data collected will inform the larger understanding and practice of space medicine.

> **Favourable Balance of Risk and Benefits**
> **Recommendation:** NASA should systematically assess risks and benefits and ensure that the benefits sufficiently outweigh the risks.[38]

A favourable balance of risk and benefit must be maintained. Assessing risks should include those to individual astronauts, as well as those to the broader mission. Benefits should consist of those expected for society, as well as future humans in space. Additionally, morally sound risk-benefit judgments should take into account both quantifiable risks and benefits, as well as those not quantifiable given current knowledge.

---

37 Institute of Medicine, 2014.
38 Institute of Medicine, 2014.

> **Principle of Respect for Autonomy**
> **Recommendation:** Astronauts' participation should be voluntary. NASA should keep astronauts informed of a mission's risks and benefits before, during, and after the mission.[39]

The principle of autonomy has been fundamental in western bioethics and is grounded in individual liberty and freedoms, including the right of non-interference and the right to self-govern. Within modern Western medicine, autonomy is among the most widely recognized ethical concepts. Informed consent is part of this principle. According to the concept of informed consent, patients must be fully informed about their diagnosis and any intervention or treatment. According to this principle, participation in space operations should be voluntary.

Obtaining voluntary consent for military operations in space presents many challenges, which will be discussed in the next section. The type of informed consent required in this context shifts towards the realm of research ethics and precisely those considerations meant to protect human subjects in that setting. However, there are also significant differences between this population and traditional research subjects. To conduct ethically responsible research, consent is an ongoing process, and participants are able to withdraw from any study at any point. The ability to revoke consent and halt participation as risks become apparent is fundamental to respect for research autonomy. In long-duration space operations, consent becomes binding and irrevocable at mission launch.[40] The irrevocability of consent significantly limits the autonomy of this group and highlights the necessity of the other principles: avoiding harm, providing benefit, and balancing those principles appropriately. According to the IOM, this irrevocability (coupled with unknown and potential long-term risk) creates an ethical imperative to define the long-term duties owed to those participating in these missions. In the context of our discussion, these would be obligations owed to military service-members by either the institution or the society it serves.

---

39 Institute of Medicine, 2014.
40 Institute of Medicine, 2014.

> **Principle of Fidelity**
> **Recommendation:** NASA should acknowledge its obligation to astronauts, who serve at significant personal risk, by providing health care and other protections not only during but also after the mission.[41]

Similar to the principle of reciprocity discussed in medical ethics, this principle recognizes what society owes to those that take on this level of risk, both certain and uncertain. Those who take on additional burdens or risks for society's benefit are entitled to fidelity or a commitment to minimize any harms that may emerge. Informed by this principle and other considerations, IOM recommends establishing a comprehensive health care system for astronauts.[42] Echoing the 2001 report's recommendations, this health care system should include not only health care but also research and training.[43] Preventive medicine should play a significant role in maximizing mitigated risk. The system should aim to optimize health and the ability to function in deep space and maintain or restore normal function in the pre and post mission phases.

> **Principle of Fairness**
> **Recommendation:** NASA should ensure fairness in its selection of astronauts and crews, distribution of the risks and benefits, and postflight support of astronauts.[44]

The principle of fairness requires like cases be treated alike and unlike differently. This is known as the principle of justice in biomedical ethics. In research, this principle applies to both individuals and groups to ensure that one group does not bear a disproportionate burden. Additionally, the crew's distribution of risks and overall crew composition must be as fair as possible.

According to the IOM, these principles dictate specific responsibilities that NASA should pursue.[45] These responsibilities are relevant to future military operations in space. They include an obligation to fully inform participants and continually work to minimize risk and promote benefit

---

41 Institute of Medicine, 2014.
42 Institute of Medicine, 2014.
43 Institute of Medicine, 2014.
44 Institute of Medicine, 2014.
45 Institute of Medicine, 2001.

(both individual and societal). The need for an ongoing process of informed consent through pre-and post-mission phases highlights another relevant principle that is not mentioned by the IOM: Transparency. As part of the principles outlined in this report, transparency will be critical to both fulfilling obligations and establishing trust with stakeholders. The military must be transparent and open about risks/benefits associated with this type of mission. Transparency is a value of public health ethics that applies to this institutional setting, as its strengths trust and bolsters the other principles.

The Military: Unique Consideration and Further Analysis

The ethical issues highlighted, and principles outlined above are not unique to military operations in space, but rather to long-term missions and habitation in that setting. The positioning of military operations and service-members in space only adds further complications to this environment.

While it is essential to first address the general ethical issues related to health and health care in space, concerns related to military medical ethics will also need to be considered. The military's hierarchical and mission-driven values have historically presented challenges to both clinically oriented medical ethics and research ethics. Physicians practicing in the military already face challenges almost unmatched in the civilian world.[46] As mentioned throughout this chapter, these differences are magnified in space; resources are exceptionally scarce, risks and uncertainty are multiplied, and specialized care is a world away. The austerity of this new environment for military operations means that priorities, values, and understandings of risk may be altered.

The principle of respect for autonomy warrants specific consideration as it is significantly complicated by military service-members' involvement. Although the degree of autonomy retained by those military service-members varies from nation to nation and service to service, it is generally understood that they relinquish a degree of autonomy upon commission or enlistment. The moral reality of military service is one of reduced autonomy and personal liberty. Service-members are not supposed to question

---

[46] Sheena M. Eagan Chamberlin, "Medicine as a Non-Lethal Weapon: The Ethics of Winning Hearts and Minds." *Ethics and Armed Forces* (2015/1): 9-15.

orders;[47] they are told how to dress, where to go, and even when to sleep, eat, and exercise. In light of this culture of obedience, coercion is a significant issue. Service-members do not say no. Beyond that, consent is generally not sought for deployment as it is a requirement of military service. This lack of autonomy extends from mandatory training and deployments to compulsory vaccinations and mandatory treatment. Individuals in military service relinquish this autonomy in a State's service, who takes their autonomy into its stewardship. Military medical ethics has long grappled with balancing respect for autonomy with military values that prioritize mission success and aggregate concerns.[48] How and when autonomy should be valued within the setting should remain a consideration, in light of the many concerns covered in this chapter.

Even if it were possible to avoid coercion, the ability to obtain genuinely informed consent is significantly limited by the inherently high levels of uncertainty and unknown risk. Still, since this type of mission will involve long-term harms not unlike any other battlespace and additional levels of consent should be considered.

Conclusion

This chapter opened with the WMA declaration, stating medical ethics in times of armed conflict is identical to times of peace.[49] Whether or not this is descriptively true on the terrestrial battlefield, medical ethics will inevitably have to change as we look upward to space. Medical decision-making in this setting will be unlike anywhere else, and ethical dilemmas will be more extreme, with many unknowns and high levels of uncertainty. These considerations present significant challenges to traditional medical ethics and should prompt continued discussion and reflection.[50]

---

47 Service-members are only under an obligation to follow lawful orders. Unlawful orders should not be obeyed.
48 Allhoff, 2008; Beam, Sparacino, Howe, Pellegrino, & Hartle, 2003; Howe, 2003.
49 World Medical Association, 2017.
50 Acknowledgements: I would like to thank Ryan Accomazzo and Tyler Brookshire for the research assistance on this project.

References

Allen, C.S., R. Burnett, J. Charles, F. Cucinotta, R. Fullerton, J. Goodman, A. Griffith, J. Kosmo, M. Perchonok, J. Railsback, S. Rajulu, D. Stilwell, G. Thomas, G and T. Tri. *Guidelines and Capabilities for Designing Human Missions*, NASA (2003).

Allhoff, F. *Physicians at War: The Dual-Loyalties Challenge*. New York, NY: Springer, 2008.

Beam, T. E., L.R. Sparacino, E.G. Howe, E.D. Pellegrino and A.E. Hartle. *Military Medical Ethics*. Washington, DC: The Borden Institute, 2003.

Beauchamp, T. and J. Childress. *Principles of Biomedical Ethics*. New York, NY: Oxford University Press, 2013.

Defense Health Board. *Ethical Guidelines and Practices for U.S. Military Medical Professionals*. United States Department of Defense, 2015.

Department of Defense. *DoD Instruction 6025.37: Medical Ethics in the Military Health System*. Office of the Under Secretary of Defense for Personnel and Readiness, 2017.

Doarn, C.R., M. Anvari, T. Low, and T.B. Broderick. "Evaluation of teleoperated surgical robots in an enclosed undersea environment." *Telemedicine journal and e-health* 15(2019): 325-335.

Eagan Chamberlin, S.M. "Medicine as a Non-Lethal Weapon: The Ethics of Winning Hearts and Minds." *Ethics and Armed Forces* (2015/1): 9-15.

Gawande, A. "Casualties of War – Military Care for the Wounded from Iraq and Afghanistan." *New England Journal of Medicine* 351 (2004): 2471-2475.

Haidegger, T., J. Sándor and Z. Benyó. "Surgery in space: the future of robotic telesurgery." *Surgical Endoscopy* 25 no. 3 (2011): 681-90.

Howe, E. G. "Mixed Agency in Military Medicine: Ethical Roles in Conflict." In *Military Medical Ethics*, eds. Beam, T. E., L. R. Sparacino, E. G. Howe, E. D. Pellegrino and A. E. Hartle. Washington, DC: Borden Institute, 2003.

Institute of Medicine. *Health Standards for Long Duration and Exploration Space Flight*. Washington, DC: The National Academies Press, 2014.

Institute of Medicine. *Safe Passage: Astronaut Care for Exploration Missions*. Committee on Creating a Vision for Space Medicine During Travel Beyond Earth Orbit, Board on Health Sciences Policy, Institute of Medicine, Washington, DC: The National Academies Press, 2011.

Madden, W. and B. S. Carter. "Physician-Soldier: A Moral Profession." In *Military Medical Ethics*, eds. Beam, T. E., L. R. Sparacino, E. G. Howe, E. D. Pellegrino and A. E. Hartle, 269-291. Washington, DC: The Borden Institute, 2003.

Moskop, J. C. "A Moral Analysis of Military Medicine." *Military Medicine* 163 no. 2 (1998): 76-79.

Novosel, M. J. *Dustoff: The Memoir of Army Aviator*. Presidio Press, 2003.

Sidel, V. W. and B.S. Levy. "Physician-Soldier: A Moral Dilemma?" In *Military Medical Ethics*, eds. Beam, T. E., L. R. Sparacino, E. G. Howe, E. D. Pellegrino and A. E. Hartle. Washington, DC: The Borden Institute, 2003.

Thirsck, R., D. Williams and M. Anvari. "NEEMO 7 undersea mission." *Acta astronautica* 60 no.2 (2007): 512-517.

USAFA Institute for Applied Space Policy and Strategy. 2019. *Military on the Moon*. Accessed November 27, 2020. https://usafaiasps.wixsite.com/space/astropolitics-copy.

USSF. *United States Space Force Mission*. Accessed November 28, 2020. https://www.spaceforce.mil/About-Us/About-Space-Force/Mission/.

Weisgerber, M. *Boots on The Moon Are Going to Have to Wait*. October 1. Accessed December 1, 2020. https://www.defenseone.com/technology/2020/10/boots-moon-are-going-have-wait-space-force-general-says/168939/).

World Medical Association. *Ethical Principles of Health Care in Times of Armed Conflict and Other Emergencies* (2017) https://www.wma.net/wp-content/uploads/2017/02/4245_002_Ethical_principles_web.pdf.

10

STAR LAWS

The Role of International Law in Regulating Civil and Military Space Activities

Cassandra Steer

Introduction

Often we hear the description that outer space is a "Wild West", or a lawless "final frontier", but nothing could be further from the truth. The 1967 Outer Space Treaty (OST),[1] and the other core space treaties, apply to all activities in outer space, whether governmental or non-governmental. Moreover, Article III of the OST states that all activities in outer space "shall be in accordance with international law", a provision which lifts large bodies of international law from their terrestrial origins into space, ensuring that in fact, most human activity is clearly governed by known laws. It is true that as technology has increased our activities in space, and as the nature of our activities have become more complex, more commercialised and more competitive, questions have started to open up about exactly what is permissible, and the core space treaties provide general principles, rather than detailed regulation of specific activities. However, together with national laws regulating space activities, these treaties still provide a very clear legal framework for both military and civilian space activities.

---

[1] "Treaty on the Principles Governing the Activities of States in the Exploration and Use of Outer Space, Including the Moon and Other Celestial Bodies," 610 U.N.T.S. 205 § (1967), https://www.unoosa.org/oosa/en/ourwork/spacelaw/treaties/outerspacetreaty.html.

This chapter will proceed in three parts. First, a background will be provided of the ways in which space-based technologies have become critical for both civil and military purposes in the twenty-first century, and why some argue there is a risk of space warfare. Second, an overview will be given of how the general legal framework ensures that space is well regulated for civilian and military uses of outer space, including how other existing branches of international law also apply to activities in space. For civilian operations, we are starting to see more questions arise as to how to govern more complex and technologically advanced space activities. For military operations, the most important ones are the law on the use of force, the law of armed conflict, human rights law, and environmental law. Despite these established bodies of law governing activities in space, we still hear assertions that space is lawless and up for grabs. The third, concluding part of this chapter will discuss the political reasons behind these assertions and clarify that space is anything but lawless. It is true that new challenges of regulation and oversight come with advances in technology and the increased commercialisation of space. But the imperative is upon States to ensure a stable, secure environment, and to ensure the rule of law prevails, just as in all terrestrial environments.

Star Laws and Star Wars: Nearing a Conflict in Space

Outer space has always been a domain in which military activity has been prevalent, however most of us are blissfully unaware of the fact that our daily activities are also highly dependent on space technologies. From telecommunications and internet, to Global Positioning Systems (GPS) for our daily navigation as well as highly complex air traffic management, and even regulating traffic lights in cities. Every time you use your bank or credit card, the transaction depends on satellites for timestamps and tracking the movement of monies, just as in the global stock market. We use satellites for weather forecasting, climate change observation and natural disaster prediction, as well as for coordinating disaster response. With satellite imaging we can track the movement of refugees, identify the sites of mass atrocities, and follow the movements of armed groups. Alongside this civilian dependency, there is an equivalent dependency of modern militaries on space technologies. GPS was originally a military invention, and is used today to track movement of troops on land, at sea and in the air, as well as aiding in remote piloting of drones, and guiding weapons to their destination. Those same satellites which we use for our daily telecommunications and internet connections are used by the military both for classified purposes and to allow troops to communicate with their families at home. Much intelligence, surveillance and reconnaissance is based on images and data gathered by "remote sensing" satellites. The First Gulf War is often referred to as the first space war[2] because of how integrated these space technologies were with the way the war was fought here on Earth, and since then this space-based dependency has only increased. This makes space a potential new domain of conflict: since the eyes and ears of most modern militaries are in space, the best way to disable an adversary with minimum risk for one's own troops would be to target these space assets. The most technologically advanced militaries are the most vulnerable should such a conflict extend into or take place in space, because their dependency on space is the greatest.

At the same time, the guiding principles of the OST are that the Moon and celestial bodies shall be used exclusively for peaceful purposes,[3] that activities in outer space shall be for the benefit and in the interest of all nations, and that space is the "province of all [hu]mankind".[4] It might be surprising that these pacifist and co-operative principles emerged from negotiations between the two superpowers competing during the Cold War, however the Soviets and the Americans both realised that if they wanted to continue to have access to and use of space, it would require mutual restraint from extending the arms race on Earth into outer space. By 1967 both the USA and the USSR had already conducted nuclear tests in outer space, the effects of which were to damage their own satellites and to impact upon radio and telecommunications from west coast USA to Australia.[5] Many other countries, though not yet active in space, recognized the need to protect this precious new domain, and thus alongside the "peaceful purposes" axiom, the OST also prohibits the placement of nuclear or other weapons of mass destruction into orbit around the Earth.[6]

---

2 Major Robert A. Ramey, "Armed Conflict on the Final Frontier: The Law of War in Space", *The Air Force Law Review* 48 (2000): 4.
3 Treaty on the Principles Governing the Activities of States in the Exploration and Use of Outer Space, including the Moon and Other Celestial Bodies Article IV.
4 Treaty on the Principles Governing the Activities of States in the Exploration and Use of Outer Space, including the Moon and Other Celestial Bodies Article I.
5 James Moltz, *The Politics of Space Security: Strategic Restraint and the Pursuit of National Interests* (Stanford University Press, 2011), 139.
6 Treaty on the Principles Governing the Activities of States in the Exploration and Use of Outer Space, including the Moon and Other Celestial Bodies, Article IV.

Despite these tenets, there has always been a military presence in outer space, and a common understanding has emerged that "peaceful purposes" only excludes aggressive activities.[7] Thus, militaries can lawfully utilise space in the ways they have always done. But in recent years, a new space arms race appears to have begun. Not with nuclear or other weapons of mass destruction, but with more covert technologies, such as the ability to interfere with a satellite by means of cyber-attacks, spoofing a signal from a satellite to a ground station, jamming a signal, blinding or dazzling the satellite, or interference on-orbit by means of another satellite.[8] There have also been more overt anti-satellite weapons (ASATs) tests, such as the catastrophic direct-ascent missile launched by China in 2007 to destroy one of its own defunct weather satellites.[9] Though China stated this was to try and remove the satellite as a piece of space debris, it is widely understood to be an ASAT test, which created an unprecedented 3,000 pieces of trackable debris and an unknown number of smaller pieces, most of which will remain in orbit for about fifty years.[10] In 2008 the USA also destroyed one if its own de-orbiting satellites, ostensibly to remove the risk of noxious gases upon burning up on re-entry, however this is read by many to be a response to China's 2007 actions, demonstrating a similar capability.[11] Russia has made many attempts at a direct-ascent kinetic ASAT, but has not been successful,[12] though it has demonstrated other capabilities with co-orbital explosives and co-orbital satellites which may potentially have the ability to interfere with a satellite.[13] In 2019, India surprised the world with a successful direct-ascent ASAT, destroying a satellite it had launched only months earlier. India immediately made many public announcements,

---

7 Carl Quimby Christol, *The Modern International Law of Outer Space* (Pergamon, 1982), 22.
8 Laura Grego, "A History of Anti-Satellite Programs," Cambridge, Mass.: Union of Concerned Scientists, January 2012.
9 "Chinese ASAT Test", CelesTrak, January 2019, https://celestrak.com/events/asat.php.
10 James Moltz, *Crowded Orbits: Conflict and Cooperation in Space* (Columbia University Press, 2014), 30.
11 Grego, "A History of Anti-Satellite Programs," 4; Laura Grego, "The Anti- Satellite Capability of the Phased Adaptive Approach Missile Defense System," 2011, 7.
12 Melissa Maday, "Russia Tests PL-19 Nudol Direct-Ascent ASAT System," SpaceWatch.Global, April 17, 2020, https://spacewatch.global/2020/04/russia-tests-pl-19-nudol-direct-ascent-asat-system/.
13 Brian Weedon and Victoria Samson, eds., "Global Counterspace Capabilities" (Secure World Foundation, April 2020), sec. 2.1, https://swfound.org/counterspace/.

stating outright that it had now joined the elite club of those with ASAT capabilities.[14]

Following the creation of the U.S. Space Force in 2020,[15] public awareness as to the importance of space for strategic purposes has been raised. China and Russia already had centralised military space branches – something akin to a "space force" – for many years.[16] Announcements from Canada, France, India and Japan[17] that they are not only increasing their budgets for military space programmes but intend to coordinate their forces into something akin to a "space force", has further increased this public awareness, but these developments have also raised tensions. As space becomes a more attractive domain for military prowess, and as more States increase their military dependency on space, the question then arises to what extent international law already governs space activities, and whether there is a need to develop further clear and explicit rules with respect to military engagement in this new domain. The next section will outline the existing legal framework and whether there is a need for new or more law.

The International Legal Framework in Space

Despite our impressive technological advances in the 60 years since the space age began, the lack of general awareness as to the extent that space technologies permeate our lives makes it easy to assert that there are no, or very few, regulations applicable to our space activities. In fact, it could

---

14 Dinakar Peri, "Successful Anti-Satellite Missile Test Puts India in Elite Club," *The Hindu*, March 27, 2019, sec. National, https://www.thehindu.com/news/national/successful-anti-satellite-missile-test-puts-india-in-elite-club/article26657024.ece; Caleb Henry, "India ASAT Debris Spotted above 2,200 Kilometers, Will Remain a Year or More in Orbit," SpaceNews.com, April 9, 2019, https://spacenews.com/india-asat-debris-spotted-above-2200-kilometers-will-last-a-year-or-more/.
15 "National Defense Authorization Act," Pub. L. No. 116–92, S. 1790 (2019) Title IX, Subtitle D.
16 Weedon and Samson, "Global Counterspace Capabilities," sec. 1.22 and 2.29.
17 Chelsea Gohd, "Everyone Wants a Space Force — but Why?," Space.com, September 11, 2020, https://www.space.com/every-country-wants-space-force.html; Mari Yamaguchi, "Japan Launches New Unit to Boost Defense in Space," Defense News, May 18, 2020, https://www.defensenews.com/global/asia-pacific/2020/05/18/japan-launches-new-unit-to-boost-defense-in-space/; "Japan Eyes New Defense Unit to Monitor Space in Fiscal 2020," *Japan Today*, 2019, https://japantoday.com/category/national/japan-eyes-new-defense-unit-to-monitor-space-in-fiscal-2020; "Glimpses of the Indian Space Program," Department of Space, Indian Space Research Organisation, 2020, https://www.isro.gov.in/sites/default/files/flipping_book/Glimpses2018/index.html.

be said that some parties have an interest in promoting this view, since the fewer regulations there are, the more permissive an environment becomes to allow almost anything to take place. Since it has not been contested in any legal forum such as a domestic or international court, this permissive approach is relatively easy to maintain.

Often in media portrayals of increased military activities in space, we hear the language of space as a lawless domain and a "wild west", in which it is up to the most powerful actors to assert their authority in the absence of any rules. Whoever has technological dominance in outer space has control over a new high ground. This picture of a "lawless frontier" rests on a narrative that space is a risky environment requiring States to be assertive, or perhaps even aggressive, in acquiring technological dominance. However, space has been governed by international law since the beginning of the space age, and that law remains just as relevant today.

International Space Law

There are five core space treaties, all of which were drafted under the auspices of the UN Committee on Peaceful Uses of Outer Space (COPUOS). All five were negotiated within 12 years in remarkably rapid succession, at the height of the Cold War, representing a feat of international cooperation. The 1967 OST can be seen as a kind of constitution for space, and the key principles determine the framework under which all human activity in space takes place. The other four space treaties emerged to expand upon certain provisions of the OST. A summary of the OST follows, with reference to those provisions which gave rise to the other treaties.

- Article I declares that all activities in space "shall be carried out for the benefit and in the interests of all countries, irrespective of their degree of economic or scientific development, and shall be the province of all mankind". Further, outer space shall be "free for exploration and use by all States without discrimination of any kind".
- Article II determines that there shall be no national appropriation, by claim of sovereignty, by use or occupation, nor by any other means. This article, combined with the obligation under Article I that space activities shall be for the benefit of all countries, is coming under contention as plans are made to mine natural resources on the Moon: is the extraction and then sale of such resources the same as "appropriation"? How can a private company own something

that is protected from appropriation? The intersection between international and domestic laws on this matter are discussed below. The last of the five core space treaties, the 1979 Moon Agreement, was negotiated mainly to ensure that an international regulatory regime would be established when the technology for mining resources in space became feasible,[18] however this treaty did not gain traction with the major space faring nations, and only has 18 signatories.

- Article III determines that all activities in space shall be in accordance with international law. This is a key provision in terms of understanding that whole existing bodies of international law apply to space, including the law of treaties, on the use of force, the law of armed conflict, international environmental law, human rights law, among others.
- Article IV prohibits the placement of weapons of mass destruction in orbit around the Earth but does not regulate the use of conventional weapons at all. It does, however, state that the Moon and all other celestial bodies shall be used for "exclusively peaceful purposes", and prohibits military bases or installations, the testing of weapons and military manoeuvres on celestial bodies.
- Article V, astronauts are deemed to be envoys of humankind and to have the right to assistance and repatriation in the event of distress no matter where they are. This article forms the basis for the 1968 Return and Rescue Agreement,[19] the second of the five core space treaties, which also regulates the obligation to return spacecraft or parts thereof which may land on the territory of another State.
- Article VI establishes the important principle that States are responsible under international law for all activities under their jurisdiction, whether governmental or non-governmental (which includes commercial, university, and other private activities). Article VI OST also requires "authorisation and continuing supervision" on the part of States over non-governmental space activities, which is what leads to the importance of national space laws.
- Article VII specifies that States which launch or procure a launch are liable for any damaged caused by that activity. This article, in

---

18 "Agreement Governing the Activities of States on the Moon and Other Celestial Bodies," United Nations Treaty Series vol. 1363, No. 23002 § (1979) Article 11(5).
19 "Agreement on the Rescue of Astronauts, the Return of Astronauts, and Return of Objects Launched into Outer Space," United Nations Treaty Series vol. 672, No. 9574 § (1968).

combination with the obligations under Article VI, are what gave rise to the 1972 Liability Convention, which provides further detail as to which States are liable and when.[20]

- Article VIII refers to the need for a national registry of objects launched into space, and determines that those objects, and the personnel thereof shall remain under the jurisdiction of that State whether the object is in space or returns to Earth. Article XI also requires States to communicate the "nature, conduct, locations and results" of their space activities. These articles were the impetus for the 1976 Registration Convention,[21] which requires States to establish a national registry and also to register all launches in an international registry deposited with the UN Office of Outer Space Affairs. This helps to track the number, type and ownership of space objects, which has become a critical necessity as space traffic management is complicated by the amount of space debris and ever-increasing number of launches each year.
- Article IX requires States to be guided by the principle of cooperation and mutual assistance in all their space activities, and to have "due regard to the corresponding interests" of all other States. This includes the obligation to undertake international consultations before undertaking a specific activity if a State has reason to believe that a planned activity would cause harmful interference to the activities of other States.
- Article X promotes the principles of cooperation and shared benefits from Article I by allowing for the possibility of States to request to observe the space activities of another State.

This brief overview demonstrates the core principles governing all human activities in space. It is worth highlighting again that, by virtue of Article III OST, general international law governs all activities in space. Specifically, we can think of the law of treaties, which governs how treaties are to be interpreted when there is a dispute, and what the general obligations are of

---

20 "Convention on the International Liability for Damage Caused by Space Objects," United Nations Treaty Series vol. 961, No. 13810 § (1972).
21 "Convention on Registration of Objects Launched into Outer Space," United Nations Treaty Series vol. 1023, No. 15020 § (1976).

a State once it has signed a treaty;[22] principles of environmental law;[23] the use of force which is prohibited under Article 2(4) of the UN Charter and is underscored specifically in Article III OST; the right to use force in self-defence as preserved in Article 51 of the UN Charter;[24] the law of armed conflict including what objects are lawful targets and the obligation to weigh up the principles of proportionality and precaution in attack;[25] and human rights law.

Apart from the core space treaties, there is also an important body of international law produced by the International Telecommunications Union (ITU); the ITU Constitution, ITU Convention and Administrative Regulations.[26] In combination, these three legal instruments provide the framework for allocating the radio frequencies necessary for satellites to transmit their data back to Earth, and for prohibiting "harmful interference" of these frequencies once allocated to specific satellites for specific purposes.[27] The most important part of the Administrative regulations are the Radio Regulations, which are updated by quadri-annual meetings of the World Radiocommunication Conference, which includes members from the commercial sector as well as government representatives. States are obliged to licence national space operations and activities according to the allocations they receive, and to require observance of all the ITU rules by any entity operating under their national jurisdiction. While military installations are exempt from adherence to the ITU Regulations, they are still obliged to prevent harmful interference and to give assistance in the case of distress, and to comply with the requirements regarding the types of

---

22 Cassandra Steer, "Sources and Law-Making Processes Relating to Space Activities," in *Routledge Handbook of Space Law*, ed. Ram Jakhu and Paul Stephen Dempsey (Routledge, 2016), 13.
23 Steer, 15.
24 Icho Kealotswe-Matlou, "The Rule of Law in Outer Space: A Call for an International Outer Space Authority," in *War and Peace in Outer Space: Law, Policy, and Ethics*, ed. Cassandra Steer and Matthew Hersch, Ethics, National Security, and the Rule of Law (Oxford, New York: Oxford University Press, 2021), 94.
25 Cassandra Steer and Dale Stephens, "International Humanitarian Law and Its Application in Outer Space," in *War and Peace in Outer Space: Law, Policy, and Ethics*, ed. Cassandra Steer and Matthew Hersch, Ethics, National Security, and the Rule of Law (Oxford, New York: Oxford University Press, 2021), 43, 48.
26 "Constitution and Convention of the International Telecommunication Union," UNTS 1825, 1826 § (1994), https://treaties.un.org/Pages/showDetails.aspx?objid=08000002800b0730&clang=_en; "ITU Regulatory Publications," International Telecommunications Union, 2021, https://www.itu.int/pub/R-REG.
27 Constitution and Convention of the International Telecommunication Union Article 6.

frequencies to be used.[28] For both civilian and military activities, therefore, space is anything but lawless.

National Space Laws

Because Art VI of the OST places international responsibility on States and requires them to "authorise and continually supervise" space activities under their jurisdiction, the legislation of national laws regulating those activities is an integral part of space law and governance. So, while space itself is beyond national jurisdiction, our human activities in space are not. It is up to each State to interpret the requirement of Art VI according to its own domestic legal system, and so there is a great difference between, for example, the many layers of regulation under various government agencies in the USA, and the very minimalistic legislation in Canada or South Africa. This also depends upon the size and activity of the space sector in each country. It is therefore logical that the USA, under whose jurisdiction more than half of all satellites and spacecraft fall,[29] and where the commercial space sector is enormous, has extensive domestic legislation to licence, oversee and regulate those activities, whereas countries with a small or no existing space agency, and only a few satellites under their jurisdiction, would have fewer laws in place.

Some space activities will fall under the jurisdiction of more than one country, and a commercial operator will have to seek licences or permits from multiple governments. An Australian satellite company will need a licence from the Australian Space Agency to operate its satellite, a licence from whichever country it wants to launch from, a contract with a launch company registered in that country, an insurance contract to cover the risks associated with launch that is valid in that country, and contracts with various clients to whom it wishes to provide its services, which may include the governments of any number of countries. As there are only about 13 countries which have launch capacity, most countries will have to seek similar licences, permits and contracts when they wish to launch a government owned satellite from the territory of another State.

---

28 Constitution and Convention of the International Telecommunication Union Article 48; Francis Lyall and Paul B. Larsen, *Space Law: A Treatise*, Second edition (London, New York: Routledge, 2018), 207.
29 "Satellite Database," Union of Concerned Scientists, August 1, 2020, https://www.ucsusa.org/resources/satellite-database; "Number of Satellites by Country 2019," Statista, 2021, https://www.statista.com/statistics/264472/number-of-satellites-in-orbit-by-operating-country/.

Most of the time there is no contention regarding the impact of domestic legislation on foreign companies or even foreign governments. To date, we also have no incidents of international litigation of disputes regarding the space treaties. However, a controversial debate is emerging over the domestic and international regulation of the extraction of natural resources in space – something the technology for which is only a few years away, and the commercial market for which will emerge in the next decade, with value estimates in the billions of dollars.[30] Humanity has a history of conflict arising over resources, and given that the resources on the Moon and asteroids of greatest value will be those that provide water or fuel to support long-term human spaceflight, it would be no surprise if the commercial and international competition for these resources becomes the centre of future conflict, whether this plays out on Earth or in space itself.

The USA and Luxembourg have passed domestic legislation which assert that such activities are in compliance with the OST, and promise to protect the rights of private entities to undertake such activities, including to protect their claims against other competing entities.[31] In 2020, a Presidential Executive Order made the surprising statement that the USA does not consider space to be a "global commons".[32] While this is not a legal term of art, the fact that the lead player in space departed from what has been a political and ethical consensus since the dawn of the space age, is concerning. Especially as this paints a questionable context for planned activities on the Moon. In 2020, NASA signed the Artemis Accords with nine countries, including Australia, which allow the signatory countries to take part in the Artemis programme to return humans to the Moon and, soon after, begin mining the Moon.[33] The Artemis Accords contain a series of general principles regarding cooperation, interoperability, and sustainability of activities on the Moon, and includes a principle that asserts that natural resource extraction can and shall take place in accordance with

---

30 "Asteroid Mining Market Value 2025," Statista, accessed February 11, 2021, https://www.statista.com/statistics/1023115/market-value-asteroid-mining/.
31 "U.S. Commercial Space Launch Competitiveness Act," Pub. L. No. 114–90 (2015); Luxembourg, "Loi Du 20 Juillet 2017 Sur l'exploration et l'utilisation Des Ressources de l'espace.," Pub. L. No. N° 674 du 28 juillet 2017 (2017), http://data.legilux.public.lu/file/eli-etat-leg-loi-2017-07-20-a674-jo-fr-pdf.pdf.
32 Donald J. Trump, "Executive Order on Encouraging International Support for the Recovery and Use of Space Resources" (2020), https://www.whitehouse.gov/presidential-actions/executive-order-encouraging-international-support-recovery-use-space-resources/.
33 "NASA: Artemis Accords," NASA, 2020, https://www.nasa.gov/specials/artemis-accords/img/Artemis-Accords-signed-13Oct2020.pdf.

the OST,[34] thus eliciting from the signatory States an agreement with this very contentious legal position. The Accords are bi-lateral agreements, Article I of which states that they represent a "political commitment", and NASA is emphatic that they do not have the status of a treaty.[35] Yet they could be considered to amount to a series of unilateral declarations made by each of the signatory States, which are themselves binding upon those States.[36] Australia is in a unique position, as it is the only Artemis signatory which is also a State party to the Moon Agreement. Australia has therefore both declared that it agrees that mining in space is in accordance with the OST, and also obliged itself to ensure an international regime is established to govern the exploitation of natural resources. It remains to be seen how this will play out in the current decade: whether we will see the international community move quickly to establish a new international regime, and whether the other Artemis signatory States will feel the need to enact domestic legislation in accordance with the principles of those Accords.

Space: Anything But A Lawless Frontier

Even with the international and domestic legal framework described above, there are still some who argue that when it comes to military activities and the potential for a conflict in space, we find ourselves in a lawless frontier. The argument that there are no laws governing a potential "battle above" is often based on the principles that there is no national sovereignty in space under Article II OST, and Article I, which provides that space shall be the province of all humankind. The assumption is that space belongs to no-one and is therefore not subject to the normal rules applicable to human or State behavior. The argument goes that it is therefore up to those actors who are able to make use of space to determine the limits of their activities. However, as is clear from the descriptions above, space is anything but lawless. Until humanity has established independent settlements on other planets, which are self-sufficient, and which will therefore become self-governing, all

---

34 "NASA," sec. 10.
35 "NASA," sec. 13(2).
36 James Crawford, *Brownlie's Principles of Public International Law*, 9th ed. (OUP Oxford, 2019), 402; "Guiding Principles Applicable to Unilateral Declarations of States Capable of Creating Legal Obligations, with Commentaries Thereto" (International Law Commission, 2006), 370, https://legal.un.org/ilc/texts/instruments/english/commentaries/9_9_2006.pdf.

of our activities in space are governed by existing international law and national laws.

It is true that the OST and the other space treaties provide much more in the way of general principles rather than detailed regulations of specific activities, which is why it is helpful to think of them as framework treaties. When it comes to military activities in particular, there are some important questions left open. Article IV of the OST prohibits the placement of weapons of mass destruction in orbit around the Earth, but it does not say anything about weapons which might be more specific in their targeting capabilities. Thus, the weaponisation of space as such is not prohibited, nor regulated in any other way, aside from the prohibition of nuclear testing in space, according to the Partial Test Ban Treaty (PTBT).[37] Even more problematic is that the OST does not define "weapons" at all, so that any satellite which has a peaceful primary purpose, for instance telecommunications, or on-orbit servicing of damaged or old satellites, for the purposes of cleaning up space debris, might be used to interfere with another satellite without being forbidden under the OST. Furthermore, the OST doesn't explicitly prohibit the use of Earth-based ASATs such as laser, or long-range missiles such as we've seen tested by the USA, Russia and China over the decades.

While the lack of more specific terminology in the OST (or any of the other four core space treaties) is problematic, this notion that space is therefore a lawless frontier ignores vast bodies of public international law that apply by virtue of Article III OST, even if they do not reference space in particular.

Such arguments echo the now outdated "Lotus" principle, that whatever is not explicitly forbidden or regulated under international law is permissible. This principle hails from the *SS Lotus* case, adjudicated by the Permanent Court of International Justice, the predecessor to today's International Court of Justice (ICJ) in 1926 between France and Turkey. The Court held that if something is not expressly prohibited in international law, it is permitted. However, at the time that this was held to be a valid principle, international law was far more centralized and entirely deferential to State action and expressions of State will. Today, international law is described as consisting of multiple self-regulating regimes, such

---

37 "Treaty Banning Nuclear Weapon Tests in the Atmosphere, in Outer Space and under Water," Pub. L. No. UNTS 480 (1963), https://treaties.un.org/doc/Publication/UNTS/Volume%20480/volume-480-I-6964-English.pdf.

as trade law, environmental law, intellectual property law, international criminal law and others.[38] As well as these self-regulating regimes, there are principles of general international law which apply across all regimes, and there are ways in which separate regimes apply to each other, such as environmental law or human rights law applying to the law of armed conflict, trade law and space law. It can therefore no longer be said that international law operates as a single legal regime in which activities are either explicitly regulated, or where silence on a specific activity can be interpreted as permitting that activity. As well, the number and types of international actors has increased significantly, such that no longer do we determine the content and applicability of international law based solely on what the most powerful States codify in treaties or express as their will through custom. Today we must take into account multiple participants in the international legal order,[39] including individuals who have human rights enforceable against States; multinational corporations which have legal personality; international organizations which can create law and act as subjects under it, such as the UN, the EU or NATO; and non-governmental organizations which are highly influential on the formation of law, such as the work of the International Committee of the Red Cross with respect to the law of armed conflict, or the emerging influence of organisations such as Moon Village Association[40] which has observer status at COPUOS and is advocating for principles to protect the Moon once mining begins there.

There is therefore a difference in today's public international law between the notion that an action does violate an applicable rule, or what could be termed a prescriptive approach, and the notion that whatever is not expressly prohibited is permissible, or what could be termed a permissive approach. *Lotus* represented a permissive approach, suitable to nineteenth century positivism, which required codification and indisputable rules of custom. In the latter half of the twentieth century it became clear in the jurisprudence of the ICJ, and in much of the leading literature in public international law, that this principle could no longer be considered a sound base for any assertion under international law.

---

38 Anne-Marie Slaughter and Jose E. Alvarez, "A Liberal Theory of International Law," in *Proceedings of the Annual Meeting (American Society of International Law)* (JSTOR, 2000), 240–53; Martti Koskenniemi, "Fragmentation of International Law? Postmodern Anxieties," *Ljil* 15 (2002): 553; Martti Koskenniemi, "Hegemonic Regimes," *Regime Interaction in International Law: Facing Fragmentation*, 2012, 305–24.
39 Cassandra Steer, "Non-State Actors in International Criminal Law," in *Non-State Participants in the International Legal Order*, ed. Jean D'Aspremont (Oxford: Routledge, 2011).
40 "Moon Village Association," 2021, https://moonvillageassociation.org/about.

Thus, even when it is not clear what specific normative rules of international law would apply in a given situation, to say that outer space is a lawless frontier – or even that it is lawless when it comes to military activities and the potential for conflict in space – is to use anachronistic language which misunderstands the fact that all human and State activity in space falls under some form of international law. Silence as to specific acts does not necessarily amount to permissiveness, rather we must ask in what ways these activities are restricted, permitted or tolerated, according to the various regimes of international law.

The notion of space as a "lawless frontier" is appealing to States who desire to be as unlimited as possible in their use of space. Since one of the freedoms guaranteed by the OST is access to and use of space for all nations,[41] some have argued that this translates to the freedom to use it in self-defense, and thus the use of force can be legitimised because it is unlimited. In fact, some have argued that the right to use force in self-defense in space is an unlimited norm, perhaps even reaching the status of *jus cogens*, trumping any other limitations international law may place on use or threat of force.[42] But the use of force in space is always limited by international law, in the same way use of force at sea, on the land or in the air is. The fact that Article III OST stipulates that all activities in outer space must be conducted "in accordance with international law, *including the Charter of the United Nations*, in the interest of maintaining international peace and security" [emphasis added] demonstrates that use of force and the correlative right of self-defence were at the forefront of the negotiations between the two super powers when drafting the OST.[43] As will be discussed in another chapter, Article 51 of the United Nations Charter provides for lawful use of force in self-defense only if it is in response to an armed attack, and the response must be proportionate and temporary, with the intention being to return to a situation of peace and security. The significant body of law that we have on the prohibition on the limits of the right to use force in self defence will apply in space and needs no further regulation.

The notion of a "lawless frontier" is in fact a dangerous lexicon to employ. It conjures up images of the eighteenth century myth of "terra

---

41 Treaty on the Principles Governing the Activities of States in the Exploration and Use of Outer Space, including the Moon and Other Celestial Bodies Article I.
42 Michel Bourbonniere, "National-Security Law in Outer Space: The Interface of Exploration and Security," *Journal of Air Law and Commerce* 70 (2005): 60.
43 Olivier Ribbelink, "Article III," in *Cologne Commentary on Space Law: Outer Space Treaty*, ed. Stephan Hobe, Bernhard Schmidt-Tedd, and Kai-Uwe Schrogl, vol. 1 (BWV Verlag, 2017), 65.

nullius" which justified the British territorial claim of Australia, the nineteenth century "wild west" in the USA, or the early twentieth century rhetoric of civilizing savage peoples in India and Africa.[44] Wherever there has been a "lawless frontier" rhetoric historically, it has been used by a powerful few States to justify claims of sovereignty, access to resources, and the use of violence as they expanded their empires. In a terrestrial context this has been at the cost of Indigenous lives and culture, as well as at the cost of the natural environment. In space there is not yet, to our knowledge, a risk of endangering Indigenous populations, however there is a risk of affecting our near-Earth environment, as we are already seeing with ever-increasing problems of space debris. The impact of this debris on our own future secure use of space is serious enough, as is the impact it can cause if satellites providing essential infrastructure services are damaged. But as we consider more human space flight in the near future, it also poses a direct threat to the safety of those traversing space, whether for commercial, recreational or scientific purposes. Moreover, if a conflict in space were to escalate beyond the ways in which space is already used in terrestrial warfare, to include the interference or destruction of satellites, there is also a high risk of direct impact on human safety and security on Earth, due to our terrestrial dependency on space technologies. We must therefore take a more responsible, forward-looking approach in selecting which rhetoric we employ when analysing international law applicable to outer space.

As Judge Manfred Lachs, former judge on the ICJ and one of the most preeminent space lawyers has said, space has always been subject to international law, it was just never tested before.[45] The fact that it has not been tested in international or domestic courts should in fact be read as a signal that States historically have been more interested in maintaining political stability and agreeing upon a base level rule of law in space, than in testing out the boundaries. In the coming decades, stability in space may not necessarily entail attempts at new treaties, it can take the form of agreement on norms of responsible behaviour in space, such as the 2019 UN Guidelines on the Long-term Sustainability of Outer Space Activities,[46] express commitment to the core principles of the OST, and unilateral

---

44 Alan Boyle and Christine Chinkin, *The Making of International Law* (Oxford University Press, 2007).
45 Manfred Lachs, *The Law of Outer Space: An Experience in Contemporary Law-Making*, Re-Issued (Martinus Nijhoff Publishers, 1972), 125.
46 Working Group on the Long-term Sustainability of Outer Space Activities, "Guidelines for the Long-Term Sustainability of Outer Space Activities" (Committee on the Peaceful Uses of Outer Space, July 17, 2019), https://undocs.org/A/AC.105/C.1/L.366.

statements underscoring the importance of the rule of law. As our activities expand in space, especially our military activities, we would be better to work towards clarifying exactly what is restricted, permitted, or tolerated, before waiting for an event that would force the law to be tested under contentious circumstances.

References

Agreement Governing the Activities of States on the Moon and Other Celestial Bodies, United Nations Treaty Series vol. 1363, No. 23002 § (1979).
Agreement on the Rescue of Astronauts, the Return of Astronauts, and Return of Objects Launched into Outer Space, United Nations Treaty Series vol. 672, No. 9574 § (1968).
Bourbonniere, M. "National-Security Law in Outer Space: The Interface of Exploration and Security." *Journal of Air Law and Commerce* 70 (2005): 3.
Boyle, A. and C. Chinkin. *The Making of International Law*. Oxford University Press, 2007.
Christol, C. Q. *The Modern International Law of Outer Space*. Pergamon, 1982.
Convention on Registration of Objects Launched into Outer Space, United Nations Treaty Series vol. 1023, No. 15020 § (1976).
Convention on the International Liability for Damage Caused by Space Objects, United Nations Treaty Series vol. 961, No. 13810 § (1972).
Crawford, J. *Brownlie's Principles of Public International Law*. 9th ed. OUP Oxford, 2019.
Department of Space, Indian Space Research Organisation. "Glimpses of the Indian Space Program", 2020. https://www.isro.gov.in/sites/default/files/flipping_book/Glimpses2018/index.html.
Gohd, C. "Everyone Wants a Space Force — but Why?" Space.com, September 11, 2020. https://www.space.com/every-country-wants-space-force.html.
Grego, L. "A History of Anti-Satellite Programs." Cambridge, Mass.: Union of Concerned Scientists, January, 2012.
Grego, L. "The Anti- Satellite Capability of the Phased Adaptive Approach Missile Defense System," 2011, 7.
"Guiding Principles Applicable to Unilateral Declarations of States Capable of Creating Legal Obligations, with Commentaries Thereto." International Law Commission, 2006. https://legal.un.org/ilc/texts/instruments/english/commentaries/9_9_2006.pdf.
Henry, C. "India ASAT Debris Spotted above 2,200 Kilometers, Will Remain a Year or More in Orbit." SpaceNews.com, April 9, 2019. https://spacenews.com/india-asat-debris-spotted-above-2200-kilometers-will-last-a-year-or-more/.
"Japan Eyes New Defense Unit to Monitor Space in Fiscal 2020." *Japan Today*, 2019. https://japantoday.com/category/national/japan-eyes-new-defense-unit-to-monitor-space-in-fiscal-2020.

Kealotswe-Matlou, I. "The Rule of Law in Outer Space: A Call for an International Outer Space Authority." In *War and Peace in Outer Space: Law, Policy, and Ethics*, edited by Cassandra Steer and Matthew Hersch, 91–105. Ethics, National Security, and the Rule of Law. Oxford, New York: Oxford University Press, 2021.

Koskenniemi, M. "Fragmentation of International Law? Postmodern Anxieties." *Ljil* 15 (2002): 553.

Koskenniemi, M. "Hegemonic Regimes." *Regime Interaction in International Law: Facing Fragmentation*, 2012, 305–24.

Lachs, M. *The Law of Outer Space: An Experience in Contemporary Law-Making*. Re-Issued. Martinus Nijhoff Publishers, 1972.

Luxembourg. Loi du 20 juillet 2017 sur l'exploration et l'utilisation des ressources de l'espace., Pub. L. No. N° 674 du 28 juillet 2017 (2017). http://data.legilux.public.lu/file/eli-etat-leg-loi-2017-07-20-a674-jo-fr-pdf.pdf.

Maday, M. "Russia Tests PL-19 Nudol Direct-Ascent ASAT System." SpaceWatch.Global, April 17, 2020. https://spacewatch.global/2020/04/russia-tests-pl-19-nudol-direct-ascent-asat-system/.

Moltz, J. *Crowded Orbits: Conflict and Cooperation in Space*. Columbia University Press, 2014.

Moltz, J. *The Politics of Space Security: Strategic Restraint and the Pursuit of National Interests*. Stanford University Press, 2011.

"Moon Village Association", 2021. https://moonvillageassociation.org/about.

NASA. "NASA: Artemis Accords," 2020. https://www.nasa.gov/specials/artemis-accords/img/Artemis-Accords-signed-13Oct2020.pdf.

National Defense Authorization Act, Pub. L. No. 116–92, S. 1790 (2019).

Peri, D. "Successful Anti-Satellite Missile Test Puts India in Elite Club." *The Hindu*. March 27, 2019, sec. National. https://www.thehindu.com/news/national/successful-anti-satellite-missile-test-puts-india-in-elite-club/article26657024.ece.

Ramey, Major R. A. "Armed Conflict on the Final Frontier: The Law of War in Space." *The Air Force Law Review* 48 (2000): 1.

Ribbelink, O. "Article III." In *Cologne Commentary on Space Law: Outer Space Treaty*, edited by Stephan Hobe, Bernhard Schmidt-Tedd, and Kai-Uwe Schrogl, 1:64–69. BWV Verlag, 2017.

Slaughter, A. and J. E. Alvarez. "A Liberal Theory of International Law." In *Proceedings of the Annual Meeting (American Society of International Law)*, 240–53. JSTOR, 2000.

Statista. "Asteroid Mining Market Value 2025." 2021. https://www.statista.com/statistics/1023115/market-value-asteroid-mining/.

Statista. "Number of Satellites by Country 2019." 2021. https://www.statista.com/statistics/264472/number-of-satellites-in-orbit-by-operating-country/.

Steer, C. "Non-State Actors in International Criminal Law." In *Non-State Participants in the International Legal Order*, edited by Jean D'Aspremont. Oxford: Routledge, 2011.

Steer, C. "Sources and Law-Making Processes Relating to Space Activities." In *Routledge Handbook of Space Law*, edited by Ram Jakhu and Paul Stephen Dempsey. Routledge, 2016.

Steer, C. and D. Stephens. "International Humanitarian Law and Its Application in Outer Space." In *War and Peace in Outer Space: Law, Policy, and Ethics*, edited by Cassandra Steer and Matthew Hersch, 23–54. Ethics, National Security, and the Rule of Law. Oxford, New York: Oxford University Press, 2021.

Treaty banning nuclear weapon tests in the atmosphere, in outer space and under water, Pub. L. No. UNTS 480 (1963). https://treaties.un.org/doc/Publication/UNTS/Volume%20480/volume-480-I-6964-English.pdf.

Treaty on the Principles Governing the Activities of States in the Exploration and Use of Outer Space, including the Moon and Other Celestial Bodies, 610 U.N.T.S. 205 § (1967). https://www.unoosa.org/oosa/en/ourwork/spacelaw/treaties/outerspacetreaty.html.

Trump, D. J. Executive Order on Encouraging International Support for the Recovery and Use of Space Resources (2020). https://www.whitehouse.gov/presidential-actions/executive-order-encouraging-international-support-recovery-use-space-resources/.

Union of Concerned Scientists. "Satellite Database," August 1, 2020. https://www.ucsusa.org/resources/satellite-database.

U.S. Commercial Space Launch Competitiveness Act, Pub. L. No. 114–90 (2015).

Weedon, B. and V. Samson, eds. "Global Counterspace Capabilities." Secure World Foundation, April 2020. https://swfound.org/counterspace/.

Working Group on the Long-term Sustainability of Outer Space Activities. "Guidelines for the Long-Term Sustainability of Outer Space Activities." Committee on the Peaceful Uses of Outer Space, July 17, 2019. https://undocs.org/A/AC.105/C.1/L.366.

Yamaguchi, M. "Japan Launches New Unit to Boost Defense in Space." Defense News, May 18, 2020. https://www.defensenews.com/global/asia-pacific/2020/05/18/japan-launches-new-unit-to-boost-defense-in-space/.

11

THE WOOMERA MANUAL

Legitimising or Limiting Space Warfare?

Cassandra Steer

Introduction

As we enter a period of history where global relations and the global distribution of power are undergoing momentous shifts, space is gaining more attention as a strategic domain where these relations and tensions are being played out, alongside the other domains of land, air, sea and cyber. Our reliance on space today for both civilian and military services is far beyond what it was at the beginning of the Space Age in the 1960s, yet in some ways we are returning to similar patterns as those which gave rise to the 1967 Outer Space Treaty (OST). Indeed, many argue that one of the core motivations behind the drafting of the OST, leading to such swift cooperation between the competing superpowers at the height of the Cold War, was nuclear arms control and de-escalation in space.[1] Because of the uncontainable effects of their weapons tests in the earliest years of the space age, both the U.S. and the USSR recognised that if they wanted continued access to space for their own purposes, they needed to restrain their competitor and agree to such restraints themselves. It is for this reason that

---

1 Olivier Ribbelink, "Article III," in *Cologne Commentary on Space Law: Outer Space Treaty*, ed. Stephan Hobe, Bernhard Schmidt-Tedd, and Kai-Uwe Schrogl, vol. 1 (Carl Heymanns Verlag, 2009), 65; Kai-Uwe Schrogl and Julia Neumann, "Article IV," in *Cologne Commentary on Space Law: Outer Space Treaty*, ed. Stephan Hobe, Bernhard Schmidt-Tedd, and Kai-Uwe Schrogl, vol. 1 (Carl Heymanns Verlag, 2009), 72; James Moltz, *The Politics of Space Security: Strategic Restraint and the Pursuit of National Interests* (Stanford University Press, 2011), 150.

these adversaries agreed to the language in Article IV of the 1967 OST that the Moon and all other celestial bodies in space shall be used "exclusively for peaceful purposes".[2]

Yet space has always been militarised, in the sense that from the earliest days of the space age, militaries have used space technologies for intelligence, surveillance and reconnaissance; in particular, the U.S. and the USSR used these technologies to survey each other's nuclear programmes. In recent decades we have seen an increase in the militarisation of space, including the establishment of the U.S. Space Force in 2020,[3] similar dedicated military branches in other countries, as well as a shift in rhetoric in national space policies and strategies towards a more offensive stance. As space becomes more critical to military operations and further integrated into terrestrial warfare, it has become a strategic domain unto itself, in which militaries can also be vulnerable to various kinds of targeting, whether temporary or permanent, and whether non-kinetic or kinetic.[4] Because of this there are some who argue that a conflict extending into space is inevitable and this in turn has set many countries on edge, seeking not only to protect their assets in space, but also to threaten those of their adversaries. Others insist that space warfare is not inevitable, but we need to undertake the right diplomatic and legal steps now. This raises some questions: what are the right legal, policy and diplomatic steps to take? And is existing international law up to the job?

By virtue of Article III OST, all activities in space must take place in accordance with international law, thereby "raising" existing bodies of law into space.[5] We therefore know that both *jus ad bellum*, or the law on the use of force, and *jus in bello*, or international humanitarian law, otherwise known as the law of armed conflict, apply to activities in outer space. Yet, because space has many unique characteristics when compared with other

---

2 "Treaty on the Principles Governing the Activities of States in the Exploration and Use of Outer Space, Including the Moon and Other Celestial Bodies," 610 U.N.T.S. 205 § (1967), https://www.unoosa.org/oosa/en/ourwork/spacelaw/treaties/outerspacetreaty.html, Article I.
3 "National Defense Authorization Act," Pub. L. No. 116–92, S. 1790 (2019), Title IX, Subtitle D.
4 For a brief overview of various anti-satellite weapons, see Cassandra Steer and Dale Stephens, "International Humanitarian Law and Its Application in Outer Space," in *War and Peace in Outer Space: Law, Policy, and Ethics*, ed. Cassandra Steer and Matthew Hersch, Ethics, National Security, and the Rule of Law (Oxford, New York: Oxford University Press, 2021), 26–30; For a more detailed overview of counterspace technologies, country by country, see Brian Weedon and Victoria Samson, eds., "Global Counterspace Capabilities" (Secure World Foundation, April 2020), https://swfound.org/counterspace/.
5 Treaty on the Principles Governing the Activities of States in the Exploration and Use of Outer Space, including the Moon and Other Celestial Bodies, Article III.

environments, and because these bodies of international law were written and have developed in relation specifically to warfare on land, at sea and, relatively recently, in the air, it is not always clear how certain aspects or specific legal provisions of *jus ad bellum* or *jus in bello* will apply to military activities in space. For example, at what point does an activity in space amount to a "threat to international peace and security", or an "armed attack", both of which would justify some form of response? Where there is ambiguity, there is a risk of different interpretations of the threshold, which can lead to escalation and the risk of space warfare, or of terrestrial warfare in response to a space activity. In an attempt to provide some clarity, the Woomera Manual on the International Law of Military Space Activities[6] (Woomera Manual) is being developed by a group of independent experts from around the world. But there is an enormous tension between seeking to clarify these questions and respecting the cardinal principle of "peaceful purposes" in the OST.

This chapter will therefore tackle the question whether such a Manual could have the effect of legitimising further militarisation of space, or even space warfare, or rather whether it might have a restraining effect on the risk of space warfare, and on the potential impacts if such warfare were to take place. First, the tension between the "peaceful purposes" principle and the existing and increasing use of space for military purposes will be considered. Second, a brief background will be provided of the Woomera Manual, including the main justifications that are made for the project, and the issues it seeks to cover. Third, a comparison is made between three competing views on the ethics of undertaking such a Manual: those who consider the project to go against the "peaceful purposes" principle; those who generally resist the furtherance of binding norms in space; and those who favour clarification and regulation of norms as a form of restraint on State behaviour. Finally, the conclusion is drawn that the Woomera Manual has an important role to play at this juncture in history, to limit rather than legitimise space warfare.

"Peaceful Purposes" and Military Activities in Outer Space

The text of the OST is based on a series of early UN General Assembly Resolutions, which were adopted soon after the launch of the first satellite,

---

[6] "The Woomera Manual", 2020, https://law.adelaide.edu.au/woomera/home.

the Soviet "Sputnik" in 1957. That launch marked the beginning of the Space Age, and with it, the extension of the Cold War into space. In 1958 the first such resolution was titled "Questions of the peaceful use of outer space", and it established an ad-hoc Committee on the Peaceful Uses of Outer Space (COPUOS).[7] In 1959, another resolution established COPUOS as a permanent body, and recognised the "common interest of mankind as a whole in furthering the peaceful use of outer space", alongside the desire to "avoid the extension of present rivalries into this new field".[8] These express concerns formed the basis for negotiations on the OST, and as U.S. Ambassador Goldberg stated during those negotiations, "the central issue was to ensure that outer space and celestial bodies were reserved *exclusively for peaceful purposes*" [Emphasis added].[9] Around the same time, negotiations were taking place on the Antarctic Treaty,[10] which also served to protect against the contested claims of territory in the southern polar region from becoming a source of conflict, by ensuring that activities in the Antarctic could be for scientific and peaceful purposes only. States were clearly using their successful negotiations on that treaty as a model for the new international legal regime applicable to space.

Importantly, the drafters of Article IV felt compelled to highlight that this "includ[es] the Charter of the United Nations, in the interest of maintaining international peace and security". The reference to the UN Charter was intended to underscore such provisions as the prohibition of the use of force,[11] but also the lawful exceptions to this prohibition, that is, in self-defence when a State has suffered an armed attack,[12] and collective self-defence when the Security Council authorises the use of force through its powers under Chapter VII of the Charter. Moreover, the language in the rest of the article obliging States to undertake all their activities "in the interest of maintaining international peace and security and promoting international co-operation and understanding", is a directive from the international community that space was not to become a domain of conflict, since the UN

---

7 UN General Assembly, "Resolution 1348 (XXIII): Questions of the Peaceful Uses of Outer Space," December 13, 1958.
8 UN General Assembly, "Resolution 1472 (XIV): International Co-operation in the Peaceful Uses of Outer Space," December 12, 1959, Resolution A, paragraphs 1 and 3 of the preamble respectively.
9 "UN COPUOS Legal Sub-Committee Fifth Session, UN Doc A/AC.105/C.2/SR.57," October 20, 1966, 6.
10 "Antarctic Treaty," Pub. L. No. 402 U.N.T.S 71 (1961).
11 "Charter of the United Nations" (1945), https://www.un.org/en/sections/un-charter/un-charter-full-text/, Articles 2(4).
12 Charter of the United Nations, Article 51.

member States attach overriding importance to the responsibility of the UN in maintaining international peace and security. It also echoes the obligation to seek peaceful resolution of disputes, embedded in the UN Charter.[13]

One might question how it is that space has always been used for military purposes in light of the peaceful purposes principle. One answer is that Article IV only determines that celestial bodies must be used exclusively for peaceful purposes – and not space itself. However, the Preamble of the OST highlights the "common interest of all [hu]mankind in the progress of the exploration and use of outer space for peaceful purposes". While the preamble of a treaty does not have the same binding force as the provisions in each of the articles, the law of treaties tells us that preambles are to be used to interpret the text of the provisions, and to identify the object and purpose of a treaty.[14] States are obliged not to act in any way that might be in breach of that object and purpose. Thus, while there is no obligation for all of our space-based activities to be "exclusively for peaceful purposes", States may be obliged not to behave in such a way as to create the circumstances that will lead to a conflict in space. By contrast, a rapid consensus emerged in the first decades of the Space Age that military activities are lawful, as long as those activities are not aggressive and therefore in contravention of the prohibition on the use of force under international law.[15] It is therefore considered to be justified to use space-based technologies to support terrestrial warfare. The increased tensions we are seeing in the strategic uses of space, including the determination by the U.S. that space is a "warfighting domain",[16] may be putting this tacit understanding under pressure, and may in fact be in breach of the object and purpose of the OST.

Further pressures may arise in the near future with regards to another prohibition under Article IV OST, namely the prohibition on the "establishment of military bases, installations and fortifications, the testing of any type of weapons and the conduct of military manoeuvres" on any celestial body. A certain ambiguity is created by the explicit allowance

---

13 Charter of the United Nations Article 33.
14 "Vienna Convention on the Law of Treaties," 1155 U.N.T.S. 331 § (1969), Article 31.
15 Carl Quimby Christol, *The Modern International Law of Outer Space* (Pergamon, 1982), 22; It should be noted that the Soviets disagreed with this interpretation for several decades, however the Russian interpretation now accords with this consensus. See Kubo Mačák, "Silent War: Applicability of the Jus in Bello to Military Space Operations," 2018, 17.
16 The U.S. Department of Defence released an unclassified summary: United States Department of Defense, "Summary of the 2018 National Defense Strategy," 2018, https://www.hsdl.org/?view&did=807329; This term is now often used by officials in public fora, for example. "US Space Force Chief: Space Is 'A Warfighting Domain,'" Forces Network, 2021, https://www.forces.net/news/head-us-space-force-space-warfighting-domain.

for the use of military personnel for "scientific research or for any other peaceful purposes". For example, it is entirely conceivable that an activity on the Moon could be fully funded by, directed by and fulfilled by a State's military, to enable that State to assert dominance by way of physical presence, and yet convincingly declare that such activities are for scientific purposes. We are likely to see such contentious interpretations of Article IV in the coming decade.

Finally, Article IV also prohibits the placement of nuclear weapons or other weapons of mass destruction (WMD) in orbit around the Earth, on a celestial body or stationed in space "in any other manner". While this comprehensive ban on WMDs was laudable at the time, and helped characterise the OST as an arms control treaty, it leaves the door entirely open for conventional weapons, and for the development of new types of weapons, which may arguably be lawfully tested or placed in space. Logically, even the testing of anti-satellite weapons (ASATs) or the placement of weapons in space goes against the peaceful purposes doctrine, however there is insufficient international political will to support this interpretation – and in fact China, India, Russia and the US have all tested ASATs, and are testing various "counterspace" technologies, whether ground-based or space-based. The gap left by Article IV is not one these powers necessarily want closed.

The international community has not remained silent in the face of these developments. Every year, the UN General Assembly adopts the PAROS (Prevention of an Arms Race in Outer Space) resolutions, with near-universal support, and with nearly identical wording, calling on States:

> in particular those with major space capabilities, to contribute actively to the goal of preventing an arms race in outer space, as an essential condition for the promotion of international cooperation in the exploration and use of outer space for peaceful purposes.[17]

Another resolution that is reiterated each year with strong support is the call for "Transparency and Confidence Building Measures in outer space activities",[18] which calls on States to undertake diplomatic and information

---

17 United Nations General Assembly, "Prevention of an Arms Race in Outer Space, A/Res/72/26," December 11, 2017, para. 4, https://digitallibrary.un.org/record/1326281?ln=en.
18 See the resolution adopted in 2020 at the 75th Session of the General Assembly, which repeats the text of resolutions from previous years United Nations General Assembly, "Transparency and Confidence Building Measures in Outer Space Activities, A/Res/75/69," December 7, 2020, https://undocs.org/en/A/RES/75/69.

sharing agreements. While these resolutions are important political expressions, UN General Assembly resolutions are non-binding,[19] and are therefore limited in their effectiveness. Nonetheless, the near-universal support is telling, as no country votes against these resolutions. But it is also telling that the U.S. and Israel consistently abstain from voting, which indicates their position against any pro-active space arms control measures.[20]

China and Russia, on the other hand, have been promoting two parallel arms control initiatives. In 2014, China and Russia co-sponsored a General Assembly resolution urging States to make a "No First Placement" statement.[21] Although neither the PAROS resolutions nor the "No First Placement" resolution are binding, any unilateral statement made by a State in pursuance of these resolutions is binding according to international law.[22] So far, only a handful of countries have made such statements, and the U.S. and its allies have remained consistently opposed.

The "no first placement" initiative was coupled with a second, longer Sino-Russian initiative to propose a Treaty on the Prevention of the Placement of Weapons in Outer Space Treaty (PPWT).[23] In 2008 and 2014 the two countries presented drafts to the Conference on Disarmament (CD), and the 2014 draft received support by countries such as the Union of South American Nations (UNASUR), and the Non-Aligned Movement. Notably none of these are very active space "powers", however the political support comes from precisely those States which have an interest in ensuring there are clear rules applicable to those larger powers who will likely be the first engaged in a conflict in space. Conversely, major powers such as the U.S.,

---

19 Charter of the United Nations Article 10, states that the General Assembly may "make recommendations". These are non-binding, unlike Security Council Resolutions, which gain their binding force by virtue of Article 25 of the UN Charter.
20 The voting record can be seen at https://digitallibrary.un.org/record/1325775?ln=en and https://www.un.org/press/en/2020/ga12296.doc.htm. The resistance of these two countries to such statements has been noted by, for example. "Federation of American Scientists: Prevention of an Arms Race in Outer Space", 2021, https://fas.org/programs/ssp/nukes/ArmsControl_NEW/nonproliferation/NFZ/NP-NFZ-PAROS.html.
21 General Assembly of the United Nations, "No First Placement of Weapons in Outer Space," December 2, 2014, https://undocs.org/en/A/RES/69/32.
22 James Crawford, *Brownlie's Principles of Public International Law*, 9th ed. (OUP Oxford, 2019), 402; "Guiding Principles Applicable to Unilateral Declarations of States Capable of Creating Legal Obligations, with Commentaries Thereto" (International Law Commission, 2006), 370, https://legal.un.org/ilc/texts/instruments/english/commentaries/9_9_2006.pdf.
23 People's Republic of China Ministry of Foreign Affairs, "Treaty on the Prevention of the Placement of Weapons in Outer Space, the Threat or Use of Force against Outer Space Objects (Draft)" (2014), https://www.fmprc.gov.cn/mfa_eng/wjb_663304/zzjg_663340/jks_665232/kjfywj_665252/t1165762.shtml.

Canada, France and others rejected these proposals altogether.[24] Reasons given for this resistance are that it is difficult to define a "weapon" in space, since so much technology can be re-purposed; it is problematic that the treaty draft only refer to space-based weapons, and not ground-based weapons that could target space systems; and that a verification procedure is difficult to implement.[25] Resistance by the U.S. and its allies continued even in the face of the UN Group of Government Experts process established in 2019 to discuss whether there is a need for a treaty on space arms.[26] The U.S. has stated that it is not interested in any binding international law measures when it comes to space arms control, preferring non-binding efforts, yet it has failed to provide any real alternatives.

Why the Woomera Manual?

In the face of this international diplomatic stand-off, and as tensions in the space domain continue to grow, other avenues are needed. The Woomera Manual was conceived as a project to clarify what appears to be a legal gap, and to assist in reducing the risk of misinterpretation or miscalculation of others' military activities in space, which could itself trigger a conflict. The Manual is developed by an international group of independent experts in the law of conflict, the use of force, space law and military space operations. It is therefore not a source of law, but it sets out to clarify how customary international law and treaty law applies in space, in such a way that decision-makers can refer to it for guidance.

International law has specific sources which make up its normative content. Article 38 of the Statue of the International Court of Justice is the main reference point, where four sources are listed:

a) treaties
b) customary international law

---

24 Jinyuan Su, "The Legal Challenge of Arms Control in Space", in *War and Peace in Outer Space: Law, Policy, and Ethics*, ed. Cassandra Steer and Matthew Hersch, Ethics, National Security, and the Rule of Law (Oxford, New York: Oxford University Press, 2021), 179–98.
25 Gilles Doucet, "Notification of Transfer of Kinetic Energy to an Object in Earth Orbit: A Proposed TCBM for Space Security", in *War and Peace in Outer Space: Law, Policy, and Ethics*, ed. Cassandra Steer and Matthew Hersch, Ethics, National Security, and the Rule of Law (Oxford, New York: Oxford University Press, 2021), 248.
26 "Group of Governmental Experts on Further Effective Measures for the Prevention of an Arms Race in Outer Space – UNODA", United Nations Office for Disarmament Affairs, 2019, https://www.un.org/disarmament/topics/outerspace/paros-gge/.

c) general principles
d) as a subsidiary source, if the other three do not provide a clear answer as to the law, judicial decisions and the writings of "the most highly qualified publicists".[27]

As discussed above, when looking to the treaty law of the OST, Article III tells us that international law in general applies in space. Therefore, with respect to military activities, the legal principles of *jus ad bellum* and *jus in bello* apply to outer space just as they apply to activities on land, at sea, or in the air. Current *jus ad bellum* and *jus in bello* laws have developed over time in response to the ways in which force is used and conflicts are fought. For the most part, the law of *jus ad bellum* is based upon the prohibition against the threat or use of force prescribed in art 2(4) of the UN Charter, as well as on customary law and some case law emerging from the International Court of Justice (ICJ).[28] *Jus in bello* has a much more detailed body of law, based in large part on the Geneva Conventions which were negotiated following World War II, and on more than eighty treaties which have been negotiated since, as well as a large body of customary law and case law.[29] But the treaties, custom and even general principles have developed in application to the other physical domains, and some treaties even contain language that specifies when it applies to land or to sea. The question, therefore, is exactly which rules apply to space, which might be excluded, and which might require more attention for development by States. The Woomera Manual sets out to clarify this by examining the treaty and customary international law developments and then testing them for their applicability to space. Essentially, unless specific rules are written or developed to apply to particular domains, the principles do not change, and it is a matter of applying them to new sets of facts.

Manuals as a Project to Clarify the Law

The Woomera Manual bases itself upon the lessons learned from previous international manuals developed by international experts independently

---

27 "Statute of the International Court of Justice," 33 UNTS 993 § (1946), https://www.icj-cij.org/en/statute, Article 38.
28 See generally Christine Gray, *International Law and the Use of Force* (Oxford University Press, 2018).
29 "ICRC Databases on International Humanitarian Law | International Committee of the Red Cross", 2021, https://www.icrc.org/en/icrc-databases-international-humanitarian-law.

of States, as a way of forming global agreement on the application of *jus ad bellum* and *jus in bello* to new technologies and different domains.

In 1995 the San Remo International Institute of Humanitarian Law recognised the need to clarify how *jus ad bellum* and *jus in bello* applied to new technologies and new forms of warfare at sea. The development of new treaties was unlikely and, in any case far too slow, and so the *Manual on International Law Applicable to Armed Conflicts at Sea*[30] was spearheaded by the Institute, bringing together military lawyers, academics, and technical experts to formulate a restatement of customary international law where there was no clear treaty law applicable to armed conflict at sea. The manual was drafted in a similar style to national military manuals, in which States set out a series of rules to guide decision-makers and included a commentary on each rule to reflect the opinions and discussions of the experts, and identify where there was consensus. The Manual has since been referred to by many States in the development of their own rules of engagement and national manuals. It reflected the position of States at the time and has since contributed to further crystallisation of customary international law norms for conflict at sea. The success of this project was what inspired the 2003 *Harvard Manual on International Law Applicable to Air and Missile Warfare*,[31] and the 2013 *Tallinn Manual on International Law Applicable to Cyber Warfare*,[32] which deals with these precarious questions in the context of cyber-attacks. Even though there has been some criticism as to the process by which the Tallinn Manual was drawn up, in that it did not include non-Western States and was funded largely by the NATO Cooperative Cyber Defence Centre of Experts, it seems that most States are keen to have an international standard by which they can operate in these new technological domains, and Tallin 2.0 was published in 2017 to cover a further set of questions.[33] Space is the next domain in which this need has become very clear.

In 2016 work began on a Manual on the International Law Applicable to Military Operations in Outer Space (MILAMOS).[34] Experts in space law,

---

30 Louise Doswald-Beck, "The San Remo Manual on International Law Applicable to Armed Conflicts at Sea," *American Journal of International Law* 89, no. 1 (1995): 192–208.
31 Claude Bruderlein, *HPCR Manual on International Law Applicable to Air and Missile Warfare* (Cambridge University Press, 2013).
32 Michael N. Schmitt, *Tallinn Manual on the International Law Applicable to Cyber Warfare* (Cambridge University Press, 2013).
33 *Tallinn Manual 2.0 on the International Law Applicable to Cyber Operations*, 2nd ed. (Cambridge: Cambridge University Press, 2017), https://doi.org/10.1017/9781316822524.
34 https://www.mcgill.ca/milamos/.

the laws of *jus ad bellum*, and *jus in bello*, space operations and cyber law were gathered, including some individuals who had been involved in the Harvard and Tallinn Manuals. The project was ambitious, and it soon became clear there was too much for one manual. Thus, the MILAMOS continues under the governance of the McGill Institute of Air and Space Law and seeks to focus on space law in times of peace. In 2018, the Woomera Manual commenced, spearheaded by Adelaide University Law School,[35] with a focus on international military and security law in space, especially the use of force during times of tension and conflict. Ultimately it is the intention that both manuals will complement each other. The Woomera Manual has experts from 11 different countries including China and India, as well two technical experts from the Union of Concerned Scientists and Secure World Foundation, and an observer from the International Committee of the Red Cross.

Issues Covered in the Woomera Manual

Part of the problem is that right now we do not have an internationally agreed definition of what amounts to a threat of force or an act of aggression in space. Does deliberate interference with a satellite that provides military intelligence or communications amount to use of force? What if this interference is temporary, and does not involve kinetic destruction of the satellite in question? Does interference amounting to sending a satellite off its intended orbital trajectory amount to use of force? If one State unilaterally interprets these actions as acts of aggression, is it justified in using force in self-defence in space or on Earth? And if so, how do the laws of *jus in bello* apply with respect to identifying which satellites or space systems are lawful targets, and making a proportionality and necessity calculation?

Another example where clear international rules would be helpful to decision-makers and operators is the paramount question of lawful targeting. One of the core legal principles of *jus in bello* is that of distinction, which requires that, "in order to ensure respect for and protection of the civilian population and civilian objects" parties to a conflict "shall at all times distinguish between the civilian population and combatants and between civilian objects and military objectives and accordingly shall direct their operations only against military objectives".[36] As discussed in the previous

---

35 "The Woomera Manual". The Woomera Manual is supported by the University of Exeter, the University of Nebraska, and the University of New South Wales in Canberra.
36 "Protocol Additional to the Geneva Conventions of 12 August 1949 and Relating to the Protection of Victims of International Armed Conflicts (Protocol I)" (1977), Article 48.

chapter, the "dual use" nature of many satellites, providing imperative services for both civilian and military purposes, may make it extremely difficult to identify whether such a satellite is a lawful target.[37] And even if it is lawful, there must be an assessment made of whether disabling or destroying such a dual-use satellite would create fallout for civilians that would be disproportionate to the military advantage gained by its targeting, for instance if it provided GPS signals for civil aviation.[38] These principles apply to a potential conflict in space as a matter of customary law and of general principle, but they were not written with the peculiarities of space technologies and the space environment in mind. The development of international regulations which clarify how these principles should apply to space can create an even ground of agreed standards, which can aid in predictability and reduce the risk of escalation.

The Woomera Manual tackles issues of right of access to space, jurisdiction, State responsibility and accountability, obligations with respect to military objects and activities in space, military space activities short of armed force, and military space activities during armed conflict.[39] It therefore provides a comprehensive guide to the most pressing questions that today's military space decision-makers and operators have. States can still apply their own interpretations of the law applicable, but the Woomera Manual will provide a neutral, central point of reference that will hopefully encourage international consistency and clarity.

But some question the role, legitimacy and impact of these manuals, which brings us to the debate at the centre of this chapter: do such projects serve to limit or rather legitimise warfare? And given that our activities in space must be for peaceful purposes and for the benefit of all nations, perhaps a manual is particularly problematic for the space domain. The fact is, there have been various attempts over the years to improve space arms control and reduce the risk of a conflict in space, but the effectiveness of these attempts could be questioned, which raises the concern of many that we may be on the cusp of conflict extending into space. In the next section, the political history of these attempts will be discussed, as a context for the different views on the ethics of a manual relating to space warfare.

---

37 Steer and Stephens, "International Humanitarian Law and Its Application in Outer Space," 40.
38 Steer and Stephens.
39 The expectation is that by the end of 2021, the Woomera Manual will have gone through a process of State engagement and final review and be ready for publication. Until then the contents are not publicly available.

Competing Views on the Ethics of Regulating Space Warfare

As we move into the third decade of the twenty-first century, the global political order is entirely uncertain, other than that we have returned to a multipolar world. There are multiple political, economic and social fronts of tension, and commercial entities are very much a part of today's international political competition. All of these terrestrial factors play out in space in exactly the same way, and because of how critical space-based technologies are, the risk of a crisis or conflict extending into space is very real – though not necessarily inevitable. Despite the geopolitical shifts we are witnessing, there remains a general commitment to a rules-based international order. It is therefore imperative to clarify the extent of this rules-based order in its application to space.

There are three main schools of thought regarding whether international law is already sufficient to cover crisis or conflict in space: first, some would assert that there is no law governing a potential conflict in outer space, and that it is a lawless frontier, which requires assertive action. Those adhering to this school of thought tend to resist all initiatives to clarify existing law or develop new legal instruments regulating military activities or an arms race in space. As the discussion in the previous chapter demonstrates, this position is simply incorrect, and in fact potentially dangerous, since it can lead to escalation. This position will not receive further consideration in this chapter. Second, some would argue that if we are to maintain the peaceful purposes principle, and to discourage further militarisation or weaponisation of space, then we should not develop rules and laws of military engagement in space, since to do so would be to not only condone it, but in fact to create legitimacy for States which may wish to justify increased military activity in space. A third school of thought takes a kind of middle road: existing bodies of international law apply to space, including on the use of force, the law of armed conflict and environmental law, however work needs to be done to clarify exactly which normative rules would apply and how. These last two schools of thought will be given more attention here.

"Peaceful Purposes" as Paramount

There is no doubt the 'peaceful purposes' directive of the OST must be upheld and respected. The question is whether the development of the Woomera Manual goes against this principle. There are some who assert that all efforts should be directed towards the prevention of an arms race in space, and that any efforts to regulate space warfare would be to tolerate or even condone its occurrence. According to this school of thought, any international declaration of the law applicable to space warfare would be a legitimisation of it, particularly if the participants involved in such a project were to include stakeholders such as military or government representatives of major powers.

The position from the perspective of these States is that any attempts to regulate potential space warfare would be in opposition to this commitment to a new binding treaty such as the PPWT or any variation and would legitimise any State wishing to use force in space. How can we have international rules determining that no weapons shall be placed in space, as well as international rules determining the lawful use of weapons in space? Surely this is contradictory.

In fact, it is not at all contradictory. There is a traditional divide between *jus ad bellum*, or the law regarding the use of force, and *jus in bello*, or the law regarding the conduct of conflict. Whereas the conditions under which force may be lawfully employed are very limited according to *jus ad bellum*, once there is a situation of armed conflict, the *jus in bello*, or law of armed conflict, is binding on all parties with the intention to clarify how a conflict may be conducted, in order to mitigate the effects of war and protect civilian objects and lives. While the laws of *jus ad bellum* may determine who is a wrongful aggressor, or when force may be used as a justified means of self-defense, the laws of *jus in bello* come into play as soon as there is an armed conflict and is neutral as to who was the wrongful actor in the first place. Thus, while we may be moving towards new norms prohibiting the placement of weapons in space, it is not beyond comprehension that a State actor might breach this norm, leading to a conflict involving space objects, and the need for clarity on the applicable *jus in bello* legal principles.

Added to this is the unique nature of the space environment and the kinds of technologies we are using; it may not even be clear exactly what amounts to a weapon, nor to a use of force. China, Russia and the U.S. are all testing various covert technologies in space, whose capabilities and purposes are difficult to divine. Some are co-orbital spacecraft, some have been "spawned" by the original spacecraft that was launched, some are operating as spaceplanes, landing and re-launching multiple times.[40] It is

---

[40] For a country-by-country overview of these capabilities, see Weedon and Samson, "Global Counterspace Capabilities".

becoming increasingly difficult to attribute intent of these capabilities as they fly near other satellites, and if there is a satellite failure or glitch in the service, it is near impossible to attribute the cause. Was it space weather, debris, or deliberate interference? And if the latter, would this amount to a threat or use of force? Could another State retaliate? At what point could a State interpret such an interference as an armed attack and assert a right of self-defense, thus escalating a situation to a state of conflict? Without clarifying what amounts to an armed conflict according to the law of *jus ad bellum*, and how the legal principles of *jus in bello* should apply in an ensuing conflict, we can find no solace in rules proclaiming the "peaceful purposes" axiom to be paramount or prohibiting the placement of weapons in space.

Resistance to Binding International Norms

Proponents of the primacy of the peaceful purposes maxim are absolutely correct that efforts should be directed towards arms control, and towards the prevention of space warfare. There is no question that this is the chief priority for all actors in the space sector. However, it is a cheerless fact that since the UN Charter was signed in 1945, prohibiting the use of force under Art 2(4), there has not been a single year in which there has not been an armed conflict of some sort somewhere in the world.[41] Proscriptive legal rules of *jus ad bellum* and *jus in bello* serve to limit and restrain States, but cannot completely prevent conflict. Some actors take this realist starting point to argue that potential conflict in space should not be the subject of international regulation, not because they favour a PPWT, but for the opposite reason: because they do not favor any binding regulation at all. Powerful space-faring nations such as the U.S.[42] and France[43] as well as others

---

41 https://www.icrc.org/eng/who-we-are/history/since-1945/history-ihl/overview-development-modern-international-humanitarian-law.htm.

42 The U.S. stated that that it would, in principle, be in favour of a binding instrument but not one without a verification procedure Delegation of the United States of America to the Conference on Disarmament, "Note Verbale Dated 2 September 2014 Addressed to the Acting Secretary-General of the Conference Transmitting the United States of America Analysis of the 2014 Russian-Chinese Draft Treaty on the Prevention of the Placement of Weapons in Outer Space, the Threat or Use of Force against Outer Space Objects, CD/1988", September 3, 2014, https://documents-dds-ny.un.org/doc/UNDOC/GEN/G15/007/57/PDF/G1500757.pdf?OpenElement.

43 Louis Riquet, "Intervention de M. Louis RIQUET Representant Permanent Adjoint de la France aupres de la Conference du Desarmement", October 27, 2014, https://unoda-web.s3-accelerate.amazonaws.com/wp-content/uploads/assets/special/meetings/firstcommittee/69/pdfs/TD_OS_27_Oct_France.pdf.

such as Canada and Australia[44] have refused to support the PPWT proposal, and some have opposed the work of the 2019 UN Group of Governmental Experts. Neither COPUOS nor the CD have been able to facilitate States towards any binding instruments with respect to space since the Cold War. If these international bodies are unable to facilitate the negotiations for new binding instruments, then, the argument goes, there is clearly no political will to create new laws. However, there is no reason why non-binding instruments and the clarification of binding regulations on specific issues cannot go hand in hand. In fact, combined efforts of this sort may have a mutually strengthening and bolstering effect, and the recent adoption of UN General Assembly Resolution on "Reducing space threats through norms, rules and principles of responsible behaviours" is indicative of this.[45]

Until very recently, the opposition of the U.S. against any binding international space arms control efforts has been the dominant stance by virtue of the status of the U.S. as the unipolar power on Earth and the dominant power in space. However, the balance of power is shifting dramatically as we move into the 2020s, and as more actors, both governmental and commercial, begin to change the dynamics of space politics, we are faced with new tension points. On the one hand, destabilising effects emerge as China and India grow more powerful and improve on their technological abilities, and as the U.S. responds with escalatory rhetoric surrounding Space Force and the desire to maintain dominance in space. On the other hand, international space diplomacy is now being led by the UK and some European nations, and there is a growing role for traditional middle powers such as Australia. The statement that rang true for the last decade that there is no desire for new treaties or more detailed international legal regimes in space is no longer true. Thus, the position opposing regulation of military space activities is becoming weaker. But new technologies and political shifts are outpacing any international law-making processes.

If we want to be able to minimise the risk of escalation during times of tension, if we want to be able mitigate the effects of a potential conflict taking place in or through space, and if we want to be able to bring such a

---

[44] Darren Hansen, "Thematic Statement: Outer Space (Disarmament Aspects)", October 27, 2014, https://unoda-web.s3-accelerate.amazonaws.com/wp-content/uploads/assets/special/meetings/firstcommittee/69/pdfs/TD_OS_27_Oct_Australia.pdf.

[45] UN General Assembly "Resolution 75/36: Reducing Space Threats through Norms, Rules and Principles of Responsible Behaviours" 16 December 2020, https://digitallibrary.un.org/record/3895440?ln=en.

potential conflict to a close sooner, we must have clarity on the rules on the use of force and the laws of armed conflict applicable to space, just as we do for conflicts on land, sea and in the air.

The Preference for Clarification and Regulation

Neither the laws of *jus ad bellum* nor *jus in bello* purport to prevent aggression or conflict. Rather we have international humanitarian law precisely because the prohibition on the use of force is sometimes flouted, and States (as well as a growing number of non-State actors) agree that it is better to have rules in place to regulate what amounts to lawful or unlawful force, and what amounts to lawful or unlawful warfare, in order to mitigate its effects. But regulating the use of force, and regulating the rules of armed conflict, does not amount to condoning their existence.[46] The same reasoning should be applied to the push now to regulate the use of force in space.

An analogy can be drawn with criminal law. Although in public international law we do not talk about State acts as being criminal, the analogy is appropriate since we are talking about why we apply the law to acts we consider wrongful, and why it is important to define these acts as unlawful. Sometimes individuals flout the moral conditions of civil society. However, regulating the limits of human interaction, including violent crimes, does not in any way condone this behavior. On the contrary, there are three main theories underpinning criminal law and why we punish; prevention, retribution, and norm expression.[47] With respect to prevention, it may be impossible to empirically prove how many premeditated crimes were aborted due to the threat of punishment. Yet by having a clear legal order in place, citizens know exactly what the lawful limits are on their behaviour, and the moral calculations made when considering committing an act of theft or a crime of violence will include the possible ramifications. It helps us all to know what is acceptable and what is forbidden behaviour. The same can be said for military behaviour in space.

The second main criminal law theory is retribution. By clarifying what amounts to a breach of the limits on acceptable behaviour, society

---

[46] International Committee of the Red Cross, "International Humanitarian Law and the Challenges of Contemporary Armed Conflicts," Report, November 13, 2015, 40, https://www.icrc.org/en/document/international-humanitarian-law-and-challenges-contemporary-armed-conflicts; Mačák, "Silent War," 37.
[47] George P. Fletcher, *Rethinking Criminal Law* (Oxford University Press, USA, 2000), 414–15.

can respond appropriately when those limits are breached.[48] Individuals know that even if they are wronged, they are not to take the law into their own hands and enact vengeance; rather they are to go through the channels provided by the State to make an arrest and prosecute and punish the wrongdoer. Though no criminal law system is perfect in terms of efficiency or answering all victims' grievances, to have such an order in place ensures greater chances of equality before the law, and of final conflict resolution through retributive justice.[49] Similarly, with a clear legal order in place regarding the use of force and the law of armed conflict in space, the avenues for pursuing a State or entity which breaches those rules can include many more conflict-resolving possibilities than just the use of force in retaliation, which would lead to escalation. When a State breaches these rules there are avenues through the law on State responsibility, including demands of cessation and reparation.[50] When an individual or non-State actor breaches these laws, there are avenues such as international criminal law and collective terrestrial responses short of the use of force.

The third main criminal law theory is norm expression. H.L.A. Hart referred to the "denunciatory" theory, that is, by choosing to criminalise certain acts, a society is making clear what behaviour it denounces as unacceptable.[51] These norms become internalised both at an individual and a collective level.[52] This expressive role of identifying wrongful behaviour by clarifying in the law exactly what is considered acceptable or unacceptable is important even with respect to actions which everyone already knows are wrong. The expression of the wrongfulness in the law galvanises the norms of a society and specifies the exact limits of acceptable behaviour. In international criminal law this expressive function is particularly important with respect to actions of actors in leadership positions during a conflict, and the breach of the laws of *jus ad bellum* or *jus in bello*.[53] It is a logical corollary of this that it is necessary to clarify the application of the legal principles of *jus ad bellum* and *jus in bello* to the new domain of outer space

---

48 Imogen Tallgren, "The Sensibility and Sense of International Criminal Law", *European Journal of International Law* 13, no. 3 (2002): 579.
49 H. L. A. Hart, *Punishment and Responsibility: Essays in the Philosophy of Law*, 2nd edition (Oxford University Press, 1970), 9.
50 James Crawford, *The International Law Commission's Articles on State Responsibility: Introduction, Text and Commentaries* (Cambridge University Press, 2002) Articles 30 and 31.
51 Hart, *Punishment and Responsibility*.
52 Tallgren, "The Sensibility and Sense of International Criminal Law," 580.
53 Cassandra Steer, *Translating Guilt: Identifying Leadership Liability for Mass Atrocity Crimes*, vol. 9 (Springer, 2017), 45.

before it is tested in the throes of a conflict. As UK's Lord Henning has put it, "the ultimate justification of any punishment is [...] that it is the emphatic denunciation by the community of a crime".[54]

Beyond this analogy with criminal law, there is anecdotal evidence that those involved in decision-making in times of tension and conflict in fact seek legal clarifications to help them make the "right" decision. More regulation means less guess work, and less risk of escalation.

For example, as mentioned above, it is not yet clear what might amount to a use of force in space. Because the technologies being employed are so specific to the space domain, and the ways in which one might interfere with or disable these technologies differ from the ways in which actions interpreted as force might be taken in a terrestrial context, there is a risk that during times of tension one State might interpret the actions of another as force, and take this to mean that a state of conflict has entered into being, or that it may justifiably use force in retaliation against the first State. Military satellite operators would prefer to know when an action can be considered an unlawful threat or use of force, and what the legal limitations are on the responses they can consider, than to leave everything up to interpretations made under the pressure of time and political tensions.

The claim made here is that by clarifying the norms applicable, space warfare may not be prevented *in toto*, however the number and kinds of cases for escalating a situation from tension to conflict could be reduced simply by having more clarity. As well, in the eventuality that conflict does take place in space, internationally recognized rules could help to mitigate its effects and bring it to a close. Multiple international instruments could contribute to this and could work in parallel rather than in competition with each other. Thus, even those States or stakeholders who advocate for the development of new binding instruments such as a PPWT, could advance their cause if they supported the clarification of international law applicable to a potential conflict in space. In the end, what is central is a shared commitment to the rule of law.

---

54 Report of the Royal Commission on Capital Punishment (London, 1953), Section 53.

The Role of the Woomera Manual in Regulating but Not Legitimising Star Wars

It has been argued here that it is not contradictory to support all efforts to prevent the weaponisation of space on the one hand and to support the regulation of armed conflict in space on the other hand, in part because the law governing armed conflict is neutral with respect to the use of force. In other words, even if we work to prevent the weaponisation of space, we may not be able to prevent an act of aggression in space by means of a space object that is not designated as a weapon, or by means of a simple breach of the emerging norms against weaponisation. We must therefore concurrently work to prevent the weaponisation of space, and to clarify the application of the law of armed conflict to space as a preventive measure; should conflict take place in space or should space assets be used in aggressive ways with respect to a terrestrial conflict, it is better to have internationally recognised rules in place, which may help to de-escalate a situation and bring a conflict to a close. A lack of regulation would leave it up to technological prowess and "might is right", with disastrous effects for us all. Internationally agreed regulation can aid in reducing the risk of a conflict in space, mitigating the effects of an eventual conflict, and providing clarity which may also lead to bringing such a conflict to an earlier close.

While prevention should be the priority of all actors, no amount of regulation, binding or non-binding, has ever been able to prevent conflict *in toto*. History paints a sad reality in this respect. But history can also teach us that having internationally agreed standards and rules can mitigate the potential for self-interest being the only guide in times of tension and conflict. Unilateral determinations of, for example, the presence of weapons of mass destruction or the justification of the use of force, have led to contentious situations of conflict and the loss of many lives in our very recent history. International regulation and the pressure to abide by it can have a softening and mitigating effect, such that it in fact can work hand in hand with efforts to clarify international standards of conduct in space, and international norms against weaponisation of space.

Rather than legitimising the use of force in space, regulating potential space warfare in fact limits, by definition, exactly when force may be used in self-defence, and how. We must not forget that the use of force in self-defence is a norm that exists in international law independently of *where* it might be employed. Regulating potential conflict in space is not to expand the reasons, means and methods of warfare, but rather to clarify and strengthen the limits placed on warfare *before* it becomes a reality in a new domain.

References

Antarctic Treaty, Pub. L. No. 402 U.N.T.S 71 (1961).
Bruderlein, Claude. *HPCR Manual on International Law Applicable to Air and Missile Warfare*. Cambridge University Press, 2013.
Charter of the United Nations (1945). https://www.un.org/en/sections/un-charter/un-charter-full-text/.
Christol, C.Q. *The Modern International Law of Outer Space*. Pergamon, 1982.
Crawford, J. *Brownlie's Principles of Public International Law*. 9th ed. OUP Oxford, 2019.
Crawford, J. *The International Law Commission's Articles on State Responsibility: Introduction, Text and Commentaries*. Cambridge University Press, 2002.
Delegation of the United States of America to the Conference on Disarmament. "Note Verbale Dated 2 September 2014 Addressed to the Acting Secretary-General of the Conference Transmitting the United States of America Analysis of the 2014 Russian-Chinese Draft Treaty on the Prevention of the Placement of Weapons in Outer Space, the Threat or Use of Force against Outer Space Objects, CD/1988," September 3, 2014. https://documents-dds-ny.un.org/doc/UNDOC/GEN/G15/007/57/PDF/G1500757.pdf.
Doswald-Beck, L. "The San Remo Manual on International Law Applicable to Armed Conflicts at Sea". *American Journal of International Law* 89, no. 1 (1995): 192–208.
Doucet, G. "Notification of Transfer of Kinetic Energy to an Object in Earth Orbit: A Proposed TCBM for Space Security". In *War and Peace in Outer Space: Law, Policy, and Ethics*, edited by Cassandra Steer and Matthew Hersch, 245–62. Ethics, National Security, and the Rule of Law. Oxford, New York: Oxford University Press, 2021.
"Federation of American Scientists: Prevention of an Arms Race in Outer Space". Accessed March 2, 2021. https://fas.org/programs/ssp/nukes/ArmsControl_NEW/nonproliferation/NFZ/NP-NFZ-PAROS.html.
Fletcher, G. P. *Rethinking Criminal Law*. Oxford University Press, USA, 2000.
General Assembly of the United Nations. "No First Placement of Weapons in Outer Space", December 2, 2014. https://undocs.org/en/A/RES/69/32.
Gray, C. *International Law and the Use of Force*. Oxford University Press, 2018.
"Guiding Principles Applicable to Unilateral Declarations of States Capable of Creating Legal Obligations, with Commentaries Thereto". International Law Commission, 2006. https://legal.un.org/ilc/texts/instruments/english/commentaries/9_9_2006.pdf.

Hansen, D. "Thematic Statement: Outer Space (Disarmament Aspects)", October 27, 2014. https://unoda-web.s3-accelerate.amazonaws.com/wp-content/uploads/assets/special/meetings/firstcommittee/69/pdfs/TD_OS_27_Oct_Australia.pdf.

Hart, H. L. A. *Punishment and Responsibility: Essays in the Philosophy of Law*. 2nd edition. Oxford University Press, 1970.

"ICRC Databases on International Humanitarian Law | International Committee of the Red Cross". Accessed February 12, 2021. https://www.icrc.org/en/icrc-databases-international-humanitarian-law.

International Committee of the Red Cross. "International Humanitarian Law and the Challenges of Contemporary Armed Conflicts". Report, November 13, 2015. https://www.icrc.org/en/document/international-humanitarian-law-and-challenges-contemporary-armed-conflicts.

Mačák, K. "Silent War: Applicability of the Jus in Bello to Military Space Operations", 2018.

Ministry of Foreign Affairs, People's Republic of China. Treaty on the Prevention of the Placement of Weapons in Outer Space, the Threat or Use of Force against Outer Space Objects (Draft) (2014). https://www.fmprc.gov.cn/mfa_eng/wjb_663304/zzjg_663340/jks_665232/kjfywj_665252/t1165762.shtml.

Moltz, J. *The Politics of Space Security: Strategic Restraint and the Pursuit of National Interests*. Stanford University Press, 2011.

National Defense Authorization Act, Pub. L. No. 116–92, S. 1790 (2019).

Protocol Additional to the Geneva Conventions of 12 August 1949 and relating to the Protection of Victims of International Armed Conflicts (Protocol I) (1977).

Ribbelink, O. "Article III". In *Cologne Commentary on Space Law: Outer Space Treaty*, edited by Stephan Hobe, Bernhard Schmidt-Tedd, and Kai-Uwe Schrogl, 1:64–69. Carl Heymanns Verlag, 2009.

Riquet, L. "Intervention de M. Louis RIQUET Representant Permanent Adjoint de la France aupres de la Conference du Desarmement", October 27, 2014. https://unoda-web.s3-accelerate.amazonaws.com/wp-content/uploads/assets/special/meetings/firstcommittee/69/pdfs/TD_OS_27_Oct_France.pdf.

Schmitt, M. N. *Tallinn Manual on the International Law Applicable to Cyber Warfare*. Cambridge University Press, 2013.

Schrogl, K, and J. Neumann. "Article IV". In *Cologne Commentary on Space Law: Outer Space Treaty*, edited by Stephan Hobe, Bernhard Schmidt-Tedd, and Kai-Uwe Schrogl, 1:70–93. Carl Heymanns Verlag, 2009.

Statute of the International Court of Justice, 33 UNTS 993 § (1946). https://www.icj-cij.org/en/statute.

Steer, C. *Translating Guilt: Identifying Leadership Liability for Mass Atrocity Crimes*. Vol. 9. Springer, 2017.

Steer, C. and D. Stephens. "International Humanitarian Law and Its Application in Outer Space". In *War and Peace in Outer Space: Law, Policy, and Ethics*, edited by Cassandra Steer and Matthew Hersch, 23–54. Ethics, National Security, and the Rule of Law. Oxford, New York: Oxford University Press, 2021.

Su, J. "The Legal Challenge of Arms Control in Space". In *War and Peace in Outer Space: Law, Policy, and Ethics*, edited by Cassandra Steer and Matthew Hersch, 179–98. Ethics, National Security, and the Rule of Law. Oxford, New York: Oxford University Press, 2021.

Tallgren, I. "The Sensibility and Sense of International Criminal Law". *European Journal of International Law* 13, no. 3 (2002): 561–95.

Tallinn Manual 2.0 on the International Law Applicable to Cyber Operations. 2nd ed. Cambridge: Cambridge University Press, 2017. https://doi.org/10.1017/9781316822524.

"The Woomera Manual", 2020. https://law.adelaide.edu.au/woomera/home.

Treaty on the Principles Governing the Activities of States in the Exploration and Use of Outer Space, including the Moon and Other Celestial Bodies, 610 U.N.T.S. 205 § (1967). https://www.unoosa.org/oosa/en/ourwork/spacelaw/treaties/outerspacetreaty.html.

"UN COPUOS Legal Sub-Committee Fifth Session, UN Doc A/AC.105/C.2/SR.57", October 20, 1966.

UN General Assembly. "Prevention of an Arms Race in Outer Space, A/Res/72/26", December 11, 2017. https://digitallibrary.un.org/record/1326281?ln=en.

UN General Assembly. "Resolution 1348 (XXIII): Questions of the Peaceful Uses of Outer Space", December 13, 1958.

UN General Assembly. "Resolution 1472 (XIV): International Co-operation in the Peaceful Uses of Outer Space", December 12, 1959.

UN General Assembly. "Transparency and Confidence Building Measures in Outer Space Activities, A/Res/75/69", December 7, 2020. https://undocs.org/en/A/RES/75/69.

UN General Assembly. "Resolution 75/36: Reducing Space Threats through Norms, Rules and Principles of Responsible Behaviours", 16 December 2020, https://digitallibrary.un.org/record/3895440?ln=en.

United Nations First Committee on Disarmament and International Security. "Reducing Space Threats through Norms, Rules and Principles of Responsible Behaviours, A/C.1/75/L.45/Rev.1", October 23, 2020. https://www.un.org/ga/search/view_doc.asp?symbol=A/C.1/75/L.45/Rev.1.

United Nations Office for Disarmament Affairs. "Group of Governmental Experts on Further Effective Measures for the Prevention of an Arms Race in Outer Space – UNODA", 2019. https://www.un.org/disarmament/topics/outerspace/paros-gge/.

United States Department of Defense. "Summary of the 2018 National Defense Strategy", 2018. https://www.hsdl.org/?view&did=807329.

"US Space Force Chief: Space Is 'A Warfighting Domain'". Forces Network. 2021. https://www.forces.net/news/head-us-space-force-space-warfighting-domain.

Vienna Convention on the Law of Treaties, 1155 U.N.T.S. 331 § (1969).

Weedon, Brian, and Victoria Samson, eds. "Global Counterspace Capabilities". Secure World Foundation, April 2020. https://swfound.org/counterspace/.

12

RESPONDING ON EARTH TO KINETIC ATTACKS IN SPACE

Geordie Jacobs[1]

Most of the literature written about kinetic attacks in space refers to actions taken by one State against another State's space based-asset, considering the situation if China or Russia, for example, were to employ a ground or space-based weapon against a United States satellite. Extraordinarily little discussion is being had regarding non-state actors carrying out kinetic attacks against satellites. In this chapter, I wish to consider exactly this sort of situation, with a particular focus on the question of when, if ever, a kinetic response on Earth would be an appropriate reaction by a State to a kinetic attack carried out on a satellite by a non-state actor.

Despite remarks made in 2006 by Robert Joseph, President Bush's Under Secretary for Arms Control and International Security, regarding the threat of non-state actor attacks in space, more than a decade later the 2019 United States Defense Intelligence Agency (DIA) report *Challenges to Security in Space* references non-state actors only from a commercial point of view and fails to address the existence of any belligerent non-state actors or terrorist organisations that could threaten the security of space

---

1 I wish to acknowledge the support and kind assistance of Squadron Leader, Reverend, Dr, Nikki Coleman (Royal Australian Air Force – Space Domain Review), Associate Professor Stephen Coleman (University of New South Wales – School of Humanities and Social Sciences), and Sergeant Amy Hestermann-Crane (Royal Australian Air Force – Directorate of Intelligence, Surveillance, Reconnaissance, Electronic Warfare & Space). The views expressed are those of the author and do not reflect the official policy or position of the Royal Australian Air Force, the Department of Defence, or the Australian Government.

based assets.² The Chief of Space Operations in the United States Space Force, General John Raymond, has spoken of Russian satellites conducting on-orbit weapons testing and speculation has been increasing regarding China's advancement in on-orbit anti-satellite capabilities however nothing has been said regarding terrorist on-orbit attacks.³ The DIA *Challenges to Security in Space* report lists six types of orbital threats including kinetic kill vehicles (KKV), radiofrequency jammers, lasers, chemical sprayers, high-power microwaves, and robotic mechanisms. However, again fails to mention the threat posed by terrorist organisations.⁴

The general response from most people, even experts, when the topic of space terrorism is raised is that it isn't possible – so why talk about it? Apart from being short-sighted, this attitude is incredibly naïve, and this naivety is exemplified using two other inventions from the twentieth century that have changed the way we live – much like the invention of satellites. Within one hundred years of their creation both inventions had been used by terrorist organisations to pursue their political or ideological aim: aircraft and the internet.

In 1903, Wilbur and Oliver Wright achieved "powered, sustained, controlled flight from level ground in their Flyer 1" for the first time in human history.⁵ Eight years after this momentous achievement, the Italian First Aeroplane Flotilla carried out the world's first instance of aerial attacks against Turkish positions during the Italian–Turkish War of 1911-12.⁶ Thirty-four years after this, much of Europe had been decimated by the aerial bombardment of World War II, and by 1975, over seven and a half million tonnes of explosives had been dropped on Vietnam, Laos and Cambodia

---

2 "Remarks on the President's National Space Policy delivered to the George C Marshall", Robert G. Joseph, last modified December 13, 2006, http://www.spaceref.com/news/viewsr.html?pid=22773, "Challenges to Security in Space", United States Defense Intelligence Agency, last modified January 2019, https://www.dia.mil/Portals/27/Documents/News/Military%20Power%20Publications/Space_Threat_V14_020119_sm.pdf.
3 "Russia conducts space-based anti-satellite weapons test", U.S. Space Command Public Affairs Office, U.S. Space Command, last modified July 23, 2020, https://www.spacecom.mil/MEDIA/NEWS-ARTICLES/Article/2285098/russia-conducts-space-based-anti-satellite-weapons-test/.
4 U.S. DIA "Challenges to Security in Space".
5 Lorraine Cavaliere, "The Wright Brothers' Odyssey: Their Flight of Learning", *New Directions for Adult & Continuous. Education*, no. 53 (1992): 51-59.
6 Michael Paris, "The First Air Wars – North Africa and the Balkans, 1911-13", *Journal of Contemporary History*, no. 26 (1911): 97-109.

during the Vietnam War.⁷ Ninety-eight years after the Wright brothers first flight, on September 11, 2001, nineteen members of the al-Qaeda terrorist organisation flew American Airlines Flight 11 and United Airlines Flight 175 into the World Trade Centre towers in New York, American Airlines Flight 77 into the Pentagon, and United Airlines Flight 93 into a field in Somerset County, Pennsylvania.⁸ This event was the most destructive terrorist attack on United States soil to-date, and used aircraft alone as the means and method of attack.

On October 29, 1969, Leonard Kleinrock and his University of California team sent the first successful host-to-host message, "Lo", a precursor to the modern internet.⁹ By 1989, computers no longer took up entire rooms and commercial dial-up access was accessible in the United States. Thirty years after the sending of "Lo", in August 1999, a year after Google's creation, Jonathan James, a fifteen year old Californian, hacked into the U.S. Department of Defence and the National Aeronautics and Space Administration (NASA) acquiring the plans for the International Space Station's (ISS) life-support system while operating undetected in the network for over a month.¹⁰ In 2019, the same year the internet had its fiftieth birthday, Cybersecurity Ventures predicted that cybercrime, a category of activity that includes cyberterrorism, would cost the global economy (U.S.) $6 trillion in 2021.¹¹

I do not believe in 1903 the Wright brothers could have predicted that within one-hundred years, the descendants of their Flyer 1 would be the cause of such devastation in global wars nor be the weapons responsible for the most destructive terrorist attack on U.S. home soil to-date. Likewise, I do not believe in 1969, Leonard Kleinrock and his team could predict that within fifty years, their half-sent message of "Lo" would eventually evolve into a system like the Internet, attacks on which would cost the global economy over $6 trillion dollars in one year. *Sputnik* 1, the first human-made satellite,

---

7 Eric Jaworski, "Military Necessity and Civilian Immunity", in *International Crime and Punishment: Selected Issues*, ed. Sienho Yee (University Press of America, 2003), 96-97; "Bombing Missions of the Vietnam War", Environmental Systems Research Institute, last modified 2017, https://storymaps.esri.com/stories/2017/vietnam-bombing/index.html.
8 Gary Solis, *The Law of Armed Conflict: International Humanitarian in War* (New York, Cambridge University Press, 2010), 527.
9 "A Crash Into the Future", UCLA History, UCLA Alumni, last modified – Unknown, https://alumni.ucla.edu/ucla-history/ucla-history-45/.
10 Michael Newton, *The Encyclopedia of High-tech Crime and Crime-fighting* (New York: Infobase Publishing, 2003), 164.
11 "Cybercrime Damages $6 Trillion By 2021", Steve Morgan, *Cybercrime Magazine*, last modified October 26, 2020, https://cybersecurityventures.com/annual-cybercrime-report-2020/.

was launched into orbit on October 4, 1957.[12] In the subsequent sixty-four years, humankind has launched hundreds of people into orbit and millions of objects into Earth orbit and beyond.[13] With the launch of the ISS in 1998, humankind has had an enduring presence in space since November 2000.[14] Since the mass proliferation of satellites in the 1970s our reliance on space for everyday life has dramatically increased. However, this reliance is not only in civilian uses such as global communication, navigation, weather prediction and global financial systems, but in critical military uses such as intelligence, surveillance and reconnaissance, strategic early warning, and precision guidance for weaponry.[15] This dual civil and military reliance on satellite systems makes them "particularly attractive to terrorist groups because of the dramatic impact such an attack would have on most of the (civilian) population of the earth"[16] not to mention the impact such an attack would have on the advanced military forces that rely on satellites to operate. Additionally, space has become increasingly vulnerable as it has become significantly more accessible as technological and cost barriers have decreased, especially the decrease in cost of launching small readily available, low-cost satellites such as Cubesats. The ability to launch objects into space is no longer limited to the world's great space powers with indigenous launch capabilities.[17] Reputable international bodies such as the European Space Agency (ESA) and several private industries including the IHI Corporation, Mitsubishi Heavy Industries, Rocket Lab, SpaceX, Sea Launch and United Launch Alliance are all capable of launching their own rockets into orbit, carrying payloads from other companies and nations. The increase in launch providers and decrease in cost of launch (the 2011 cost to launch NASA's space shuttle into Low Earth Orbit (LEO) was approximately US$54,500 per kilogram, in 2018, the price quote for a SpaceX Falcon launch was US$2,720 per kilogram) has resulted in a dramatic increase in the number

---

12 "Timeline of Space Exploration", *Australian Geographic*, last modified August 27, 2012, https://www.australiangeographic.com.au/topics/science-environment/2012/08/timeline-of-space-exploration/.
13 U.S. DIA, "Challenges to Security in Space"; "The Forgotten Cold War Plan That Put a Ring of Copper Around the Earth," Joe Hanson, *Wired Magazine*, last modified August 13, 2013, https://www.wired.com/2013/08/project-west-ford/.
14 "International Space Station Facts and Figures", NASA, last modified May 14, 2021, https://www.nasa.gov/feature/facts-and-figures.
15 Remarks on the President's National Space Policy by Robert G. Joseph, December 13, 2006.
16 Nikki Coleman and Stephen Coleman, "Terrorism and Space Security", 69th *International Astronautical Congress*, Bremen, Germany (2018): 1-5.
17 U.S. DIA, "Challenges to Security in Space".

of satellites being launched into orbit.[18] Cataloguing and Monitoring what each of these organisations is launching into orbit in order to maintain Space Situational Awareness (SSA) is an increasingly expensive and momentous task and made all the more difficult by the increase in the amount of space debris, which threatens satellites in every orbit.[19]

It is this threat of space debris that could be exploited and weaponised by terrorist organisations in order to overcome the technological advantage which adversarial advanced military forces have gained through their use of satellites. Without a significant amount of expertise and at a relatively low cost, it is possible for a terrorist organisation to launch a Suicide Kinetic Kill Vehicle (SKKV), created with the sole purpose of creating a cloud of debris that would impact with satellites and create more debris – eventually impacting and damaging or possibly destroying the satellites in that and nearby orbits, or forcing those satellites to change their orbit in order to avoid damage or destruction.

The most likely device used by a terrorist organisation for a SKKV is an improvised explosive device (IED) aimed at a specific target or direction – effectively a directional fragmentation charge (DFC) concealed within an easily sourced and affordable cube-satellite. DFC are easy to construct, have a range of trigger mechanisms (time, proximity, pressure/impact and command signal), can be loaded with various manner of projectile (including small non-metallic projectiles which would be near impossible to track in space, such as gravel or golf balls) and have been used terrestrially by terrorist organisations to great effect, exemplified in countless attacks in Afghanistan where the Taliban and other terrorist organisations have employed DFC in targeted killings against officials, public figures and Afghan security and coalition force personnel.[20] Given that a DFC/IED would be the simplest form of SKKV, and thus the most likely for a terrorist group to use, for the rest of this discussion I will use the term SKKV to refer to such a device, though other devices are possible.

---

18 Harry Jones, "The Recent Large Reduction in Space Launch Cost", *48th International Conference on Environmental Systems Albuquerque*, New Mexico, July 8-12, 2018, last modified July 12, 2018, https://ttu-ir.tdl.org/handle/2346/74082.
19 "Bluestaq wins $280 million contract for Space Situational Awareness Library", Nathan Strout, C4ISRNET, last modified March 24, 2021, https://www.c4isrnet.com/battlefield-tech/space/2021/03/24/bluestaq-wins-280-million-contract-for-space-situational-awareness-library.
20 Ancil Adrian-Paul et al. *Afghanistan's Security Sector Governance Challenges – the Geneva Centre for the Democratic Control of Armed Forces* (Sofia: Procon Ltd. 2011), 34; "The Afghan War: Why the Kandahar Campaign Matters", Jason Motlagh, Time, last modified October 18, 2010, http://content.time.com/time/world/article/0,8599,2026158,00.html.

As also exemplified in recent conflicts in the Middle East, one of the most challenging aspects of asymmetric warfare is positively identifying who the threat actor is. Without positive identification (PID) you cannot hope to satisfy the *Jus in Bello* principles of discrimination and proportionality when responding to the threat. However, we have already seen the extraordinary lengths threat actors will go to obfuscate their actions to maintain plausible deniability. Attacks in Iraq and Afghanistan are often claimed or denied by multiple groups to gain or limit attribution, depending on the organisation's desired outcome.[21] Senior terrorist organisations and State security forces, such as the Haqanni Network, Taliban, Iranian Revolutionary Guard Corps and conventional Iranian military, often facilitate attacks carried out by other terrorist organisations providing them with training, weapons and equipment.[22] This facilitation increases the difficulty of determining the identity of the threat actors. In a similar vein, a terrorist act in space would also, possibly unknowingly, be facilitated by a larger organisation as terrorist organisations currently lack the capacity to launch their own satellites into space.

While many would still believe it would not be possible for a terrorist organisation to covertly place a SKKV in orbit, we have seen civilian organisations go to extreme lengths in order to achieve their desired space outcomes. One example that has occurred in space was perpetrated by the U.S. company Swarm.

On 12 January 2018, Swarm a U.S. based start-up communications company defied the Federal Communications Commission (FCC) by launching four pico-satellites into orbit as secondary payloads on an Indian rocket. The FCC had already ruled these pico-satellites could not be placed into orbit because they measure one-quarter the width of a cube satellite unit and were too small to be reliably tracked.

Swarm circumvented the FCC by having the pico-satellites launched from India and were caught when signals from the four pico-satellites were detected by ground stations in the U.S.

Swarm was fined $900,000, were retrospectively given authorisation for these satellites, and were then allowed to launch a further three

---

21 "With the U.S.-Taliban Deal in Place, IS-K Seeks to Build A Reign of Terror in Afghanistan", Ahmad Shah Kataw, Global Security Review, last modified, August 7, 2020, https://globalsecurityreview.com/with-the-us-taliban-deal-in-place-is-k-seeks-new-reign-of-terror-afghanistan/.

22 "In Afghanistan, a new 'great game' with ISIS, ISK and Pakistan is on with a vengeance", Tara Kartha, *The Print*, last modified March 11, 2020, https://theprint.in/opinion/in-afghanistan-a-new-great-game-with-isis-isk-and-pakistan-is-on-with-a-vengeance/378365/.

pico-satellites in December 2018 (onboard the SpaceX Falcon 9 rocket) – with the FCC's endorsement.[23]

This case demonstrates several important factors:

1. Those wishing to launch satellites can (and will) circumvent strict launch protocols to achieve their goal
2. The launcher of the satellite is not necessarily the owner of the satellite
3. The launcher of the satellite may be unaware of what they are launching as a secondary payload
4. Those responsible for launching satellites may not care about what they are launching as secondary payloads, and
5. Satellites can be launched and put into orbit without proper authority.

However, if an attack was launched using an SKKV, determining how the SKKV was placed into orbit is a necessary step that assists in determining its ownership and the identity of the threat actor. To determine how the SKKV was put into orbit one must determine:

- When was it launched / put into orbit?
- Where was it launched from?
- Was it launched by a State enterprise (for example ESA) or private enterprise (such as SpaceX)?
- This would indicate whether the SKKV is sponsored by a State (unlikely)
- An assessment must also be made as to whether the SKKV was launched as the primary payload or as a part of the secondary payload.

This final assessment must be made because if the SKKV was launched as part of the primary payload, this implies launching the SKKV was a deliberate action by the launcher, who is thus likely the threat actor. If the SKKV was launched as part of the secondary payload, this implies the SKKV was a source of income for the launcher. Being part of the secondary payload is more likely as it is highly unlikely an organisation capable of launching bodies into space would knowingly jeopardise the future use of reliable orbits.

---

23 "FCC fines SWARM $900,000 for unauthorized satellite launch", David Shepardson, Reuters, last modified December 21, 2018, https://www.reuters.com/article/us-usa-satellite-fine-idINKCN1OJ2WT.

Legitimately launched satellites are registered and are generally marked with symbology to show their ownership, such as a national or European Union flag or a corporate symbol. While the SKKV is unlikely to be registered or have identifiable markings or symbols on it, its identity may be deduced by other means. Where was it placed in orbit? What does it immediately threaten? What does it threaten long-term? Determining what is threatened in the long-term is a difficult assessment to make due to the fact debris caused in the first instance by the SKKV will likely damage multiple other systems. Determining the long-term effects would require sophisticated modelling software which is probably unavailable to terrorist organisations at this time. Therefore, it is likely the satellites that are immediately threatened by the SKKV are the intended target. It is possible the terrorists are targeting satellites indiscriminately, however this seems unlikely considering the effort taken to put a SKKV into orbit. Additionally, the threat actor's identity may be determined using signals analysis. Is the SKKV emitting or receiving any signals, such as repositioning commands to better position it to destroy its target? Could it be remotely commanded by signal to detonate? Locating the source or receiver of these signals would assist greatly in determining PID of who is controlling the satellite. However, satellite control (repositioning) is often outsourced to third party corporations who are responsible for deconflicting the path of a satellite with other satellites and space debris. If the satellite is being controlled, determining who is controlling the satellite may also assist in determining the identity of the threat actor.

Once both PID and the level of threat posed by the SKKV have been determined, the next question that must be asked is whether a kinetic reaction against the threat actor on Earth will prevent the use of the SKKV in space and overcome the threat. This is likely only possible if the SKKV is command detonated by remote signal, as a SKKV that detonates based off proximity, impact or time would not be affected by a kinetic response against the operator on Earth. Additionally, the question must be asked whether a kinetic response on Earth for an attack in space is worth the ramifications. Kinetic actions on Earth are far more visible to the international community than attacks in space. The State responding to the threat in space will likely face public backlash for a kinetic response on Earth, and the State would be required to respond publicly in order to make other terrorist organisations aware of the ramifications of threatening their satellites. Additionally, the threat actors are unlikely to be on the sovereign territory of the State they are attacking or indeed in the territory of a State allied to the State they are attacking, so extradition to face criminal charges is unlikely. From a western point of view, the threat actors are more likely to be situated in a country that does not have extradition treaties with the State they are attacking, for example China, Iran, North Korea or Russia, incidentally all countries where cyber-attacks against the west have occurred without generating a kinetic response.[24] Future attacks in space can be compared to ongoing attacks in cyber-space: the attacks have been conducted by non-state actors, occur out of the public eye, have the potential to cause civilian death and/ or destruction of infrastructure and are often unable to be conceptualised by the general public. Conversely, responding kinetically to cyber-attacks, for example using long range ballistic missiles against the source of the cyber-attack, is not out of the public eye, would likely result in immediate human death that can be shown on the evening news and as a result can be conceptualised by the public and would therefore likely result in public and international condemnation. In addition to this public condemnation, a kinetic response would likely violate the sovereign territory of another State and threaten an escalation of violence potentially leading to inter-State war. So as with the question of how to respond to cyber-attacks, the question of whether it is ethically permissible to kinetically respond on Earth to the threat of kinetic attacks in space then must be broken down to scenarios to consider situations where it is not only the threat actor that is being responded to, but also others who may be responsible for the SKKV, as well as considering whether it is worth responding kinetically against the threat actor when that actor could be physically located in another country where a kinetic response risks an escalation in violence and may violate international law. In answering this question, to be an ethical response, all kinetic actions must satisfy the *Jus in Bello* criteria of discrimination and proportionality. The principle of discrimination asserts that the only appropriate targets are those concerned with the enemy's war effort, and the principle of proportionality claims that the damage that is done in prosecuting such targets needs to be in line with the actual military value of the target itself.[25]

In the event a SKKV is launched into space by a terrorist group there are two definite, and multiple possible other, persons or organisations that a State could respond against; different groups would need to be

---

24 Quentin E. Hodgson et al. *Fighting Shadows in the Dark – Understanding and Countering Coercion in Cyberspace* (Santa Monica: RAND Corporation, 2019).
25 Stephen Coleman, *Military Ethics: An Introduction and Case Studies* (New York: Oxford University Press, 2013), 150.

responded to in different ways, and not all situations would justify a kinetic response. As Robert Joseph stated, "The United States views the purposeful interference with its space systems as an infringement on our rights, just as we would view interference with U.S. naval and commercial vessels in international waters. If these rights are not respected, the United States has the same full range of options – from diplomatic to military – to protect its space assets as it has to protect its other critical assets".[26] In the event a terrorist organisation has successfully placed a SKKV in space it is not only the threat actor that could, or should, be responded to. The following groups must all be considered.

1. The organisation responsible for launching the SKKV: the Launcher.
2. The organisation responsible for the SKKV: the Threat Actor.
3. The person or organisation responsible for creating the SKKV (who may be separate to the Threat Actor): the Creator.
4. The person or organisation responsible for positioning the SKKV: the Current Controller.
5. The organisation responsible for sponsoring or providing the training and expertise to the Threat Actor and/or Creator: the Sponsor.

Groups 1-3 will exist in all cases, though it is possible that one group could fulfil two or more categories, especially categories 2 and 3. Group 4 will only exist if the SKKV requires maneuvering in space. Group 5 may exist in some cases but not in others.

The Launcher

It is highly unlikely the Launcher is also the Threat Actor. Those able to launch satellites into orbit are probably unwilling to threaten the use of space orbits through the creation of debris. However, the Launcher is partially responsible for the destruction caused, or the threat of destruction posed, by the SKKV since they were integral in enabling that destruction. This raises the question, is it ethically permissible to respond kinetically against the Launcher to ensure they never act this way again? – the men and women working at the responsible launch facilities contributed to placing a weapon in space that could, in a worst-case scenario, render whole

---

26 Remarks on the President's National Space Policy by Robert G. Joseph, December 13, 2006.

orbits unusable due to the cascading effect of "debris begats debris".[27] As previously discussed, this could lead to hundreds of thousands of people being adversely affected as more and more satellites are damaged or destroyed. Millions of people could realistically die as a result of the loss of these satellites, not only through loss of navigation, communications and financial collapse, but due to a decrease in food production since much of the world's food supply is farmed using satellite technology.[28] The Launcher can be relatively easily identified, however any kinetic response that leads to the death of those responsible for the launch of the SKKV would be unjustified as it is unlikely they launched the SKKV with the intent of causing harm. The Launcher would however still be subject to other punishments such as criminal prosecution or sanctions.

The Threat Actor

The argument for an ethical kinetic response is clearest when discussing the Threat Actor. As the group with the most responsibility for the SKKV they are the most appropriate to be responded to kinetically and with lethal intent. However, the response must remain discriminate and proportional. For a kinetic response to be discriminate it must, to the greatest extent, only effect the Threat Actor and have minimal to no effect on non-combatants or structures with protected status. For example, if the Threat Actors were operating from an apartment building that also contained non-combatants, or in a building next to a structure with protected status such as a hospital, a response that destroyed the entire apartment building or that caused unnecessary damage to the hospital would not be discriminate and is also likely to be deemed a non-proportional response. While some levels of collateral damage can be approved by high-ranking defence officials and national leaders (this level of collateral damage varies between officials and nations), to be deemed discriminate any kinetic response would need to limit the collateral damage as much as possible and only directly target the Threat Actors. Terrorist groups are aware of the intent by most military forces to limit non-combatant death, and often position themselves in areas inhabited by, or frequented by, non-combatants, or near culturally significant

---

[27] General William Shelton in "How to Collect Space Junk", *The Economist*, last modified November 21, 2018, https://www.economist.com/the-economist-explains/2018/11/21/how-to-collect-space-junk.
[28] Remarks on the President's National Space Policy by Robert G. Joseph, December 13, 2006.

buildings such as hospitals or places of worship. An indiscriminate large ton inter-continental ballistic missile therefore would not be an appropriate response weapon of choice, however a smaller munition aimed at the specific apartment the Threat Actors are in or a discreet mission carried out by Special Operations personnel may be appropriate. The reasonableness of such an attack, however, is rather more problematic. The SKKV may be fired indiscriminately, and the initial damage caused may not affect a "live" satellite but simply collide with other space debris. So, the immediate result of the SKKV may not actually cause significant destruction or cost human lives. However, as debris begats debris, and as the Threat Actor has deliberately threatened the peaceful use of space, the second and third order effects of the detonation of the SKKV may not be immediately apparent but are likely significant. These second and third order effects may cause the death of people on earth and substantially impact the use of those orbits for a considerable period.[29] Carrying out an appropriate response to an attack that rendered sections of orbits unusable and acted as a significant contributor to the Kessler syndrome is extremely difficult especially as any kinetic response would likely violate the sovereignty of another nation.[30] Despite this, a kinetic response may be necessary, both to eliminate the Threat Actor and to serve as a deterrent to other terrorist organisations who may consider conducting their own attack in space.

The Creator

While the creation of DFC is proven to be within the capabilities of multiple terrorist organisations, the ability to create a reliable SKKV is likely beyond their current capabilities. An impact, proximity or time detonated SKKV could potentially be within their grasp, however an impact or proximity detonated SKKV could unintentionally detonate while the rocket is launched, and a timed SKKV does not offer the certainty that it will detonate near or in the path of other satellites. A signal commanded SKKV is almost certainly too technologically advanced for terrorist organisations,

---

29 "Space Debris Threatens Our Continued Use of Near-Earth Space for the Benefit of Humankind", European Space Agency, last modified March 11, 2021, https://scitechdaily.com/space-debris-threatens-our-continued-use-of-near-earth-space-for-the-benefit-of-humankind/.

30 Marit Undseth et al. "Space sustainability: The economics of space debris in perspective", *OECD Science, Technology and Industry Policy Papers*, No. 87 (2020), 0–63. https://doi.org/10.1787/a339de43-en.

currently, however a satellite (and therefore SKKV) capable of receiving and emitting signals is not beyond the current technological expertise of countless universities and schools around the world, so it is a group like this that a terrorist organisation would likely turn to.[31]

Both Hamas and Hezbollah are responsible for the education of hundreds of thousands of students in Palestine and Lebanon respectively.[32] These students may spend their entire school life under the tutelage of anti-western, anti-Israeli teachers, funded by their patron organisation, and eventually become engineers and scientists after university. When asked by Hamas or Hezbollah, their societal benefactors, to design or build a SKKV, potentially under threat of losing their life, livelihood, home or loved ones, it is unlikely they will hesitate or indeed have much of a choice. The question then becomes, is it ethically permissible to respond kinetically against the Creator of the SKKV? This is a much harder question to answer and effectively relies on an assessment of the level of coercion applied to the Creator. Bomb makers who are part of terrorist organisations or who help arm terrorists are generally viewed as legitimate targets for kinetic response as their actions directly contribute to the loss of civilian and security forces lives. During the Second Intifada in September 2000 for example, the Israeli Defence Force intentionally targeted Palestinian suicide bombers, bomb makers and strategists alike, making no distinction between the three groups despite only the suicide bombers being directly involved in attacking Israeli forces.[33] This was because all three groups were seen to be complicit in the devastation caused by the explosive devices. However, if the bomb makers were *forced* to create their weapons of devastation, killing them would not have been an appropriate response, and the same would be true if the Creator of the SKKV had been coerced into constructing it. This however does not mean the Creator is not liable to other responses, such as criminal prosecution, even if forced to create the SKKV.

---

31 "UCS Satellite Database", Union of Concerned Scientists, last modified January 1, 2021, https://www.ucsusa.org/resources/satellite-database.
32 Matthew Levitt, *Hamas: Politics, Charity, and Terrorism in the Service of Jihad* (New Haven CT: Yale University, 2006) p. 237 quoted in Coleman, Military Ethics, 189. "Hezbollah's Martyrs Foundation: Purpose, Mode of Operation and Funding Methods", Hezbollah: The Israeli Intelligence Heritage and Commemoration Centre, last modified March 14, 2019, https://www.terrorism-info.org.il/app/uploads/2019/04/E_058_19.pdf; Counter Extremism Project, "Hezbollah's Influence in Lebanon", December 2020, https://www.counterextremism.com/sites/default/files/2021-02/Hezbollah%20Influence%20in%20Lebanon_121120.pdf.
33 Coleman, *Military Ethics*, 188.

The Current Controller

Due to the number of organisations and States that have launched satellites into orbit, satellite collision deconfliction (with other satellites and space debris) and positioning (in the correct location) is often an outsourced independent business carried out by a third party: the Current Controller, often located in Satellite Operations Centers or Ground Stations.[34] It is highly unlikely the Threat Actor will have the ability to reposition the SKKV and so unless the SKKV has been placed directly into its desired location by the Launcher, the Threat Actor will be reliant on the Current Controller to position the SKKV in the correct location in order to have the greatest effect. It is unlikely the Current Controller is complicit in the attack and is probably unaware of the nature of the SKKV, believing it to be just another satellite launched by a small organisation for research, communications or other non-violent purposes. Therefore, like the Launcher, while a kinetic attack against the Current Controller isn't likely to be ethically justified, the Current Controller would still be subject to other punishments such as criminal prosecution and would likely face international condemnation for facilitating terrorism.

The Sponsor

Groups like Hamas in Palestine, Hezbollah in Lebanon, and the Shia Militia Groups in Iraq are only able to survive with support from a sponsor, in their cases their largest sponsor is the Iranian government and military.[35] This support is largely monetary, but also includes the provision of weaponry, training, direction and protection by their benefactor. When the Iranian Qods Force Commander, Major General Qasam Soleimani was killed by a U.S. airstrike in Iraq on January 3rd, 2020, also killed was the deputy chairman of Iraq's Popular Mobilisation Committee and commander of the Iranian-backed Kata'ib Hezbollah (a group that originated from Hezbollah), Abu Mahdi al-Muhandis, who had been designated a terrorist

---

34 "AWS Ground Station", Amazon, https://aws.amazon.com/ground-station/; "Satellite Ground Stations – Everything You Wanted to Know," Earth Sciences, last modified October 22, 2020, https://www.essearth.com/satellite-ground-stations/.
35 Bureau of Counterterrorism, "Country Reports on Terrorism 2019", https://www.state.gov/wp-content/uploads/2020/06/Country-Reports-on-Terrorism-2019-2.pdf.

by the United States.[36] This attack was conducted in part "to deter Iran from conducting or supporting further attacks against United States forces and interests, to degrade Iran's and Qods Force-backed militias' ability to conduct attacks..."[37] This attack shows that, in the view of the United States at least, the Sponsors of terrorist activities are considered legitimate targets that can be prosecuted using kinetic means. Following this argument, it is evident that a kinetic strike against the Sponsor to a SKKV attack in space could also be considered appropriate. However, this would likely be problematic in regard to *Jus ad Bellum* as well as *Jus in Bello*, in that any kinetic response that violates the sovereignty of another State that kills State and military leaders or destroys government buildings would likely lead to a declaration of war. It is possible that senior members of the Sponsoring group or strategic facilities owned by the Sponsoring group could be targeted and prosecuted using kinetic means, however, any kinetic response would be subject to the *Jus in Bello* principles of discrimination and proportionality as well as the principles of *Jus ad Bellum*.

Conclusion

While there are several groups that can be plausibly be thought to be legitimate targets for a kinetic response to a kinetic attack in space, it can be seen that in practice the majority cannot be responded to kinetically under *Jus in Bello* for one reason or another. Thus, while legal or other sanctions may be appropriately applied to the Launcher, Current Controller and possibly the Creator of a SKKV, in practice a kinetic response against these three actors is, at best, highly unlikely to be justifiable. On the other hand, a kinetic response against the Threat Actor and/or the Sponsor might be justifiable in some circumstances, but particularly in the case of the Sponsor there are other considerations that will need to be taken into consideration including the principles of *Jus ad Bellum*. How these considerations and competing principles will be weighed and justified is yet to be seen, however, the statement by Robert Joseph, "No nation, no non-state actor,

---

36 "Treasury Designates Individual, Entity Posing Threat to Stability in Iraq", U.S. Department of the Treasury, last modified – Unknown, https://www.treasury.gov/press-center/press-releases/Pages/tg195.aspx.
37 "White House Releases Report Justifying Soleimani Strike", Elliot Setzer, Lawfare, last modified, February 14, 2020, https://www.lawfareblog.com/white-house-releases-report-justifying-soleimani-strike.

should be under the illusion that the United States will tolerate a denial of our right to the use of space for peaceful purposes" likely stands as true today as when it was first stated in 2006.[38]

References

Adrian-Paul A. et al. *Afghanistan's Security Sector Governance Challenges – the Geneva Centre for the Democratic Control of Armed Forces* (Sofia: Procon Ltd. 2011).

Amazon, "AWS Ground Station", last modified 2021, https://aws.amazon.com/ground-station/.

Australian Geographic, "Timeline of Space Exploration," last modified August 27, 2012, https://www.australiangeographic.com.au/topics/science-environment/2012/08/timeline-of-space-exploration/.

Bureau of Counterterrorism, "Country Reports on Terrorism 2019", https://www.state.gov/wp-content/uploads/2020/06/Country-Reports-on-Terrorism-2019-2.pdf.

C4ISRNET, Strout N. "Bluestaq wins $280 million contract for Space Situational Awareness Library", last modified March 24, 2021, https://www.c4isrnet.com/battlefield-tech/space/2021/03/24/bluestaq-wins-280-million-contract-for-space-situational-awareness-library.

Cavaliere, L. "The Wright Brother's Odyssey: Their Flight of Learning", *New Directions for Adult & Continuous Education*, no. 53 (1992), 51-59.

Coleman, N. and S. Coleman. "Terrorism and Space Security", *69th International Astronautical Congress*, Bremen, Germany (2018), 1-5.

Coleman S. *Military Ethics: An Introduction and Case Studies* (New York: Oxford University Press, 2013), 150.

Counter Extremism Project, "Hezbollah's Influence in Lebanon", December 2020, https://www.counterextremism.com/sites/default/files/2021-02/Hezbollah%20Influence%20in%20Lebanon_121120.pdf.

Cybercrime Magazine, Morgan, S. "Cybercrime Damages $6 Trillion By 2021", last modified October 26, 2020, https://cybersecurityventures.com/annual-cybercrime-report-2020/.

Earth Sciences, "Satellite Ground Stations – Everything You Wanted to Know", last modified October 22, 2020, https://www.essearth.com/satellite-ground-stations/.

Environmental Systems Research Institute, "Bombing Missions of the Vietnam War", last modified 2017, https://storymaps.esri.com/stories/2017/vietnam-bombing/index.html.

European Space Agency, "Space Debris Threatens Our Continued Use of Near-Earth Space for the Benefit of Humankind", last modified March 11, 2021, https://

---

38 Remarks on the President's National Space Policy by Robert G. Joseph, December 13, 2006.

scitechdaily.com/space-debris-threatens-our-continued-use-of-near-earth-space-for-the-benefit-of-humankind/
Global Security Review, Kataw A.S. "With the U.S.-Taliban Deal In Place, IS-K Seeks to Build a Reign of Terror in Afghanistan", last modified, August 7, 2020, https://globalsecurityreview.com/with-the-us-taliban-deal-in-place-is-k-seeks-new-reign-of-terror-afghanistan/.
Hodgson Q.E. et al. *Fighting Shadows in the Dark – Understanding and Countering Coercion in Cyberspace* (Santa Monica: RAND Corporation, 2019).
Jaworski, E. "Military Necessity and Civilian Immunity", in *International Crime and Punishment: Selected Issues*, edited by Sienho Yee, Chapter 3. University Press of America, 2003.
Jones H. "The Recent Large Reduction in Space Launch Cost", *48th International Conference on Environmental Systems*, Albuquerque, New Mexico, July 8-12, 2018, last modified July 12, 2018, https://ttu-ir.tdl.org/handle/2346/74082.
Joseph R.G. "Remarks on the President's National Space Policy delivered to the George C Marshall", last modified December 13, 2006, http://www.spaceref.com/news/viewsr.html?pid=22773.
Lawfare, Setzer, E. "White House Releases Report Justifying Soleimani Strike", last modified, February 14, 2020, https://www.lawfareblog.com/white-house-releases-report-justifying-soleimani-strike.
NASA, "International Space Station Facts and Figures", last modified May 14, 2021, https://www.nasa.gov/feature/facts-and-figures.
Newton, M. *The Encyclopedia of High-tech Crime and Crime-fighting*. New York: Infobase Publishing, 2003.
Paris, M. "The First Air Wars – North Africa and the Balkans, 1911-13", *Journal of Contemporary History*, no. 26 (1911), 97-109.
Reuters, Shepardon D. "FCC fines SWARM $900,000 for unauthorized satellite launch", last modified December 21, 2018, https://www.reuters.com/article/us-usa-satellite-fine-idINKCN1OJ2WT.
Space.com, "Salyut 1: The First Space Station", last modified July 27, 2012, https://www.space.com/16773-first-space-station-salyut-1.html.
Solis, G. *The Law of Armed Conflict: International Humanitarian in War*. New York: Cambridge University Press, 2010.
The Economist, "How to Collect Space Junk", last modified November 21, 2018, https://www.economist.com/the-economist-explains/2018/11/21/how-to-collect-space-junk.
The Israeli Intelligence Heritage and Commemoration Centre, "Hezbollah's Martyrs Foundation: Purpose, Mode of Operation and Funding Methods", last modified March 14, 2019, https://www.terrorism-info.org.il/app/uploads/2019/04/E_058_19.pdf.
The Print, Kartha T. "In Afghanistan, a new 'great game' with ISIS, ISK and Pakistan is on with a vengeance", last modified March 11, 2020, https://theprint.in/opinion/in-afghanistan-a-new-great-game-with-isis-isk-and-pakistan-is-on-with-a-vengeance/378365/.

Time, Motlagh J. "The Afghan War: Why the Kandahar Campaign Matters", last modified October 18, 2010, http://content.time.com/time/world/article/0,8599,2026158,00.html.

Undseth, M., C. Jolly and M. Olivari, "Space sustainability: The economics of space debris in perspective", *OECD Science, Technology and Industry Policy Papers*, No. 87 (2020), https://doi.org/10.1787/a339de43-en.

Union of Concerned Scientists, "UCS Satellite Database", last modified January 1, 2021, https://www.ucsusa.org/resources/satellite-database.

United States Defense Intelligence Agency, "Challenges to Security in Space", last modified January 2019, https://www.dia.mil/Portals/27/Documents/News/Military%20Power%20Publications/Space_Threat_V14_020119_sm.pdf.

United States Department of the Treasury, "Treasury Designates Individual, Entity Posing Threat to Stability in Iraq," last modified – Unknown, https://www.treasury.gov/press-center/press-releases/Pages/tg195.aspx.

United States Space Command Public Affairs Office "Russia conducts space-based anti-satellite weapons test," last modified July 23, 2020, https://www.spacecom.mil/MEDIA/NEWS-ARTICLES/Article/2285098/russia-conducts-space-based-anti-satellite-weapons-test/.

University of California, Los Angeles History, "A Crash Into the Future", last modified – Unknown, https://alumni.ucla.edu/ucla-history/ucla-history-45/.

Wired Magazine, Hanson J. "The Forgotten Cold War Plan That Put a Ring of Copper Around the Earth", last modified August 13, 2013, https://www.wired.com/2013/08/project-west-ford/.

13

ROCKET CARGO

The Vanguard of the U.S. Space Force

Nathan Phillips[1]

> Logistics speed is at the heart of military supremacy. If a commercial company is in advanced development for a new capability to move materiel faster, then DOD needs to promptly engage and seek to be early adopters. That is the fundamental motivation for initiating the Rocket Cargo program.
>
> Air Force Research Laboratory (AFRL)[2]

On June 4 2021, the U.S. Space Force (USSF) received its first designation as a project lead. The Vanguard project designated Rocket Cargo is the fourth in the high priority, accelerated development program, and the first for which the USSF bears responsibility.[3] The simple premise of the project is to leverage off developments in the commercial rocket sector to create more agile logistical support for the DoD. The statements from the Department of the Air Force (DAF) and the Air Force Research Lab (AFRL), list a number of requirements; up to 100 tonne cargo capacity; the ability to land on a wide range of traditional and non-traditional surfaces (including

---

[1] The views expressed are those of the author and do not reflect the official policy or position of the Royal Australian Navy, the Department of Defence, or the Australian Government.
[2] "Rocket Cargo for Agile Global Logistics", *Air Force Research Laboratory* 2021. https://afresearchlab.com/technology/successstories/rocket-cargo-for-agile-global-logistics/.
[3] "Department of the Air Force announces fourth Vanguard program", *U.S. Air Force*, https://www.af.mil/News/Article-Display/Article/2646703/department-of-the-air-force-announces-fourth-vanguard-program/.

remote locations); a rapid load and unload cargo bay; and the ability to air drop after re-entry to the atmosphere.[4]

As of August 2021, the project is in the very early stages of existence, with little more than putting forward an effect that is trying to be achieved. By the time this book is published, and even more so by the time this chapter is read, it is likely that a number of developments that give specificity to the project have progressed, and indeed the project may have already been deemed a success or been divested. Therefore, the purpose of this chapter is not to make comment on the project itself, but to examine questions that demonstrate what must be examined in such a project. At times, this may seem negative. After all, this project is proposing a significant paradigm shift to the way in which logistics are considered. This chapter examines the theory behind the benefits of logistical speed, whether the theory is supported by historical example, and raises some of the questions than need to be answered before the program commenced. After all, should the program be a success, the benefits are inherent in the requirements being met. But should the program not be able to deliver what is expected, it is better to know the limitations early, and either work around them, accept them, or divest the program before it absorbs too much of the funding that could be better used elsewhere.

History and Logistical Theory

The first question is more of a historical theory and example. The AFRL has stated that the overall aim of the project is to achieve faster "[l]ogistical speed".[5] In more common terminology, rapid logistics. Historically logistics have been considered critical, though there are disagreements on the way they impact the overall success of a military operation. Perhaps one of the most quoted strategists in the Western world, Major General Carl von Clausewitz saw logistics as a separate element of warfare – critical, but as

---

4 "Department of the Air Force announces fourth Vanguard program", *U.S. Air Force*, https://www.af.mil/News/Article-Display/Article/2646703/department-of-the-air-force-announces-fourth-vanguard-program/; "Rocket Cargo for Agile Global Logistics", *Air Force Research Laboratory* 2021. https://afresearchlab.com/technology/successstories/rocket-cargo-for-agile-global-logistics/.
5 "Rocket Cargo for Agile Global Logistics", *Air Force Research Laboratory* 2021. https://afresearchlab.com/technology/successstories/rocket-cargo-for-agile-global-logistics/.

a support enabling a combatant to fight in the right place at the right time.[6] It could be argued, noting that Clausewitz spoke of minimising the reliance on logistics through supplementation through the use of natural resources, that Clausewitz saw logistics as critical, but also as a key weakness. From this perspective, rapid logistics assists timeliness (enabling the military force to act in the "right place, right time"), and reduces fragility (reducing unprotected transit time, and therefore reducing the length of time a supply vehicle is susceptible to attack). Utilising natural resources in place removes the supply chain as a single source, and therefore a single point of failure, and can be replicated in effect by utilising local resources rather than relying on long supply chains.

Henri Antoine Jomini, a Napoleonic General, presented a different view, namely including logistics as an inherent part of fighting unable to be removed from the art of warfare. In Jomini's view, a rapid logistical chain equated to a rapidly moving military.[7] While the two have decidedly different takes, both recognised the benefits of rapid logistics. With rocket cargo aiming for a 60-minute delivery window from launch to receipt, both theories seem to support the principal aim of the project when compared to the current time frames expanded upon later in this chapter.

History suggests that rapid logistics are congruent with a success in the practical application of military force as well. Classical leaders such as Xenaphon[8] and Alexander the Great sought to reduce reliance on long chains to increase their own speed of advance. In the case of Alexander, such was his dedication to rapid logistics, that after the introduction of carts seemed to slow his advance, he burned his own cart as an example, and ordered the remainder of the army do the same to their own.[9] Even with long supply chains, the ability to manoeuvre and march at great pace – with the logistics to support that – assisted in the success of both leaders, supporting the inherent view of Jomini.

In the following centuries, the Roman empires used foraging and local market economies to minimise reliance on long, exhaustive supply chains, ensuring the tyranny of distance would not slow the advance of

---

6 Carl von Clausewitz, *On War*, eds. and trans. Michael Howard and Peter Paret (Princeton: Princeton University Press, 1984), 95.
7 Antoine-Henri Jomini, *The Art of War* (New York: Greenwood Press, 1971), 32–46.
8 JF Lazenby "Logistics in Classical Greek Warfare," *War in History* 1 no 1 (1994): 6.
9 Donald W. Engels, *Alexander the Great and the Logistics of the Macedonian Army* (Los Angeles, CA: University of California Press, 1980), 13-17.

their military units.¹⁰ Further to this, the widespread construction of roads and control of the entire Mediterranean coastline meant that should their militaries be unable to fully resupply in situ – which was a reasonable assumption – resupply from great distances was much quicker than it would be without these features.¹¹

In more recent conflicts, both Napoleon Bonaparte and Gen William T. Sherman's successes were largely reliant on their ability to apply rapid logistical support to their troops. Sherman's in particular was impressive due to the sheer distance of his supply lines compared to those of his adversaries. Due to the use of rail transport, and denying such use to opposing forces, Sherman was able to use supremely rapid logistics to press advantage on his adversary.¹² Conversely, a large part of Napoleon's well-documented 1812 defeat in Russia was due to the adversary denying Napoleon's army the rapid logistics he had made available for previous campaigns,¹³ demonstrating the corollary of both Jomini and Clausewitz – that a lack of sufficiently rapid logistics will inevitably weaken the effectiveness of a military force.

Interestingly, Napoleon had not initially had such rapid and reliable supply chains, and yet still emerged victorious from a number of battles as he did not rely on logistics.¹⁴ This could perhaps be interpreted as supporting the Clausewitzian corollary of logistics equally being a weakness as much as a strength, and also supporting the idea that logistics is not inherently required as Jomini argues; however there are few, if any, studies able to demonstrate that rapid logistics has a negative effect on a military's effectiveness.

Therefore, the answer to the initial question of whether the U.S. military is justified in pursuing the utilisation of space to improve logistical speed, is that there is a historically demonstrable advantage to rapid logistics. Both theoretically, and by practical example, there is a clear benefit to rapid logistics and a demonstrable effect on successful campaigns. At the very

---

10 Peter Kahne, "War and Peacetime Logistics," *A Companion to the Roman Army* ed Paul Erdkamp (Carlton, Victoria: Blackwell Publishing, 2007), 327.
11 Justin Leidwanger, "Maritime Networks and Economic Regionalism in the Roman Eastern Mediterranean", *Les nouvelles de l'archéologie*, 135 (2014): 32-38.
12 Richard Alan Hardemon, *General Logistics Paradigm: A Study of the Logistics of Alexander, Napoleon, and Sherman* (Tannenberg Publishing,1998).
13 Dominic Lieven, *Russia Against Napolean: The Battle for Europe, 1807-1814* (United Kingdom: Penguin Books Limited, 2009).
14 Sean W Toole, *Logistics and the Fight – Lessons from Napoleon* (USMC Command and Staff College, 2011), 1.

least, should the Rocket Cargo program prove to potentially increase the logistical speed (that it, meet the overarching desired effect of the program) it is worth investigating for viability.

Testing Assumptions and Viability

Whether or not the program is viable comes down to a number of factors, including which requirements are absolutely essential – that is, the requirements that if not met, will mean the program cannot be successfully completed. Whether delivery within an hour can be achieved is framed as an essential requirement, however it should not be taken as a hard limit. As any introduction to a project management course will cover, project costs, the time to complete a project, and the defined quality of the outcome are all interconnected in the "iron triangle" of project management.[15] In the case of Rocket Cargo, one hour delivery is considered a defined quality of outcome, and any adjustment to it will impact both time to complete the project (schedule) and the cost. If the defined quality is simply the requirement that it be faster than current logistics, then there is significant scope to adjust cost and schedule to achieve a successful outcome. Should a hard-line approach of one hour be enforced as the defined quality, then the USSF is at risk of increasing both cost and time to achieve this outcome. In summary, the USSF has the option of achieving either a quality loosely defined as "better than the current system", or risk delays and budget blowout to achieve what appears to be an arbitrary time requirement. As a new service and an early responsibility, there may be a temptation for USSF leadership to want to prove itself by delivering to the highest quality. Already there is evidence of USSF leadership enthusiasm to demonstrate the utility of the newest service.[16] This is not inherently a bad trait, however Vanguard projects are specifically chosen to take advantage of industry to deliver quickly, and without undue pressure on funding.[17] Therefore, with pressure on time and cost being inherent in the Vanguard setup, and

---

15 Pollack, J., Helm, J. and Adler, D., "What is the Iron Triangle, and how has it changed?", *International Journal of Managing Projects in Business*, 11 no. 2 (2018): 527-547.
16 '… our crew … detected and missiles than launched and provided warning … Had that not happened, we might be talking about folks that died in that attack as opposed to injury. That's Space Force' Lt General David Thompson, USSF. See the chapter *'The United Stated Space Force and Space as a Military Domain'* in this volume for full quote and discussion on whether the capability existed independent of the USSF.
17 United States Air Force, *Science and Technology Strategy* (2019) 11.

with cost previously being prohibitive for rocket delivery of logistics,[18] defined quality is the only vertex of the iron triangle that has flexibility.

However, if the USSF want to succeed with this program, negotiating quality is not as disastrous as it sounds. In the context of time frame as the measure of quality, the USAF budget request indicates a requirement for delivery in 'less than one hour'. However, in all the documentation released by the United States Air Force (USAF) and the USSF, justification for this time frame appears to be missing.[19]

The lack of justification – or, to give benefit of the doubt, an as yet unreleased justification – for a one-hour limit is potentially a benefit to the program. As discussed earlier, if there is no specific justification for this time limit in delivery this makes the defined quality more flexible. Furthermore, a basic analysis of the current time frame gives significant scope for *over* an hour that would still provide a significant benefit to U.S. military logistics.

Referring to the AFRL statement prioritising logistical speed, if a hard limit of one hour is ignored, then the benefit becomes a sliding scale of what speed (as a measure of quality) can be achieved within the acceptable cost. Should it be better than the current efficiency, then there is an argument that program success would be achievable. For comparison, the primary cargo aircraft of the USAF, the Boeing C-17 Globemaster III,[20] can carry up to about 76 tonnes with an unrefuelled range of 4,400 km at a cruising speed of 400 kts (740.8 km/h). To use a recent operational base as an example, consider Al Minhad Air Base which is approximately 11,160 km from the east coast of the United States, which means that a C-17 would need refuelling twice (either in air or on the ground) and take a minimum of fifteen hours flight time to deliver a payload. That is, the current system would take fifteen hours to deliver just over three quarters of the proposed delivery load of a rocket cargo system, and it would need support along the way. Should the proposed Rocket Cargo program solution meet a timeframe that is under the fifteen hours of a C-17, then it would simply come down to whether the speed was worth the cost, rather than meeting a seemingly arbitrary constraint.

---

18 "Department of the Air Force announces fourth Vanguard program", *U.S. Air Force*, https://www.af.mil/News/Article-Display/Article/2646703/department-of-the-air-force-announces-fourth-vanguard-program/.
19 USAF, *Air Force Budget Request for Research, Development, Test & Evaluation*, 5.
20 The following data comes from the U.S. Air Force fact sheet, https://www.af.mil/About-Us/Fact-Sheets/Display/Article/1529726/c-17-globemaster-iii/.

However, determining the relationship between cost and defined quality is not necessarily the defining factor either. For simplicity, the cost/quality relationship will be referred to as efficiency in the following section.

When determining logistical speed, it appears that the program is measuring the time from launch to receipt. However, referring to Clausewitz's thoughts on logistics; if the "right place, right time" concept is about military effect, then the arguably the "right time" is the datum from which to measure logistical speed, rather than measuring from time of launch. In an urgent/worst case scenario, the right time is the time of request, and therefore the full process should be included to understand logistical speed, rather than just the vehicle travel time.

An analogy for this is evident in any online store. The delivery time, whether ordering meals, goods, supplies, or any other product, refers to time from order to receipt. It takes into account the processing of the order, packing of the items, post schedules (if necessary) and vehicle travel time. This becomes relevant to the Rocket Cargo program when examining the limited locations of launch platforms and the time taken to prepare a rocket for launch. Both factors will depend on the preferred solution of the USSF, should there be one, but regardless of the specific details, logistical speed will be determined by the full scope of the supply chain. That includes analysing the full C-17 supply chain, beyond the flight time presented in this chapter, to give an accurate comparison. An in-depth discussion of the C-17 supply chain is beyond the scope of this chapter; however, the established network of C-17 may offer flexibility and that is not yet available for a Rocket Cargo program. If the early components of a Rocket Cargo supply chain are unable to match the flexibility of a C-17 to load and launch at a wide network of possible locations (effectively reducing the distance and speed of the preparation of a payload), then a one-hour flight time is even more arbitrary. If it takes longer to deliver to a launch site and load up a rocket than it does to deliver via C-17 (or any other methodology), then a Rocket Cargo flight time is irrelevant when considering the overall logistic speed.

Using the internal supply chain to get cargo to the launch site is not the only challenge either. Other factors include minimum launch requirements, as well as whether alternate technologies may be overlooked through the focus on rockets. In particular, a minimum loadout impacts urgent or emergency delivery, in which one-hour delivery may seem an attractive option. For example, if a critical part for a system or medical supplies are needed urgently, one-hour delivery sounds like an optimum way to ensure items are received as soon as possible. However, the minimum loadout requirement is currently listed as 30 tonnes.[21] If the domestic supply chain is unable to provide 30 tonnes of cargo in a timely fashion, then the flight time is arbitrary.

Another key part of the program – and one of the other vertices of the iron triangle – is cost. Regardless of the time constraints on the program, if the efficiency is lost either by high costs or the discussed challenges with delivery time (as defined quality) then the program will not be successful. High costs undermine efficiency, which means that any delivery system will need to ensure the finances are kept to a minimum, particularly if the delivery time is not considered overly beneficial.

Cost is another factor that is hard to define without knowing the exact solution the program will select, however the USSF has indicated it will utilise industry contracting rather than develop organic capability. The current industry capability only has a single option at this stage that is capable of meeting the tonnage of the USSF requirement – SpaceX Starship. Elon Musk's latest space-related endeavour claims to have space and power to deliver 100+ tonnes,[22] whereas the closest competitors are designed for much smaller payloads; Blue Origin's New Glenn claims 45 tonnes as a maximum,[23] and United Launch Alliance's Vulcan Centaur carrying 26.2 tonnes in its optimum configuration.[24] Using Starship as a guide, Musk has indicated that each launch will cost approximately USD$2m.[25] The cost is interesting and will need verification should Starship be the chosen solution. While significantly higher than the operational costs of a C-17, the factor by which it is lower than NASA launches is far more significant, coming in at 0.1 percent of the cost.[26] This doesn't mean that Musk is necessarily incorrect or that his claim is untrue – the entrepreneur has made and backed up extraordinary claims in the past – but it does require breakdown and verification to be considered accurate. Whether this includes recovery, use of landing platforms, booster and rocket transport,

---

21 Elizabeth Howell, "US Air Force wants a commercial Rocket Cargo Vanguard to fly stuff anywhere on Earth", *www.space.com*, 2021, https://www.space.com/air-force-rocket-cargo-vanguard-commercial-rockets/.
22 "Flight Test: Starship SN15", SpaceX, https://www.spacex.com/vehicles/starship/.
23 "New Glenn," Blue Origin, https://www.blueorigin.com/new-glenn.
24 "Vulcan," ULA Launch, https://www.ulalaunch.com/rockets/vulcan-centaur.
25 Mike Wall, *SpaceX's Starship May Fly for Just $2 Million Per Mission, Elon Musk Says*, space.com, https://www.space.com/spacex-starship-flight-passenger-cost-elon-musk.html.
26 Anna Domanska, *SpaceX Starship at 0.1% cost per launch of Anazza's $2 Million* https://www.industryleadersmagazine.com/spacex-starship-at-0-1-cost-per-launch-of-nasas-2-million/.

staffing of facilities, storage and maintenance of rockets, or other costs is unclear. Adding confusion is the figure quoted by Christopher Couluris, SpaceX Director of Vehicle Integration, closer to USD$28 million.[27] Such a cost may be untenable to continuous operations. If a program is able to provide a quicker deliver but less often, the inefficiency of such a system may be such that it *decreases* the overall speed of logistics.

Another challenge for the Rocket Cargo program is the more efficient and on-demand availability of alternate technology pathways such as additive manufacturing. While printing 100 tonnes of cargo (likely requiring different printers for different materials) may prove too big a challenge, the capability provides significant agility and almost immediate responsiveness to urgent needs.[28] Additive manufacturing has the benefit of being produced in situ, and by printing with raw materials, it also allows the possibility of a more flexible supply network. In a sense, additive manufacturing draws a parallel to foraging and the Roman system of networks and local purchase, providing a robust supplementary supply chain. There are drawbacks to additive manufacturing as well though – it still requires the availability of raw materials, printing can be a slow process, and unless one is already held it takes time to create an appropriate design from which to print. My point here is not to make an argument in favour of additive printing, but rather to point out that there are options not covered by the Rocket Cargo program that may deliver the same desired effect. Alternatively, the program may benefit from being used in conjunction with other technologies. The challenge for the Rocket Cargo program is to recognise whether the efficiency of the full logistical effort supports the military need, whether that means advancing the program, reverting to more traditional methods, or integrating with alternate technologies.

There are a number of other challenges to the program; costs that are not necessarily captured within the usual scope of a project, but which are relevant to the specifics of this program and could arguably be a chapter in and of themselves. One such challenge is that of potential misinterpretation. Strategically, a large rocket launching from mainland U.S. – particularly if any stealth measures are applied – risks being misidentified. The location of launch and size of a logistics rocked would, hopefully, give enough

---

27 Caleb Henry, *Falcon 9 reaches new reusability record during Starlink, SkySat launch* https://spacenews.com/falcon-9-reaches-new-reusability-record-during-starlink-skysat-launch/.
28 Hayley Everett, "Department of Defence Unveils Additive Manufacturing Strategy", *3D Printing Industry*, https://3dprintingindustry.com/news/department-of-defense-unveils-additive-manufacturing-strategy-183832/.

evidence to suggest the benign nature of the mission; however, unless actively addressed, there is still a risk of misidentification of the rocket as a more hostile entity, opening the possibility of an aggressive reaction. Such misidentifications are rare but disastrous, as demonstrated by the USS Vincennes tragedy in 1988.[29] A like reaction for a misinterpreted hostile act then opens the possibility of escalation as well. This is a challenge that needs to be addressed by the program prior to operating civilian rockets as a military system.

Another factor is the engineering parameters of the requirements. Again, this could be a chapter on its own, and would require in depth analysis of the specific systems being considered for operation; however the current situation is that a safety margin of several kilometres is required for rocket launches.[30] While this would reduce once the rockets are no longer considered experimental, the significant safety exclusion zone and environmental impact on a landing site would be have ramifications on the requirement to land "on a wide range of non-traditional materials and surfaces, including at remote site" and "near personnel and structures".[31]

Again, none of these are sufficient reason not to pursue the program, and neither are they intended as arguments against the progression of the program. History has shown that rapid logistics are a significant factor in the success of a military campaign, and, as has been investigated in the past, the potential logistical speed offered by a rocket delivery system demand its investigation. However, while the enthusiasm previously shown by USSF leadership is commendable, it must be tempered with a balanced and mature approach to the project. Should the challenges – many of which are engineering or systematic, and therefore manageable, if only at cost – not prove to be practicable in their solutions, then the pursuit of the program would not be in the interests of the U.S. military. Similarly, if pursuit of the program proves to be financially untenable or beyond the technology of the near future, it would be counterproductive to the tenets of the Vanguard program. As the first Vanguard program to be under the remit of the USSF, there may be temptation to succeed at all costs. While due diligence would demand investigation of all pathways to success, it

---

29 Gene I. Rochlin, *Compex, Large Scale Military Systems and the Failure of Control* (Berkeley: University of California, 1990).
30 Using the formulas available under U.S. Department of Transportation Federal Aviation Administration Guide 437.52-1 of 2011.
31 "Department of the Air Force announces fourth Vanguard program", *U.S. Air Force*, https://www.af.mil/News/Article-Display/Article/2646703/department-of-the-air-force-announces-fourth-vanguard-program/.

may be that the option of most benefit to the U.S. joint force is not to pursue the program, or to instead pursue a combination of systems. Whichever system is selected though, the program promises to be an exciting one, with a genuine possibility that should it succeed, it would change logistics not just for the military, but for humanitarian organisations, global trade routes, and likely many areas not even thought of yet.

References

Air Force Research Laboratory, "Rocket Cargo for Agile Global Logistics", 2021, https://afresearchlab.com/technology/successstories/rocket-cargo-for-agile-global-logistics/.

Blue Origin "New Glenn," n.d., https://www.blueorigin.com/new-glenn.

Domanska, A. *SpaceX Starship at 0.1% cost per launch of Anazza's $2 Million, Industry Leaders Magazine*, 2019, https://www.industryleadersmagazine.com/spacex-starship-at-0-1-cost-per-launch-of-nasas-2-million/.

Engels, D. *Alexander the Great and the Logistics of the Macedonian Army*, Los Angeles, CA: University of California Press, 1980.

Everett, H. "Department of Defence Unveils Additive Manufacturing Strategy", *3D Printing Industry*, https://3dprintingindustry.com/news/department-of-defense-unveils-additive-manufacturing-strategy-183832/.

Hardemon, R. *General Logistics Paradigm: A Study of the Logistics Of Alexander, Napoleon, and Sherman*, Tannenberg Publishing, 1998.

Henry, C. "Falcon 9 reaches new reusability record during Starlink, SkySat launch," *spacenews.com*, 2020, https://spacenews.com/falcon-9-reaches-new-reusability-record-during-starlink-skysat-launch/.

Howell, E. "US Air Force wants a commercial Rocket Cargo Vanguard to fly stuff anywhere on Earth", *www.space.com*, 2021, https://www.space.com/air-force-rocket-cargo-vanguard-commercial-rockets/.

Jomini, A. *The Art of War*. New York: Greenwood Press, 1971.

Kahne, P. "War and Peacetime Logistics," *A Companion to the Roman Army* ed Paul Erdkamp. Carlton, Victoria: Blackwell Publishing, 2007.

Lazenby, J. "Logistics in Classical Greek Warfare," *War in History*, 1 no 1 (1994): 232-338.

Leidwanger, J. "Maritime Networks and Economic Regionalism in the Roman Eastern Mediterranean" *Les nouvelles de l'archéologie*, 135 (2014): 32-38.

Lieven, D. *Russia Against Napolean: The Battle for Europe, 1807-1814*, United Kingdom: Penguin Books Limited, 2009.

Pollack, J., J. Helm and D. Adler. "What is the Iron Triangle, and how has it changed?", *International Journal of Managing Projects in Business*, 11 no. 2 (2018): 527-547.

Rochlin, G. *Compex, Large Scale Military Systems and the Failure of Control*, Berkeley: University of California, 1990.

SpaceX "Flight Test: Starship SN15", https://www.spacex.com/vehicles/starship/.

Toole, S. *Logistics and the Fight – Lessons from Napoleon*, USMC Command and Staff College, 2011.

ULA Launch "Vulcan," ULA Launch, https://www.ulalaunch.com/rockets/vulcan-centaur.

United States Air Force, *Air Force Budget Request for Research, Development, Test & Evaluation*, 2020.

United States Air Force, *Science and Technology Strategy* 2019.

U.S. Air Force, "Department of the Air Force announces fourth Vanguard program", 2021, https://www.af.mil/News/Article-Display/Article/2646703/department-of-the-air-force-announces-fourth-vanguard-program/.

von Clausewitz, C. *On War*, eds. and trans. Michael Howard and Peter Paret. Princeton: Princeton University Press, 1984.

Wall, M. *SpaceX's Starship May Fly for Just $2 Million Per Mission, Elon Musk Says*, space.com, 2019, https://www.space.com/spacex-starship-flight-passenger-cost-elon-musk.html.

14

WHERE SPACE IS NOT AN OPTION

African Ethics and the Options of Non-contenders in Space Warfare

Ibanga B. Ikpe

Introduction

In 2016 a Namibian schoolboy, Simon Petrus, with little or no training in cellular phone technology, confounded the tech world by cobbling together a functional cellular phone from discarded cellular phone spares and other odds and ends.[1] The pride of his achievement was not in the fact that the phone could send and receive messages like any regular functional equipment but that it did so using radio frequencies, thus bypassing cell phone towers and the charges that come from using them. It may appear a little odd to headline a discourse on space warfare with the antics of a Namibian schoolboy, but the story is important because it serves as a bulwark against the idea that technological innovations can only flow from the technologically savvy north to the technology challenged south. It shows that a race is not won by an early start but that people who join a race later may still end up as victors if they persevere and do their best, not only to catch up, but also to set a new pace for the race. It shows that new technology can grow out of the most unexpected places and that those that

---

1 News24. 2021. Namibian Schoolboy Invents Cellphone That Doesn't Use Airtime | News24. [online] https://www.news24.com/News24/namibian-schoolboy-invents-cellphone-that-doesnt-use-airtime-20160729.

we usually look upon as technology giants do not have a monopoly of new ideas. This, of course, is not new, as it has been demonstrated many times in the course of human history.

The above opening makes even more sense in the light of a discourse that I had the privilege of engaging in with student-officers in an African military college on the ethical issues that arise from the militarization of space and the possibility of space becoming a war fighting domain. There was a general apathy among student-officers towards such a discourse with some of them questioning the relevance of such a discourse to officers whose military neither possesses nor controls military assets in space. This was despite their acknowledging the strategic advantage of space assets in contemporary warfare and the likelihood of a space conflict arising from the deployment of such assets. This was also despite their acknowledgement that countries with assets in space have the right to protect such assets from hostile action and this may lead to the militarization of space. For most of them, an ethical discourse on the deployment of military assets in space and the likely implication of conflicts in space should be left either to those who have strategic investments in space or such others that have the capacity to influence events in space by generating good or bad-will for those with such strategic assets. Specifically, these students identify the United States, China, Russia and peripherally India as those with significant strategic space assets as well as reasonable space warfare capabilities. They identify the largely pacifist countries of Europe and welfarist nations of Scandinavia as those capable of generating good or bad will for countries with such assets through advocacy. They maintained their apathy towards all discourse on the militarization of space despite acknowledging that the assets do not merely target other nations with similar capacities but may also target other nations, including those with pacifist constitutions. Their position that a discourse on the militarization of space should be left to those who have space assets and those that can influence opinion on the issue should not be seen as an indication that they are willing to roll over and surrender to any force that claims to have space assets. On the contrary, it simply means that they are prepared to perform their duties with the facilities that are available to them.

A Space Outside the Planet

A discourse on conflict in space is not only a discussion of something literally outside our planet but also figuratively outside the planet for those without space assets. While a handful of countries in Africa have or share satellites for civil use, only South Africa has a designated military satellite, and it is unclear that having such a satellite accords them space warfare capabilities. Anecdotal evidence from the continent suggests that most Directing Staff and student-officers in military colleges do not regard space warfare as something within their purview and there are several reasons for this. First, many such officers express the view that, given the basic needs issues that confront African states, any investments in space technology goes beyond the realm of luxury to the realm of a white elephant. This, however, flies in the face of reason given the many less important things those African countries have spent monies on, sometimes with little to show for it. A second commonly held view is that African countries are already disadvantaged since the space race already started without them. However, even new entrants into the space race have a chance to catch up, and even surpass others, either in the current race or in the next phase of the global power contest. Proof of this comes from the current position of China, which despite launching its space programme behind the United States and Russia[2] is now aspiring to having an independent space station. Their third argument for not taking space conflict seriously is that there are already enough players in space and, as such, joining their ranks will turn space contests into a melee where friends and foes are indistinguishable. It is, however, not clear how true this is since the hi-tech knowledge required for maintaining space assets should ensure that each asset is properly identifiable by all. The fourth reason that is often cited for disinterest in space is that engaging in space contestation may draw unnecessary attention to African countries and thereby expose them to security problems that their current indifference protects them from. This position is supported by Achebe's view on cowardice:

> It is praiseworthy to be brave and fearless, my son, but sometimes it is better to be a coward. We often stand in the compound of a coward to

---

[2] Smith, M., 2003. China's Space Program: An Overview. Report for Congress. [online] Congressional Research Service, pp.1-6. https://www.everycrsreport.com/files/20031021_RS21641_9ca49050ff378051e258f8685747d4e35ae6747f.pdf.

> point at the ruins where a brave man used to live. The man who has never submitted to anything will soon submit to the burial mat.[3]

In other words, it is not always wise to enter contests where one hopes to prevail, and prevailing is a matter of life and death. On the contrary, it is better to stand back and risk being labelled a coward while those who claim to be brave fight and kill themselves. It is an argument for conflict avoidance, especially in situations where one judges oneself to be technically ill-equipped to prevail. Although this idea is attractive on paper, the history of military conflicts has shown that militarily weak countries and those that refuse to get involved in major conflicts, rarely succeed in remaining neutral and unscathed as such conflicts unfold.

Although student-officers were generally disinterested in discourses on space warfare, a Socratic dialogue with them revealed that they were not totally oblivious of the perils of non-participation in such a discourse. For instance, some of the officers opined that by not making their views known, countries without space assets deny others knowledge of their concerns about the militarization of space and this has implications beyond the immediate discourse. One such implication is that nonparticipants end up being affected by decisions that are influenced by such a discourse. Also, student-officers pointed out that failing to participate in such a discourse denies nonparticipants the opportunity of gaining knowledge of the intricate problems of space as well as the understanding of how space assets and militarization affect the defence capacity of each country. Without such knowledge it will be impossible to relate space asset capacities to their current defence capabilities and impossible to articulate defence plans that may be needed in the event of confrontation with a country with space assets. Again, student-officers believed that such disengagement is a disservice to future generations who could have benefited from the insights that could have been gained through such participation. Student-officers also contended that a discourse on the ethical issues that arise from the use of space as a war fighting domain should be of interest to everyone since the placement of military assets in space renders the territorial boundaries of countries ineffective. Yet despite these insights, there was a general disinterest in space as a warfighting domain as well as reluctance in discussing related issues.

---

3 Achebe, C., *Arrow of God*, 1969, London: Heinemann.

The apathy of the student-officers is strange when we consider that the constant threat of insurgency, rebellion and cross-border violence has fuelled the desire in African militaries to develop skills and character that approximate the best armies in the world. Also, the involvement of African militaries in peacekeeping around the world should be an incentive for their interest in discourses that broaden their horizon on issues such as the militarization of space. Again, their avoidance of debate and discourse is strange given the fact that debate and intellectual engagement is a common feature of many African cultures. Specifically, most of the student-officers involved were from cultures where the *Pitso* system is practiced. The system, which was originally a meeting of all adult males in the community but was later democratized to include women, is a forum where the chief consults the citizenry on issues of governance and welfare.[4] At *Pitso*, a thorough discussion of issues is encouraged and participation in such discourses is considered a civic duty, to which citizens should aspire. One would have expected that products of such a culture would be interested in a discourse on the militarization of space, since it borders on issues of life and death. The rest of this chapter is dedicated to exploring this question as it relates to the African ethics of *Ubuntu*, an operating moral principle upon which many Sub-Saharan Africans base their decision making. *Ubuntu* also happens to be the operating moral principle of the communities from which the staff college class under discussion was drawn. The attempt to understand the curious stance of this military college class is also based on what has become a common practice of the college. Being acutely aware of the disconnect between civil and military culture, the practice of the college has been to relate issues of military ethics to traditional ethics, with the intention of determining how the two relate to each other on moral issues. The belief has been that if it can be shown that both ethical approaches complement each other, the disconnect between civil and military morality could be resolved. On the other hand, if such a disconnect is allowed to persist, it might result in indecision or hesitation which at a crucial time may negatively affect efficiency and command decision-making.

---

4 Wallman, S., 1968. Lesotho's Pitso: Traditional Meetings in a Modern Setting. *Canadian Journal of African Studies / Revue Canadienne des Études Africaines*, 2(2), 167.

A Light into the Tunnel

The primary question of this chapter is whether within the context of *Ubuntu*, the disinterest of student-officers in discourses surrounding the militarization of space and by extension, the participation of African countries in space, is morally justified. This question is important, not only because it explains the attitude of the student-officers but also because it lays the foundation for understanding their moral disposition. The choice of *Ubuntu* as a moral framework is based on the realization that it is the default moral principle for many sub-Saharan African communities. Thus, while the use of various analytic decision-making instruments like the Military Decision-Making Process are encouraged, there is a growing realization that such instruments need to be grounded on student-officers' default principle of moral reasoning. This will avoid a situation where decisions arrived at through such analytic decision-making instruments are mediated or rejected by this default moral principle. The importance of this is exemplified in the frustrations of Placide Tempels, whose African converts "return to their former ways of behaviour whenever they are overtaken by moral lassitude, danger or suffering".[5] Tempels, who was working as a Catholic missionary among the Luba people of the Congo, found out that conversion to Christianity did not obscure the traditional belief system of his converts and that this system had a more powerful hold on them than the new religion. According to him, they seek solutions in their traditions "because their ancestors left them their practical solution of the great problem of humanity, the problem of life and death, of salvation or destruction".[6] Thus, despite their conversion to Catholicism, it was common for his parishioners to consult traditional oracles and seek non-Christian solutions to the problems of life. Tempels' experience supports the assumption of the said military educators that a default moral principle from traditional culture is likely to affect command decision-making. Student-officers are, therefore, more likely to make unencumbered decisions if they can justify the decision by this traditional moral principle. But what is *Ubuntu*?

The idea of *Ubuntu* is encapsulated in the Sotho expression, *motho ke motho ka batho* which in the Nguni languages is *umuntu ngumuntu ngabantu*. This expression has been variously translated into English to

---
5 Tempels, P. and Read, M., 1959. *Bantu Philosophy*. (Paris): Presence Africaine, 13.
6 Tempels, P. and Read, M.

mean "a person is a person through other persons", "a human being is a human being because of other human beings," and "I am because we are; and since we are, therefore I am". As an aphorism, it "articulates our inter-connectedness, our common humanity and the responsibility to each that flows from this connectedness".[7] According to Metz, "this maxim has descriptive senses to the effect that one's identity as a human being causally and even metaphysically depends on a community. It also has prescriptive senses to the effect that one ought to be a mensch, in other words, morally should support the community in certain ways".[8] Although the Sotho and Nguni languages are limited to a few southern African countries, many other sub-Saharan African communities have versions of the expression in their own languages. Thus, one could argue, like Mbiti, that *Ubuntu* is a "cardinal point in the African view of man"[9] and consists in "the capacity to express compassion, justice, reciprocity, dignity, harmony and humanity in the interests of building, maintaining and strengthening the community".[10] It follows from this principle that "if one harms others, for example, by being exploitive, deceptive or unfaithful, or even if one is merely indifferent to others and fails to share oneself with them, then one is said to be lacking '*Ubuntu*,' literally lacking in personhood or humanness".[11] This idea constitutes the default moral principle which the said student-officers and others in command and staff roles are likely to resort to when they are confronted with decisions that have moral implications. It is also the moral principle which may ultimately determine whether their apathy towards a discourse on the militarization of space constitutes a moral failure.

The Socratic dialogue with student-officers in the college did not only focus on space assets and the militarization of space, but also touched on their attitude towards the discourse. Primarily, the question was whether their apparent disinterest in engaging in such a discourse was morally justified within the context of *Ubuntu*. Their consensus opinion on this was that since *Ubuntu* is essentially a pacifist moral principle, any discourse on the preservation and perfection of weapons of war was antithetical to the principle and is therefore not supported by it. They held this view

---

[7] Mbiti, J., *African Religions and Philosophy*, 1969, 108-109.
[8] Metz, T., Toward an African Moral Theory, *Journal of Political Philosophy*, 2007, 15(3), 323.
[9] Mbiti, J., 1969. *African Religions and Philosophy*, 1969, 109.
[10] Bessler, J., In the Spirit of Ubuntu: Enforcing the Rights of Orphans and Vulnerable Children Affected by HIV/AIDS in South Africa. *Hastings International and Comparative Law Review*, 2008, 31(1), 48.
[11] Metz, T. and Gaie, J., The African Ethic of Ubuntu/Botho: Implications for Research on Morality. *Journal of Moral Education*, 2010, 39(3), 275.

despite noting the difference between a discourse on space assets and the militarization of space and the actual possession of space assets. They also maintained their stance despite noting that the discourse is not an endorsement of military adventures in space but may actually sensitize the public to the dangers of space conflicts and thereby influence opinion on the militarization of space. Also, the idea that such a discussion could lead to the codification of an ethics of space warfare did not change their stance. As far as they were concerned *Ubuntu* supports warfare and everything pertaining thereto only in circumstances where war is pursued in response to an initial military aggression. For them, wars of aggression are inevitably immoral because, by its nature, war generally violates the dignity of persons within the war zone. It violates the dignity of civilians who are forced to flee their homes and endure the indignities that such a displacement brings. It also violates the dignity of wounded combatants who carry with them the physical scars of war and henceforth live with the knowledge that they are less than who they were prior to the war. Again, it violates the dignity of combatants without physical scars as they must live with the knowledge that they have taken the lives of others and in so doing robbed parents of their children, spouses of their partners, children of their parents and many others of close familial or fraternal relationships. But even if, for some reason a combatant returns from war without taking a life or having noticeable injuries, the witnessing of death, injury, and general human misery violates the peace of mind and therefore the dignity of those involved. Furthermore, war violates the dignity of those who make political decisions about warfare as they ponder the human and material loss foisted on humanity by the war they authorized. Under *Ubuntu*, therefore, warmongering, as well as extensions of warfare beyond defensive action, is indefensible and morally reprehensible.

The above stance may not be readily appreciated by those who have no in-depth understanding of the culture, so it is necessary to explain specific concepts that reflect the students' positions and show how they relate to *Ubuntu* as a moral principle. Reference in the above to the notion of human dignity provides a starting point for this insight into their belief that space assets, the militarization of space, and offensive war in general, violates human dignity. This is because human dignity is closely tied to traditional views of personhood and as such any violations thereof are likely to reduce the capacity of those involved to achieve full personhood. Since personhood is closely tied to all spheres of human existence, including moral, legal, political, and social concepts of identity, responsibilities, privileges, rights, citizenship, equality, liberty, etc, it is important in determining whether a person has or lacks *Ubuntu*. Perhaps the best way to understand this interplay between *Ubuntu* and personhood is to look at the interplay between the moral and metaphysical view of personhood. According to the moral view, personhood is "a status earned by meeting certain community standards, including the ability to take on prescribed responsibilities that are believed to define personhood".[12] Metaphysical personhood refers to "'transmissible life essence", "innate power" or "life-breath" that enables one to "live on after death and become ancestors themselves".[13] Good deeds which constitute the community standards for moral personhood do not only contribute to moral personhood but also to metaphysical personhood. In other words, moral uprightness, as defined by the community, is not only important in determining the honourable (person) from the dishonourable (non-person) but also plays a role in determining those that are venerated as ancestors by the living and accepted, after their death, into the ancestral world by the dead. It is important to note that achieving moral and metaphysical personhood is not only important to the individual but also plays a role in determining whether the descendants of the individual are likely to achieve personhood. This is because part of what constitutes a new individual is inherited from his/her forebears and if one's forebears lacked personhood, the task of becoming a person becomes tougher for the progeny. Good deeds enhance the life essence of an individual, thereby protecting that individual from illness and death while bad deeds diminish their essence and render them vulnerable to malevolent forces. Warfare, as a harbinger of suffering and death can only be seen as limiting people's capacity for personhood and should therefore be avoided as much as possible.

The above shows that under *Ubuntu*, both combatants and non-combatants are likely to suffer diminution of their personhood in situations of war. Thus, since military adventures in space aim at giving some nations superiority and control over others, they cannot be said to enhance the capacity to express compassion, justice, reciprocity, dignity, harmony, and humanity. One could therefore argue that under *Ubuntu* any discourse on space assets and the militarization of space can be justified only if it consists of an outright condemnation of the practice.

---

12 Ikuenobe, P., The Idea of Personhood in Chinua Achebe's Things Fall Apart, *Philosophia Africana*, 2006, 9(2),117.
13 Jindra, M., Christianity and the Proliferation of Ancestors: Changes in Hierarchy and Mortuary Ritual in the Cameroon Grassfields. *Africa*, 2005, 75(3), 358.

Ubuntu and Space Warfare

The justifications by student-officers for their disinterest in space warfare and the militarization of space reflect the arguments that are sometimes used to justify the non-participation of African countries in space. As shown above, the arguments are heavily influenced by the pacifist nature of *Ubuntu*, its general abhorrence of war and its bias towards community interests which inevitably raises the profile of criticisms of military spending. However, this current understanding of *Ubuntu* is strange given the fact that pacifism never featured prominently in the African past. On the contrary, war and conflict have been part of the history of Africa, as in the history of other peoples of the world. The great African empires of yore, such as Songhai, Mali, Mutapa, Kanem-Bornu, Benin, Kongo, Kush, and Zulu, had been forged by wars of expansion which would have been unjustified under a pacifist view of *Ubuntu*. Many of the empires forged weapons of war and developed war strategies which, in terms of the technology of the time, could have been as revolutionary as the location of military infrastructure in space. The question therefore is how the protagonists of those wars justified their actions in a context where such actions could lead to the diminution of their life force and their capacity to attain personhood. The only circumstance under which such wars of expansion could have been pursued alongside an acceptance of the metaphysical and moral dimensions of *Ubuntu* would consist in some creative understanding of *Ubuntu* that excused such wars of expansion. Such creative understanding may be enough not merely to justify current participation of Africans in discourses on space warfare, but also to justify their possession and deployment of space assets, alongside the capacity to protect those assets from hostile interests.

Perhaps our understanding of *Ubuntu* and how its pacifist principles could be reconciled with the wars and conflicts of the African past would be better understood using Thaddeus Metz's rational reconstruction of *Ubuntu*. Metz posits that under *Ubuntu*, "an action is right just insofar as it produces harmony and reduces discord; an act is wrong to the extent that it fails to develop community".[14] According to him, this formulation of *Ubuntu* presents the preservation of relationships as an important moral goal of *Ubuntu*, which ought to be promoted by every moral agent. It also identifies harmony and togetherness as consisting in a common sense of self

---

14 Metz, T., Toward an African Moral Theory. *Journal of Political Philosophy*, 2007, 15(3), 324.

(shared identity) and a certain caring or supportive relationship (goodwill). Metz however admits that shared identity cannot be morally important in itself and that it is only when it "wishes another person well (conation); believes that another person is worthy of help (cognition); aims to help another person (intention); acts so as to help another person (volition); acts for the other's sake (motivation); and, finally, feels good upon the knowledge that another person has benefited and feels bad upon learning she has been harmed (affection)"[15] that one could be said to have *Ubuntu*. Thus, for the attitude of the student-officers to be just under *Ubuntu*, it must be seen not only to promote a shared identity that is grounded on goodwill, but to do so with appropriate conation, cognition, intention, volition, motivation, and affection. But what does this mean in real terms?

Following Metz, the attitude of the student officers would promote a shared identity if:

- It enables the individual to see himself or herself as part of a group with whom he or she shares identity
- It creates an environment within which the group of which they consider themselves a member, also considers them a member of that group
- Members of the group have common ends, even if they do not have the same motives or reasons that underlie them
- People in the group coordinate their activities in order to realize their ends, even if they do not use the same means or make the same amount of effort.[16]

It is immediately apparent that the apathy of the student-officers does not contribute to any of the above. Indeed, it appears doubtful whether such a shared identity is at all possible within the context of the military. This is because although soldiers share counterpart or fraternal relationship with other soldiers, by which military personnel recognize and accord courtesies to personnel of other militaries, they are not commonly seen as sharing an identity in the same way that members of a homogeneous African community do. Thus, even though they all go by the name "soldiers", there is no universal group to which they see themselves as belonging, neither is there a universal group of officers nor enlisted that see them as a part of itself. The absence of such a group means that the student-officers

---

15 Metz, T., 326.
16 Metz, T., 335.

do not coordinate their activities with anyone. On their contrary, their loyalty to the constitution of their various countries mean that they should always look at other military formations (even friendly ones) as potential adversaries with whom they should constantly be wary and as such deny access to their strategic secrets. Thus, while there may be intergroup cooperation towards some short-term goals, they are no universal goals that militaries are jointly committed to, either in the short-or long-term. It would therefore be a category mistake to discuss the attitude or actions of soldiers within the context of a shared identity.

Although the sharing of identity is not something that is attributable to soldiers under *Ubuntu*, it should still be possible to justify the participation of the student-officers in a discourse on the militarization of space. Soldiers may not share an identity in the sense envisioned by Metz but they do share a common humanity, which is a legitimate relationship under *Ubuntu*. This shared humanity urges individuals to consider themselves as part of the human race and to believe that other humans will see them as such. A shared humanity also makes it apparent that humanity has a common end, which is what Wiredu refers to when he argues that African moral thought seeks to "postulate the harmonization of interests as the means, and the securing of human wellbeing as the end of all moral endeavour".[17] Despite the fact that there may be no universal agreement on how this could be achieved, there is evidence of attempts by humanity to coordinate activities towards achieving human wellbeing. Such universal initiatives as the Millennium Development Goal, the Kyoto Protocol on climate change, the Treaty on the Non-Proliferation of Nuclear Weapons, reflect coordinated human actions towards a common goal. This notion of a shared humanity is supported by what Metz refers to as the "others-regarding" account of basic moral reasoning within the concept of *Ubuntu*. Under this tenet, individuals have a basic moral duty to others, irrespective of whether they have community relations with them. This basic moral duty flows from the idea that "others are worth acting for without ultimate reference to one's self-interest".[18] A failure to participate in such a discourse either sets one apart as only engaging in conversations that revolve around oneself or as ignoring those that one disagrees with or has differences with. Either way, such behaviour does not promote a shared identity but creates divisions that

---

17 Wiredu, K., Abraham, W., Irele, A. and Menkiti, I., *A Companion to African Philosophy*, Malden, Mass.: Blackwell Publishing, 2004, 398.
18 Metz, T., 336.

may result in one harbouring ill-will against others. Under this tenet, even where individual and national interests make participation in a discourse on space ethics unappealing, student-officers still have a moral obligation under *Ubuntu* to participate in such a discourse.

There is a sense in which participation in the discourse is also justified under the more traditional understanding of *Ubuntu*. This would consist in assessing the attitude of the student-officers constituting indifference to humanity and a failure to share oneself with others.[19] It is a negation of a fundamental principle of *Ubuntu*, that, "my humanity is caught up, is inextricably bound up, in yours". "We belong in a bundle of life".[20] The idea here is that despite the differences between the peoples of the world and the possible absence of a shared identity between them, a shared humanity that transcends national, regional and community divisions is conceivable under *Ubuntu*. The notion of a shared humanity becomes more apparent when the human race faces a global emergency which no subsidiary human group can effectively control. Such maladies as outer-terrestrial threats, pandemics, climate change, and sea and environmental pollution, frighten everyone equally and show that we share the same vulnerabilities even though individual circumstances may render some more susceptible to their impact than others. This common humanity imposes certain obligations on everyone, the least of which is avoiding appearing indifferent to the concern of others. Leaving the discourse on the militarization of space to those with space assets and to others that can influence opinion on the issue appears to reflect this indifference, and as such is immoral under *Ubuntu*.

It could be argued that the presentation of individual or group opinion in a discourse on the militarization of space does not approximate the pursuit of space technology and as such it would be inappropriate to jump from justifying the involvement of student-officers in such a discourse to justifying ownership of space assets by African countries. This is because presenting one's views in a discourse neither contributes to an arms race nor increase the likelihood of a global apocalypse in the way that the possession of space assets and the capacity to deploy them both for offensive and defensive purposes does. Thus, whereas one can use the notion of shared identity or a shared humanity to justify taking part in an

---

19 Metz, T. and Gaie, J., The African Ethic of Ubuntu/Botho: Implications for Research on Morality. *Journal of Moral Education*, 2010, 39(3), 275.
20 Tutu, D., *No Future without Forgiveness*. New York, NY 2000: Image/Doubleday, 31.

ethical discourse, one may not use it to justify ownership of weapons of war. A closer examination, however, shows that there are similarities between the two and that *Ubuntu* may not be diametrically opposed to ownership of space assets. For instance, just as *Ubuntu* enjoins us to participate in discourses to enter our unique perspective into the universal knowledge pool, *Ubuntu* also enjoins us to share our technologies with humanity in order increase the practical and theoretical knowledge that is available to the world. For the technology to be shared, it must be acquired first. Thus, even where national priorities inform restrictions on seeking space technology, total inaction is not an option. Elementary steps should still be taken in the hope that such steps could move research in a new direction or provide a foundation for future knowledge development in the area. In other words, one's limitations should not deter the seeking of knowledge, since the next big technology may emanate from very humble beginnings, as was demonstrated at the beginning of this discussion through the case of Simon Petrus and his phone. Becoming fully human entails honing all our knowledge and skills and becoming the best that we can be, even where the path to such excellence is riddled with difficulties.

The argument that each person approaches a discourse from a unique perspective and as such modifies and streamlines the issue based on that perspective, could also be applied to ownership of space assets and its technology. Technology is often said to be value neutral which means that whether it is seen to be good or bad depends on how the technology is acquired and deployed. A good technology may be acquired through dubious means or through a process that degrades both human beings (those who labour to ensure the fruition of the technology and those who live within the vicinity that the technology is being developed) and the physical environment. Similarly, a technology that is acquired through an ethical process may be deployed in ways that harm human beings and the physical environment. Since a person with *Ubuntu* is always conscious of the moral and metaphysical constrains of personhood, his or her development and deployment of technology is guided by these tenets, and this will ensure an ethical interaction of the technology. Thus, whereas someone that lacks *Ubuntu* may seek to acquire technology without caring about the human and material cost of the acquisition, one who has *Ubuntu* will ensure that the process of acquiring does not harm humans and the environment. Also, whereas those without *Ubuntu* may use technology to intimidate, pressure oppress and ultimately destroy others, a person with *Ubuntu* is more likely to consider our shared humanity in the deployment of such assets. In other words, space assets in the hands of those with *Ubuntu* may be used in ways that serve humanity, such as protecting the vulnerable by checkmating the aggressive intentions of those lacking *Ubuntu* and generally growing faith in our shared humanity. It could lead others to accept that the wealth of one is the wealth of all and that "I am because we are". Thus, failing to acquire space technology amounts to denying the world the balance which a person with *Ubuntu* could bring. This may consist in using one's capacity to do good with space assets to mediate the capacity of others to cause harm using space assets. It also denies the country the enormous good that space technology could bring to non-military areas of human life.

Conclusion

The responsibility of agents, both for their action and non-action within the context of duty introduces a different perspective to the ethical evaluation of the attitude of student-officers towards engaging in discourses concerning space warfare. Keeping in mind that:

1. Student-officers are military personnel and have sworn to take whatever action they deem necessary to protect their country from enemies both foreign and domestic
2. Student-officers are aware of the importance of space assets and would jump at the opportunity to have such facilities available to them
3. Student-officers are aware that there are ongoing discussions concerning the use of space-based military assets and the militarization of space
4. Student-officers are aware that their participation in discourses on space assets and the militarization of space offers the world their own unique perspective on the subject; a perspective which cannot otherwise be part of the discourse
5. Student-officers are aware discussions of this type sometimes influence public opinion and thereby public policy.

It would seem to follow that the failure of student-officers to participate in a discourse on space assets and the militarization of space constitutes an omission, the type of which carries moral responsibility. Thus, the student-officers have a moral responsibility to participate in such discussions. A failure to participate may constitute a moral failing.

When *Ubuntu* enjoins us to share ourselves with others, the sharing is not merely physical but also mental and intellectual. A physical sharing entails helping those who are overwhelmed by responsibilities or chores of a physical nature, such as would happen when one is bereaved, building a house, moving house, farming the field, searching for a lost item, repelling an attack by bullies, burglars, armed combatants, etc. A mental or intellectual sharing on the other hand consists in giving advice, sharing skills and expertise, interpreting proverbs and aphorisms, sharing oral histories, and generally helping to enhance the intellectual state of others. Apathy towards a discourse on the militarization of space is tantamount to failing to share one's intellectual self with others and as such cannot be justified under *Ubuntu*. Similarly, failure to develop one's intellect, capacities, and abilities and through this, technologies that could be of help to humanity, is also such a failure. Space is an "open wound" that can only fester if left to flies and parasites. It is only when those with *Ubuntu* are part of it that its vast potentials will benefit mankind.

References

Achebe, C., *Arrow of God*, London: Heinemann, 1964.

Berger, S., B. Dicks and M. Fontaine. 'Community': a useful concept in heritage studies? *International Journal of Heritage Studies* (2019), 26(4), 325-351.

Bessler, J. In the spirit of Ubuntu: Enforcing the rights of orphans and vulnerable children affected by HIV/AIDS in South Africa, *Hastings International and Comparative Law Review* (2008), 31(1), 31(1), 48.

Gianan, A.N. Delving into the Ethical Dimension of Ubuntu Philosophy, *Cultura* (2011), 8(1), 63-82.

Ikuenobe, P. The Idea of Personhood in Chinua Achebe's Things Fall Apart, *Philosophia Africana* (2006), 9(2), 117-131.

Jindra, M. Christianity and the Proliferation of Ancestors: Changes in Hierarchy and Mortuary Ritual in the Cameroon Grassfields, *Africa* (2005), 75(3), 356-377.

Kaphagawani, D. 2004. African Conceptions of a Person: A Critical Survey. In: K. Wiredu, ed., *A Companion to African Philosophy*. Malden: Blackwell Publishing Ltd., 2004.

Mbiti, J. *African Religions & Philosophy*, Oxford: Heinemann, 1997.

Metz, T. Toward an African Moral Theory, *Journal of Political Philosophy* (2007), 15(3), 321-341.

Metz, T. and J. Gaie. The African ethic of Ubuntu/Botho: Implications for Research on Morality, *Journal of Moral Education* (2010), 39(3), 273-290.

News24. Namibian Schoolboy Invents Cellphone that Doesn't Use Airtime | News24. [online], 2021, https://www.news24.com/News24/namibian-schoolboy-invents-cellphone-that-doesnt-use-airtime-20160729.

Onah, G. The Meaning of Peace in African Traditional Religion and Culture, [online] *Afrika World*, 2021, http://www.afrikaworld.net/afrel/goddionah.htm.

Smith, M. 2003. China's Space Program: An Overview. Report for Congress. [online] Congressional Research Service, 2003, https://www.everycrsreport.com/files/20031021_RS21641_9ca49050ff378051e258f8685747d4e35ae6747f.pdf.

Tempels, P. and M. Read. *Bantu Philosophy* (Paris): Presence Africaine, 1959.

Tutu, D. *No Future without Forgiveness*, New York, NY, 2000.

Wallman, S. Lesotho's Pitso: Traditional Meetings in a Modern Setting. *Canadian Journal of African Studies / Revue Canadienne des Études Africaines* (1968), 2(2), 167.

15

THE UNITED STATES SPACE FORCE AND SPACE AS A MILITARY DOMAIN

Nathan J. Phillips[1]

The U.S. Space Force is potentially the first branch of the military to be satirised before it had even set up an operational command. However, while Netflix's *Space Force* series plays loosely with the truth, the fictional head of the new branch, General Mark R. Naird, sums up the real-world challenges accurately and succinctly – "space is hard".[2] Over half a century since the U.S. first landed Apollo 11's Eagle Lunar Module, it is still expensive, logistically difficult, and inherently dangerous to return. So much so that only ten people have set foot on the moon since (twelve total, including Armstrong and Aldrin). Compare that to powered flight, where following the 1903 success of the Wright brothers, just over a decade later large-scale use of inhabited aircraft had a major impact on global conflicts.[3]

Perhaps a better comparison is submarine services throughout history. Both submariners and astronauts work in hostile environments not conducive to survivability, both require highly specialised training, and while many nations have military submarine capabilities, only the extremely wealthy have the private variety. Even four centuries after the first "submarine" sailed under the Thames in the 1620's,[4] they are only now

---

[1] The views expressed are those of the author and do not reflect the official policy or position of the Royal Australian Navy, the Department of Defence, or the Australian Government.
[2] *Space Force*, "Mark and Mallory Go to Washington", Season 1 Episode 3, Directed by Tom Marshall. Written by Shepard Boucher. Netflix, 29 May 2020.
[3] John H Morrow Jr., "The First World War, 1914-1919", in *A History of Air Warfare*, 1st ed. (Dulles: Potomac Books, 2010), 3-25.
[4] Jennifer Speake, "The Wrong Kind of Wonder: Ben Jonson and Cornelis Drebbel." *The Review of English Studies*, 66, no. 273 (2015): 60-70.

becoming commonplace among militaries of smaller states. Submarines are hard because of the environment they work in, the logistics of deploying them, and the sheer expense of purchase and training (both for crew and support personnel). Four hundred years after the first underwater journey, submarines are still hard. Just over fifty years since the first inhabited journey to the moon, space is still hard too.

That does not mean that it is impossible though, and in March 2018, the U.S. President announced the formation of the United States Space Force (USSF) to work specifically in that "hard" environment.[5] The USSF is the first new branch of the U.S. Department of Defence since the introduction on the U.S. Air Force (USAF) in 1947, but contrary to some claims it is not the first military branch dedicated to warfare in space. In fact, in some areas it is not even the most advanced. While still the dominant nation regarding space capability, foreign competitors have seen the U.S. as setting the standard, and aimed to surpass them.[6] From an international relations perspective, this is the realist idea of pushing a nation's power as far as possible before inducing unnecessary risk maximising both power and security.[7] In the military domain of outer space, physical destruction of a nationally owned satellite has only occurred in a test environment, and even then only four nations have proven the capability (Russia, the U.S., China, and India).[8] In plain speak, no-one has destroyed someone else's property yet. That means the *risk* of it occurring (from a realist perspective) is minimal as there is no precedent, along with liability for destruction being clearly laid out in the United Nations Outer Space Treaty (UNOST).[9] Therefore, with little risk to space development, there is a genuine opportunity for nations to expand their influence arguably without threat. As competitor nations built their own capabilities with the aim of surpassing U.S. capability, the U.S. focus

---

[5] "President Trump Establishing America's Space Force", *The White House*, 2019. https://www.whitehouse.gov/briefings-statements/president-trump-establishing-americas-space-force/.
[6] Jayan Panthamakkada Acuthan, "China's Outer Space Programme: Diplomacy Of Competition Or Co-operation?", *China Perspectives* 63 (2006), https://journals.openedition.org/chinaperspectives/577.
[7] Brian C Rathbun, *Reasoning of State: Realists, Romantics and Rationality in International Relations*, 1st ed. (Cambridge: Cambridge University Press, 2019).
[8] Niall Firth, *How to Fight a War in Space (and Get Away with It)*, MIT Technology Review (2019), https://www.technologyreview.com/2019/06/26/725/satellite-space-wars/.
[9] Articles VII of General Assembly resolution 2222 (XXI) *Treaty on Principles Governing the Activities of States in the Exploration and Use of Outer Space, including the Moon and Other Celestial Bodies,* available from https://unoosa.org/oosa/en/ourwork/spacelaw/treaties/outerspacetreaty.html.

shifts more to space capabilities and competitiveness, and the reasons for the creation of the USSF become clearer.

This chapter will cover a little on these other military space forces and why they have necessitated the creation of the USSF, before focussing on the current state and outlook for the USSF, along with the promises the U.S. have made for it, and the obstacles stopping that becoming a reality.

Space Forces Around the World

While several nations have space capabilities, there are limited nations with a dedicated space force. The USSF is currently the only independent branch, though Russia and China have their own forces as well which operate largely with the same level of independence. China and Russia are also the major challenges identified In the U.S. Defense Space Strategy, which states that both nations "… have weaponized space as a means to reduce U.S. and allied military effectiveness and challenge [U.S.] freedom of operation in space'".[10] Nations such as the UK, Australia, France, Iran, and Japan have units within other services, however the three predominant forces are the U.S., Russia, and China, for which reason they will remain the focus of this chapter.

Russia

The U.S. are the predominant nation in space – but they are not the only ones there. With Sputnik and Yuri Gargarin, Russia is accustomed to several space "firsts". As early as 1992, while still recovering from the fall of the Soviet Union, Russia reached another first by forming a new combat branch – its own independent Space Force.[11] It has undergone some changes since, being renamed the Russian Aerospace Defence Forces in 2011[12] before being merged with the Russian Air Force to create the Russian Aerospace Force in 2015,[13] hence why it is no longer considered independent, but it remains the oldest military Space Force in existence.

---

10 U.S. Department of Defense, *Defense Space Strategy Summary* (Washington, 2020), 1.
11 Hung P. Nguyen, "Russia's Continuing Work On Space Forces", *Orbis* 37, no. 3 (1993): 413-423, doi:10.1016/0030-4387(93)90154-5.
12 "History: Ministry of Defence of the Russian Federation", *Eng.Mil.Ru*, http://eng.mil.ru/en/structure/forces/cosmic/history.htm.
13 Nicholas Myers, "The Russian Aerospace Force", *Security Forum* 2, no.1 (2018): 91-103.

In modern times, Russia holds a political position seemingly at odds with their internal space strategy. Since early in the twenty-first century, Russia and China have jointly drafted and lobbied for a ban on space weapons[14] and continue to do so through the UN General Assembly.[15] In 2018, Vladamir Putin even made a personal appeal to then-U.S. President Donald Trump requesting support on a ban for both nations to station weapons in space[16] after Russian Foreign Minister Sergey Lavrov accused the U.S. of "nurturing plans to militarise outer space" earlier that year.[17]

There are speculatively peaceful reasons to pursue such a ban. When the ban was first drafted and leading into the second decade of the twenty-first century, the Russian space industry was booming.[18] No weapons is certainly a benefit while leading certain areas of space industry; however, Russia's ten-year Federal Space Plan approved in 2016 shows a different perspective. The Russian investment into the space program was only U.S.$20.5 billion, down from the U.S.$56.4 billion drafted in 2014 due to international incidents and financial pressures.[19] Such a large slashing of the space budget might indicate an inability to compete, and give good reason to support a ban. If unable to invest themselves, it is a logical approach to ensure others cannot as well.

However, regarding military capability, Russia has remained highly advanced. In late 2019, the Russian satellite Kosmos-2542 appeared to split, as a high-speed projectile – now known as Kosmos-2543 – was released. The smaller satellite was then accused of harassing U.S. satellites n Low Earth

---

14 "International Legal Agreements Relevant To Space Weapons", *Union of Concerned Scientists*, 2004, https://www.ucsusa.org/resources/legal-agreements-space-weapons.
15 Alexey Arbatov, "Arms Control In Outer Space: The Russian Angle, And A Possible Way Forward", *Bulletin of the Atomic Scientists* 75, no. 4 (2019): 151-161, doi:10.1080/00963402.2019. 1628475. 156.
16 Bryan Bender, "Leaked Document: Putin Lobbied Trump on Arms Control", *POLITICO*, 2018, https://www.politico.eu/article/leaked-document-vladimir-putin-lobbied-donald-trump-on-arms-control-helsinki-summit/.
17 "Выступление И Ответы На Вопросы СМИ Министра Иностранных Дел России С.В.Лаврова В Ходе Пресс-Конференции По Итогам Деятельности Российской Дипломатии В 2017 Году, Москва, 15 Января 2018 Года", *Mid.Ru*, 2018, https://www.mid.ru/ru/press_service/minister_speeches/-/asset_publisher/7OvQR5KJWVmR/content/id/3018203?p_p_id=101_INSTANCE_7OvQR5KJWVmR&_101_INSTANCE_7OvQR5KJWVmR_languageId=en_GB.
18 "The Future Of Russian Space Strategy", *Stratfor*, 2013, https://worldview.stratfor.com/article/future-russian-space-strategy.
19 Antoly Zak, "Russia Approves Its 10-Year Space Strategy", *The Planetary Society*, 2016, https://www.planetary.org/articles/0323-russia-space-budget.

Orbit.[20] In July, Kosmos-2543 then released *another* projection claimed to have been a "non-destructive test of a space-based anti-satellite weapon".[21] Furthermore, U.S. Defense Intelligence Organisation claims that Russia had been developing weapons specifically aimed at weakening the U.S.[22] and Russian doctrine itself sees space as a warfighting domain with space supremacy essential to victory in modern and future conflict.[23]

Kosmos-2543 tests directly targeting U.S. space systems and the overt expression of requiring space supremacy do not support the ongoing efforts towards a ban of all weapons in space, yet there is a strategic dilemma there as well. If Russia truly believes the U.S. is nurturing its own intent to create space-based weapons, then abandoning defence against such weapons would, from a strategic perspective, create a vulnerability. Yet from an international politics perspective, maintaining its own pursuit of such weapons appears contrary to the very ban it is trying to pursue. However, Russia's foreign policy and the processes thereof are not defined by such perspectives. Russia is concerned with action, rather than justification as demonstrated in 2014 with their annexation of Crimea, "justified" by drawing an equivalency with an event different in the extreme – the U.S. recognition of Kosovo.[24] There are clear questions around the legitimacy of this claim that are not in the scope of this chapter. However, the example of drawing equivalencies between Crimea and Kosovo demonstrates a willingness to use vague and broad justification towards action. Using similarly broad reasoning, Russia has argued that while the U.S. continues to develop space capabilities, Russia is also justified despite the combined efforts with China to implement the ban. The refusal of the U.S. to sign it gives them an apparent, if superficial, reason to continue their program while pushing for the international community – targeting the U.S. – towards a treaty that would leave only those nations with already developed weapons at a distinct advantage.

---

20 Beyza Unal and Mathieu Boulègue, "Russia's Behaviour Risks Weaponizing Outer Space", *Chatham House – International Affairs Think Tank*, 2020, https://www.chathamhouse.org/2020/07/russias-behaviour-risks-weaponizing-outer-space.
21 "Russia Conducts Space-Based Anti-Satellite Weapons Test", *United States Space Command*, 2020, https://www.spacecom.mil/MEDIA/NEWS-ARTICLES/Article/2285098/russia-conducts-space-based-anti-satellite-weapons-test/.
22 Defense Intelligence Agency, *Challenges to Security in Space* (Washington: United States Federal Government, 2019).
23 "The Military Doctrine of the Russian Federation", Посольство России В Великобритании, 2014, https://www.rusemb.org.uk/press/2029.
24 Samuel Charap & Cory Welt (2015) Making Sense of Russian Foreign Policy: Guest Editors' Introduction, *Problems of Post-Communism*, 62, no. 2, 67-70.

The back and forth is an interesting observance from an international relations standpoint, almost replicating a Space or Arms race attitude. The U.S. has a gained a clear military advantage due to its space systems. These remain unprotected, so competitors build systems to threaten these systems without doing any actual damage. The U.S. is obliged to protect their systems and infrastructure and react accordingly.[25] However without overt action, threat, or damage to this infrastructure, the U.S. is limited in how they can respond. One sideways step – technically not creating weapons but making a clear international statement – is the creation of the USSF. A defence that ensures any threat to U.S. space systems are a considered an attack on its military, which changes the consequences and perceived ethics of any response yet does not overtly threaten any foreign space systems in and of itself. Furthermore, the creation of the Russian Space Force almost two decades earlier makes it difficult for Russia to credibly attack the formation. The formation of the USSF is a powerful, yet passive move that on the surface creates political obstacles for adversaries targeting the U.S. in space. Whether the USSF can back up that threat remains to be seen.

China

As the co-drafter of Russia's proposed treaty to ban weapons in space, China takes a similar approach – promoting de-weaponisation, while preparing its own arsenal of space-based weapons, including directed-energy, and jamming capabilities, capable of targeting U.S. satellites.[26] Space technology is also a key element of China's grand strategy as a nation, not just in the military domain. China believes in the primacy of space technology and recognising the U.S. use of space to strengthen its status as a global superpower, sees this as its own path forward.[27]

With a grand strategy aiming to exploit space operations, it is unsurprising that the People's Liberation Army (PLA), encompassing all arms of the Chinese military forces, established the Strategic Support Forces (SSF) in late 2015. While the name may seem innocuous, the actual

---

25 Vishnu Anantatmula (2013) U.S. Initiative to Place Weapons in Space: The Catalyst for a Space-Based Arms Race", *Astropolitics* (2013) 11, no. 3 132-134.
26 Harsh Vasani, "How China Is Weaponizing Outer Space", *thediplomat.com*, 2017, https://thediplomat.com/2017/01/how-china-is-weaponizing-outer-space/.
27 Kevin Pollpeter et al., *China Dream, Space Dream; China's Progress in Space Technologies and Implications for the United States* (Washington: U.S.-China Economic and Security Review Commission, 2015).

functions include space, cyber, and electronic warfare along with satellite-satellite attacks and counter-space operations, negating the adversary's use of space. In function, it is a Space and Cyber Force, and while not technically independent, it is as independent as the PLA Navy or PLA Air Force. While no PLA arm is completely independent, the practicality of the PLA SSF is that it is as independent as the USSF.

There is a key difference here between the formation of Russia's Space Force and that of China's. Russia built their forces as an extension of the Soviet Union's space program and did so in the years following the end of the Cold War. There were many strategic unknowns, and it used an existing structure and history to establish the ability to respond to a potential threat. The SSF, however, came from the highest level possible. In 2013, at the Third Plenum of the 18th Party Congress, the Chinese government placed space as a priority among Chinese strategic policy.[28] Instead of a response to a dynamic external situation, China sees space as being inevitably weaponised and is preparing for what it sees as both a path forward to its own power and preparing for future conflicts.

That does not mean China has forgotten about the present though. While focussing on the future, China has made great strides in the launch and use of satellites for more tactical purposes such as targeting and surveillance.[29] While this gives advantage in many domains, there is a particular benefit to the maritime environment. The consistent feedback loop of surveillance from and to units is a critical part of sea control, and the use of space assets to detect weather, to use electronic intelligence to identify and track unfriendly units, and to provide early warning of long-range weapon launches makes control of the sea far more complete.[30] Considering the expanded Chinese claims to maritime areas such as the South China Sea, which facilitates over U.S.$3 trillion in trade, is believed to contain vast amounts of natural resources, and now hosts several Chinese military outposts,[31] the ability to survey and protect those outposts and claims is enhanced by the improvement in space capability. Considering

---

28 Kevin Pollpeter "Space, the New Domain: Space Operations and Chinese Military Reforms", *Journal of Strategic Studies* (2016) 39, no. 5-6, 709-727, DOI: 10.1080/01402390.2016.1219946.
29 Eric Hagt & Matthew Durnin "Space, China's Tactical Frontier", *Journal of Strategic Studies* (2011) 34, no 5, 733-761, DOI: 10.1080/01402390.2011.610660, 734-735.
30 Zaeem Shabbir, Ali Sarosh and Mahhad Nayyer, "Space Technology Applications For Maritime Intelligence, Surveillance, And Reconnaissance", *Astropolitics* 17, no. 2 (2019): 104-126, doi:10.1080/14777622.2019.1636634.
31 "Territorial Disputes In The South China Sea | Global Conflict Tracker", *Global Conflict Tracker*, 2020, https://www.cfr.org/global-conflict-tracker/conflict/territorial-disputes-south-china-sea.

the international condemnation, shows of force, and International Court of Arbitration ruling against Chinese claims in the area,[32] there is little to prevent justified action to remove such assets except China's own capabilities. While this is not a chapter on maritime strategy and conflict, the space capabilities of China play a large role in monitoring and the ability to react to situations in the South China Sea, giving both a significant and immediate advantage, while maintaining China's focus on becoming a space superpower.

A large part of this, like Russia, is due to Chinese military strategists perceiving space operations to be the critical to U.S. military primacy – a view shared by U.S. space strategists as well[33] – which also makes it a critical vulnerability.[34] With China also developing and testing anti-satellite weapons and the U.S. DoD reporting the existence of operational anti-satellite weapons in the Chinese arsenal,[35] the space architecture of the U.S. appears to be both critical and undefended. With the gap now closing between U.S. and Chinese space capabilities, and with vulnerabilities potentially exposed, the USSF has been created at an opportune time. Whether deliberately timed or not, the independence of the USSF to pursue its own strategic agenda vice existing as a service to others allows it the freedom to seek protective measures, be they kinetic or procedural, rather than exist as a unit subservient to a separate strategy. This is in no means a criticism of the USAF Space Command, who have largely been the same personnel and held the same capabilities. But whether intentional or not, a unit exists to serve a philosophy and strategy of the main force. The independence of the USSF enables a freedom to focus wholly on its own domain.[36]

---

32 Stefan Talmon, "The South China Sea Arbitration and the Finality of 'Final' Awards", *Journal of International Dispute Settlement* 8, no. 2 (2017): 388-401
33 Everett C. Dolman "A Debate about Weapons in Space: For U.S. Military Transformation and Weapons in Space." *The SAIS Review of International Affairs* 26, no. 1 (Winter, 2006): 163-174. doi:http://dx.doi.org.ezproxy.cqu.edu.au/10.1353/sais.2006.0006.
34 Elsa B Kania, "China Has A 'Space Force.' What Are Its Lessons For The Pentagon?", *Defense One*, 2018, https://www.defenseone.com/ideas/2018/09/china-has-space-force-what-are-its-lessons-pentagon/151665/.
35 Military and Security Developments Involving the People's Republic of China 2020 Annual Report to Congress, 65.
36 Everett C. Dolman, "Victory through Space Power: SSQ." *Strategic Studies Quarterly* 14, no. 2 (Summer, 2020): 3-15.

The United States Space Force

The USSF may not be the first, and it may only be the sole independent Space Force by technicalities, but it is arguably the most momentous in its formation. The U.S. not only leads the world in satellite launches and government space-based infrastructure,[37] but is also a hub of private industry entrepreneurship, with companies such as SpaceX and Blue Origin now competing with traditional Defence partners for contracts and airspace.[38]

Understandably then, there are questions over what exactly Space Force will do and what impact it will have.[39] It is not a new branch entirely, but rather a service within a service, with administrative control remaining with the USAF. It has a Chief of Space Operations and a seat at the Joint Chiefs,[40] but in the short time between the President's announcement and the first field command in October 2020 – also on a USAF base – what has actually changed since the functional roles were part of the USAF?

Some critics are expected. China, for example, has decried the formation of Space Force as militarising space despite the existence of its own space focussed service.[41] But there are plenty of critics within the U.S. as well, including former U.S. Secretary of State, General Jim Mattis (Ret'd). As a Colonel, Mattis was giving the tongue in cheek name of "CHAOS" or "Colonel Has Another Outstanding Solution", noting his tendency and enthusiasm towards innovative ideas.[42] With an expansive attitude towards options and after 44 years as part of another lodger service, the U.S. Marines, it may seem a given that Mattis would support the bold idea of a

---

37 Liane Zivitski, "China Wants To Dominate Space, And The US Must Take Countermeasures", *Defense News*, 2020, https://www.defensenews.com/opinion/commentary/2020/06/23/china-wants-to-dominate-space-and-the-us-must-take-countermeasures/.
38 Sandra Erwin, "Pentagon Picks Spacex And ULA To Remain Its Primary Launch Providers – Spacenews", *Spacenews*, 2020, https://spacenews.com/pentagon-picks-spacex-and-ula-to-launch-national-security-satellites-for-next-five-years/.
39 Brad Townsend, "Space Power And The Foundations Of An Independent Space Force", *Air & Space Power Journal* (2019): 11-24.
40 *National Defense Authorization Act for Fiscal Year 2020*, vol. 952-954 (Washington, 2019).
41 Geng Shuang, "Foreign Ministry Spokesperson Geng Shuang's Regular Press Conference On December 18, 2019", *Sydney.Chineseconsulate.org*, 2019, http://sydney.chineseconsulate.org/eng/fyrth/t1725764.htm.
42 Mackenzie Wolf, "The Origin Of Mattis' Call Sign, 'Chaos'", *Marine Corps Times*, 2017, https://www.marinecorpstimes.com/news/your-marine-corps/2017/09/21/mattis-explains-the-origin-of-the-call-sign-chaos/.

new service, yet he was reportedly cautious about the concept.[43] Strategists have consistently raised concerns about the formation of a force without a Grand Theory of Spacepower as well.[44] Such a theory, it is argued, is needed to ensure a clear and enduring vision not necessarily of what the USSF can do, but to understand what space warfare is. That is, what characteristics define space warfare, what factors influence control of space in both peace and conflict, the conditions space warfare and control impose on states, and the principles determining primacy, influence, and reach of an authority in space. Much like Mahan and Corbett did for maritime forces, a Grand Theory of Spacepower would set the conditions and guide the strategies of a Space Force.[45] The USAF itself fought against the formation, in part responsible for the "service within a service" approach adopted by the NDAA. There have also been legal and ethical concerns raised about violation of the UNOST.[46]

Some of the criticisms are, in hindsight, simple to dismiss. General Mattis was responsible for the fiscal state of the forces, and his opposition was based on the contemporary practical and financial situation. Since the formation the Space Force, such arguments are moot and as a result he has been supportive of the USSF as the concerns had been overtaken by events.[47] Similarly, the USAF had concerns about funding, though had the additional concerns about the loss of other resources, and of a capability critical to their operations. When a highly technical Air Force is reliant on GPS and satellite communication for example, losing the control of the capability may seem a loss, and there is also the doctrinal and cultural view that this may be seen (regardless of whether this is an accurate view) as a failure of the USAF to properly care for the capability.[48] Again, these are practical and legitimate concerns – but once the Space Force was

---

43 Robert Burns, "Pentagon Chief Mattis Defends His Reversal On Space Force", *Military.com*, 2018, https://www.military.com/daily-news/2018/08/13/pentagon-chief-mattis-defends-his-reversal-space-force.html.
44 Michael Martindale and David A. Deptula, "Organizing Spacepower: Conditions For Creating A US Space Force", *Mitchell Institute Policy Papers* 16 (2018).
45 Bledden E. Bowen, "From the sea to outer space: The command of space as the foundation for spacepower theory", *Journal of Strategic Studies*, 43:3-4 (2019), 532-556.
46 James James Fukazawa, "Does The U.S. Space Force Violate The Outer Space Treaty?", *Denver Journal of International Law & Policy* (2020), https://djilp.org/does-the-u-s-space-force-violate-the-outer-space-treaty/.
47 Jacqueline Thomsen, "Pentagon: Mattis Originally Opposed Creating Space Force Over Budgetary Concerns", *The Hill*, 2018, https://thehill.com/policy/defense/budget-appropriations/401283-pentagon-mattis-originally-opposed-creating-space.
48 James M Smith, "Air Force Culture And Cohesion Building An Air And Space Force For The Twenty-First Century", *Airpower Journal* Fall (1998): 40-53.

announced, the Air Force has similarly been supportive. While opposed in the decision-making process, once the USSF had been formed, the USAF took the position to welcome the new service and will continue to oversee the development.

The concerns about a Grand Theory and violation of the UNOST remain valid though. These are persistent matters that will fundamentally impact future leadership and the way in which military space operations are managed. A Grand Theory would guide the mission, vision, and methodology of the USSF not only throughout formation, but to use the Liddell Hart concept of a grand theory, through future conflict and more importantly, through peace.[49] Regarding the UNOST, the international expectations for interaction in space are reliant on this treaty, which was created in a vastly different era.[50] This makes it somewhat delicate in modern times, when nuclear weapons and weapons of mass destruction are no longer the key threats, and when nations are no longer the only entities operating in space. The UNOST is very clear about liabilities of damage to satellites (regardless of whether such damage is the result of military action), neutralities of individual personnel, and the denial of sovereignty on celestial bodies, along with placement of military bases on them,[51] all factors that impact potential kinetic conflicts in the space domain. While they might not be of immediate impact, at some point the USSF will need to deal with the UNOST along with any updates and decide on a Grand Theory of Spacepower to guide it forward.

Before tackling either of these issues, there is an important question to ask – what exactly does Space Force do? Understanding this is fundamental to how a Grand Theory or the UNOST will interact with each other and other actors in space.

---

49 B.H. Liddell Hart, *Strategy* (New York: Plume, 1991).
50 Jill Stuart, "The Outer Space Treaty Has Been Remarkably Successful – But Is It Fit For The Modern Age?", *The Conversation*, 2017, https://theconversation.com/the-outer-space-treaty-has-been-remarkably-successful-but-is-it-fit-for-the-modern-age-71381.
51 Articles VII, V, II, and IV respectively of General Assembly resolution 2222 (XXI) *Treaty on Principles Governing the Activities of States in the Exploration and Use of Outer Space, including the Moon and Other Celestial Bodies*, available from https://unoosa.org/oosa/en/ourwork/spacelaw/treaties/outerspacetreaty.html.

Esse Quam Videri — To Be Rather Than To Seem

The USSF has arguably accepted one of the most relied upon services in the U.S. military system. When discussing a missile attack on Iraqi airbase Ayn al-Asad a few weeks prior, USSF's Deputy Chief of Space Operations, Lt. General David Thompson emphasised this reliance in the following quote; "[O]ur crew... provided warning to those Americans and our friends and allies at al-Asad, which put them all in protective shelters. Had that not happened, we might be talking about folks that died in that attack as opposed to injury. That's Space Force".[52] In fact, the U.S. military – and the nation as a whole – is almost totally reliant on its space-based infrastructure.[53] The USSF has therefore come into existence with the weight of the nation already on its shoulders, though in a context far distinct from the burdens of the other services. Yet the role it fulfils is not yet unique from the former services of the USAF Space Command. There is a good reason for this, and it is the same reason that enabled such a swift move from announcement, to foundation, to the first field command in less than three years; the USSF is being built directly from the history of a pre-existing unit. Using the former USAF Space Command as a base for the USSF gives a history that dates back into the Cold War era, and uses the training, doctrine, facilities, skillsets, and equipment that have been proved effective from then right through to current conflicts.[54]

The advantage to this background is that Space Force has nearly four decades of experience, skillsets, and structure to rely on. By October 2020 – just ten months after inauguration – over a thousand qualified USAF personnel had already transferred to the USSF,[55] giving a pre-skilled, ready to use force, and the Space Mission Force White Paper already set out much

---

[52] Leigh Giangreco, "Space Force's Second-In-Command Explains What The Hell It Does", *Exo News*, 2020, https://exonews.org/space-forces-second-in-command-explains-what-the-hell-it-does/.
[53] Defence Space Strategy, *US Department of Defense*, 2020: Washington.
[54] Reid Barber, "The Purpose And Mission Of The Space Force", *American University*, 2020, https://www.american.edu/sis/centers/security-technology/the-purpose-and-mission-of-the-space-force.cfm.
[55] Jade Clark, "More Than 1,000 Air Force Cyber Security Operators To Transfer To Space Force – SNN", *Satellite News Network*, 2020, https://satellitenewsnetwork.com/2020/10/08/more-than-1000-air-force-cyber-security-operators-to-transfer-to-space-force/.

of the groundwork for structure in 2016.[56] However, it also means that from an immediate perspective, it is replicating an existing capability rather than creating a new one. Lt. General Thompson rightly extols the value of his fledgling service when discussing the Ain al-Asad attacks. *That's Space Force*, he states after discussing injuries that may otherwise have been far worse. However had the USSF not existed yet, the capability would still have remained. Given the number of staff crossing from the USAF to the USSF, it may have even been the same individuals providing the capability. The outcome, while never certain, would likely have been the same. The problem is that this raises a simple question; what is the point? In reference to the Latin phrase *esse quam videri*, will it *be* what it is touted to be, or simply *seem* independent, while remaining unchanged in function from the USAF unit it replaced?

For all the commotion surrounding the introduction of the USSF, what the conservative approach shows is two-fold. First of all, while it may have been a pre-existing capability, by accepting responsibility for that capability the USSF is able to apply an immediate utility. Holistically, there is little need to alter structures or organisational lines outside of the USSF, which means there is little difference to the other services, ensuring low levels of interruption to the provision of capability to them. While this may not be an overly exciting approach, it is sensible, and will provide a solid foundation from which to build space capability.

From a government oversight perspective, the focus on limiting the USSF activities to those of the former Space Command seems deliberate. It is focussed on terrestrial effects from space, rather than identifying space as a domain that contains combat in and of itself. In fact, the House Committee's comments on the formation of Space Force specifically defines the primary mission to "capabilities from space in support of all other combatant commands".[57] That is, it exists to support the other services. There is no primacy in Space as a military domain, rather any conflict or defence of space infrastructure occurring as "a necessary, but derived, mission to assure space capabilities are available to support all combatant

---

56 John E Hyten, "Space Mission Force: Developing Space Warfighters For Tomorrow", *Afspc. Af.Mil*, 2020, https://www.afspc.af.mil/Portals/3/documents/White%20Paper%20-%20Space%20Mission%20Force/AFSPC%20SMF%20White%20Paper%20-%20FINAL%20-%20AFSPC%20CC%20Approved%20on%20June%202029.pdf?ver=2016-07-19-095254-887.
57 US House of Representatives, *Report of the Committee on Appropriations Together with Minority Views* (Washington: US House of Representatives, 2020), https://www.congress.gov/116/crpt/hrpt453/CRPT-116hrpt453.pdf.

commands during conflict".⁵⁸ Similarly, while a supportive document overall, Lt Gen David Thomson, USAF, writes on Space as a Warfighting Domain, but in doing so demonstrates how the war in space acts more as a force multiplier than a force in and of itself.

With that in mind, the approach to creating a space force seems to have been framed as a cautious, conservative, and overall sensible starting point. The mission of the USSF is to "organize, train, and equip space forces to protect U.S. and allied interests in space and *to provide space capabilities to the joint force*. USSF responsibilities include developing military space professionals, acquiring military space systems, maturing the military doctrine for space power, and *organizing space forces to present to our Combatant Commands*" [italics added]. Compare this to the mission of the former Air Force Space Command: "Provide resilient, defendable and affordable space capabilities for the Air Force, Joint Force and the Nation".⁵⁹

If the service level responsibilities are removed from the USSF mission, then the output of both is the same—to provide space capabilities to the joint forces, or to present to the Combatant Commands.

Compare that with the USAF mission – "To fly, fight, and win in air, space, and cyberspace".⁶⁰ The U.S. Marine Corps mission likewise focusses on combat effectiveness, including statements such as "to win our Nation's battles swiftly and aggressively in times of crisis", and likewise the Army⁶¹ and Navy⁶² mission discuss combat effectiveness.

The commonality is that all discuss deploying forces in their specific domains and should the need arise, the ability to prevail in direct combat. The USSF, however, has a mission more aligned with the former Air Force unit it was born from, one which aligns it as a service provider rather than a direct combatant.

Similarly, the USSF capabilities are described as spacelift operations, command, and control of Department of Defence satellites, and influencing the battlespace through secure communications, weather and navigation

---

58 US House of Representatives, *Report of the Committee on Appropriations*.
59 "AFSPC Mission", *Afspc.Af.Mil*, https://www.afspc.af.mil/Home/AFSPC-Mission/.
60 "About Us", *Af.Mil*, https://www.af.mil/About-Us/.
61 "The Army's Vision And Strategy | The United States Army", *The Army's Vision And Strategy | The United States Army*, https://www.army.mil/about/.
62 "Our Navy'S Mission: How The Surface Forces Fit In", *Public.Navy.Mil*, https://www.public.navy.mil/surfor/swmag/Pages/Our-Navy%27s-Mission-How-the-surface-forces-fit-in.aspx.

data, and threat warnings.[63] There are further ground-based capabilities listed, but the key point is that it is almost verbatim the same capabilities of the Air Force Space Command.

It may not be exciting, but in terms of immediate functional effect, the USSF is starting from what it knows and using proven capabilities. It is not a simple rebranding of a unit – there is a future that would not be available to a subordinate unit – but gives a "business as usual" approach enabling the creation of a new service to occur as smoothly as possible. The training, equipment, and expertise already exists to fulfil the mission, and on the day of creation, Space Force was already providing capability to the combatant services. The burden of creation is reduced, while the output is unaffected, and internally the USSF is put to immediate effect. It is a solid base to build from.

From Inception to Infinity

The USSF has started with a modest, yet practical level of capability, though there are clearly plans to go further. The U.S. National Space Strategy significantly changes the tone of previous Strategies by adopting policies such as "Peace through Strength" in outer space, establishing "Space Pre-eminence through the American Spirit", and rhetoric that continues to emphasise a nationalistic and strength-based approach.[64] While guidance has not been particularly clear, there is also a definite desire to place humans on Mars. This, obviously, will not be done by communications and GPS satellites alone.

To build a force that can do this though, there is a key ingredient missing. While the Capstone Doctrine is a good guidance on policy, it lacks the ubiquitous and nationally agnostic approach of a Grand Theory. Such a theory was a critical component of giving the USAF guidance before it came into being, and without such a document, the USSF is without a basis to build that path forward, and without clear direction in how to grow.

There have been attempts and contribution towards a Grand Theory ever since former U.S. President George W Bush's administration

---

63 "About Us: About Space Force: Space Capabilities", *Spaceforce.Mil*, accessed 2 November 2020, https://www.spaceforce.mil/About-Us/About-Space-Force/Space-Capabilities/.
64 "President Donald J. Trump Is Unveiling An America First National Space Strategy | The White House", *The White House*, 2018, https://www.whitehouse.gov/briefings-statements/president-donald-j-trump-unveiling-america-first-national-space-strategy/.

commissioned a "theoretical framework for examining the fundamental aspects of spacepower and its relationship to the pursuit of national security, economic, information, and scientific objectives",[65] however no attempts as yet have reached the "fundamental" stage of being a Grand Theory. Part of this is the history, or lack thereof, behind spacepower. To use another technologically dependent service, the USAF, there was 154 years between the first manned free flight in 1783 and the formation of the Air Force in 1947. That's 154 years of refining and advancing technology and understanding the scientific principles to move from a hot air balloon to the Wright Brothers' flyer, to the war machines of the First and Second World Wars. By contrast, at the time of inception, manned space flights had only been occurring for 58 years, and even then spacepower has been developed without significant presence in space.[66] Airpower still had warfare application throughout the 1800's, but even then, it wasn't until 1910 – 127 years post first flight – that a separate force dedicated to flight (the French *l'Armee de l'Air*) was formed.

This lack of history makes it difficult to examine the idea of spacepower from first principles and develop the "fundamental aspects" President Bush requested. There is still a lack of broad experience, expertise, and historical analysis from which to draw, making a Grand Theory nigh on impossible right now in 2021. However, there are other obstacles as well; the view that "control of space is a lynchpin upon which national spacepower depends".[67] On face value, this is logical, however the issue is with the term "control" in comparison with the UN Outer Space Treaty. Like the "Peace through Strength" mentality, there is an impression that the USSF is being set up as an authoritive force in outer space. Breaches of the UNOST have occurred previously – the Israeli release of tardigrades on the moon breaches section IV of the Outer Space Treaty,[68] for example, and while technically not breaches of international space law, the use of Chinese and Russia anti-satellite weapons certainly breach the intent. While the Chinese simplified the matter by shooting their own satellite, thus not directly impacting any other nations, and therefore being deemed "legal"

---

65 Thomas G. Behling, Deputy Under Secretary of Defense (Preparation and Warning), Memorandum to the President, National Defense University, Subject: Space Power Theory, February 16, 2006.
66 James E Oberg et al., *Space Power Theory* (Colorado Springs: US Air Force Academy, 1999).
67 James E. Oberg et al., *Space Power Theory*.
68 Christopher D. Johnson et al., "The Space Review: The Curious Case Of The Transgressing Tardigrades (Part 1)", *Thespacereview.com*, 2020, https://www.thespacereview.com/article/3783/1.

by the UNOST, the Russian examples of trailing foreign satellites in 2020 could be construed as limiting freedom of movement through intimidation. While giving the impression of an authoritative and militarily strong force in space, here in 2021 there is little the USSF can do to impact other nations in space. To reference the Tsiolkovsky quote, it in expressing these desires that the USSF demonstrates it is still in the cradle – as would be expected as this stage – and has much to achieve before it can live up to the strategy it has been created to fulfil.

This is where the formation of the USSF becomes both necessary and problematic. As it stands, the intent, according to the strategy and doctrine, is to preserve freedom of action and support the joint force.[69] At the moment, it cannot do both. It can track and target on behalf of the joint services, but regarding the ability to preserve freedom of movement, its ability to influence the domain is unclear at best. That is not such an issue in the immediate future – the stable role of a service provider to the Joint Force ensures a smooth inception and immediate utility. However now that the USSF has come into being and the aggressive, nationalistic strategy has been set, it may not be long before it is tested. China, Russia, and Israel have already begun pushing the limits and enforceability of the UNOST, as have the U.S. in some ways.

Conclusion

The existence of the USSF has been opposed by other nations as it is almost an open challenge. It is the most critical of the U.S. services, and as it stands, the most vulnerable as well. It has boldly stated what it wants to be but has no clear pathway or Grand Theory to get there yet.

It is, at this early stage, an enigma, and only in the coming years will it become evident whether the USSF will find a way to expand into the strategy and Grand Theory that presents or remain in the metaphorical cradle for a time to come.

---

69 United States Space Force, "Spacepower Doctrine for Space Forces", *US Department of Defense* (2020): Colorado.

References

"About Us: About Space Force: Space Capabilities", *Spaceforce.Mil*, https://www.spaceforce.mil/About-Us/About-Space-Force/Space-Capabilities/.

"About Us", *Af.Mil*, https://www.af.mil/About-Us/.

Acuthan, J. "China's Outer Space Programme: Diplomacy Of Competition Or Co-operation?", *China Perspectives* 63 (2006), https://journals.openedition.org/chinaperspectives/577.

"AFSPC Mission", *Afspc.Af.Mil*, https://www.afspc.af.mil/Home/AFSPC-Mission/.

Anantatmula, V. "U.S. Initiative to Place Weapons in Space: The Catalyst for a Space-Based Arms Race", *Astropolitics* (2013) 11, no. 3 132-155.

Arbatov, A. "Arms Control In Outer Space: The Russian Angle, And A Possible Way Forward", *Bulletin of the Atomic Scientists* 75, no. 4 (2019): 151-161, doi:10.1080/00963402.2019.1628475.

Barber, R. "The Purpose And Mission Of The Space Force", *American University*, 2020, https://www.american.edu/sis/centers/security-technology/the-purpose-and-mission-of-the-space-force.cfm.

Behling, T. Deputy Under Secretary of Defense (Preparation and Warning), Memorandum to the President, National Defense University, Subject: Space Power Theory, February 16, 2006.

Bender, B. "Leaked Document: Putin Lobbied Trump On Arms Control", *Politico*, 2018, https://www.politico.eu/article/leaked-document-vladimir-putin-lobbied-donald-trump-on-arms-control-helsinki-summit/.

"Выступление И Ответы На Вопросы СМИ Министра Иностранных Дел России С.В.Лаврова В Ходе Пресс-Конференции По Итогам Деятельности Российской Дипломатии В 2017 Году, Москва, 15 Января 2018 Года", *Mid.Ru*, 2018, https://www.mid.ru/ru/press_service/minister_speeches/-/asset_publisher/7OvQR5KJWVmR/content/id/3018203?p_p_id=101_INSTANCE_7OvQR5KJWVmR&_101_INSTANCE_7OvQR5KJWVmR_languageId=en_GB.

Bowen, Bledden E. "From the sea to outer space: The command of space as the foundation for spacepower theory", *Journal of Strategic Studies*, 43:3-4 (2019), 532-556.

Burns, R. "Pentagon Chief Mattis Defends His Reversal On Space Force", *Military.Com*, 2018, https://www.military.com/daily-news/2018/08/13/pentagon-chief-mattis-defends-his-reversal-space-force.html.

Charap, S. and C. Welt. Making Sense of Russian Foreign Policy: Guest Editors' Introduction, *Problems of Post-Communism* (2015), 62, no. 2, 67-70.

Clark, J. "More Than 1,000 Air Force Cyber Security Operators To Transfer To Space Force – SNN", *Satellite News Network*, 2020, accessed 05 November 2020, https://satellitenewsnetwork.com/2020/10/08/more-than-1000-air-force-cyber-security-operators-to-transfer-to-space-force/.

Defense Intelligence Agency. *Challenges to Security in Space* (Washington: United Stated Federal Government, 2019).

Defence Space Strategy, *US Department of Defence*, 2020: Washington.

Dolman, Everett C. "A Debate about Weapons in Space: For U.S. Military Transformation and Weapons in Space." *The SAIS Review of International Affairs* 26, no. 1 (Winter, 2006): 163-174.

Dolman, Everett C "Victory through Space Power: SSQ." *Strategic Studies Quarterly* 14, no. 2 (Summer, 2020): 3-15.

Erwin, S. "Pentagon Picks Spacex And ULA To Remain Its Primary Launch Providers – Spacenews", *Spacenews*, 2020, https://spacenews.com/pentagon-picks-spacex-and-ula-to-launch-national-security-satellites-for-next-five-years/.

Firth, N. *How to Fight a War in Space (and Get Away with It)*, MIT Technology Review (2019), https://www.technologyreview.com/2019/06/26/725/satellite-space-wars/.

Fukazawa, J. "Does the U.S. Space Force Violate the Outer Space Treaty?", *Denver Journal of International Law & Policy*, 2020, https://djilp.org/does-the-u-s-space-force-violate-the-outer-space-treaty/.

Giangreco, L. "Space Force's Second-In-Command Explains What The Hell It Does", *Exo News*, 2020, https://exonews.org/space-forces-second-in-command-explains-what-the-hell-it-does/.

Hagt E, and M. Durnin. "Space, China's Tactical Frontier", *Journal of Strategic Studies* (2011) 34, no 5, 733-761, DOI: 10.1080/01402390.2011.610660.

"History: Ministry Of Defence Of The Russian Federation", *Eng.Mil.Ru*, accessed 12 November 2020, http://eng.mil.ru/en/structure/forces/cosmic/history.htm.

Hyten, J. "Space Mission Force: Developing Space Warfighters For Tomorrow", *Afspc.Af.Mil*, 2020, https://www.afspc.af.mil/Portals/3/documents/White%20Paper%20-%20Space%20Mission%20Force/AFSPC%20SMF%20White%20Paper%20-%20FINAL%20-%20AFSPC%20CC%20Approved%20on%20June%2029.pdf?ver=2016-07-19-095254-887.

"International Legal Agreements Relevant To Space Weapons", *Union of Concerned Scientists*, 2004, https://www.ucsusa.org/resources/legal-agreements-space-weapons.

Johnson, C., D. Porras, C., Hearsey and S. O'Sullivan. "The Space Review: The Curious Case Of The Transgressing Tardigrades (Part 1)", *Thespacereview.Com*, 2020, https://www.thespacereview.com/article/3783/1.

Kania, E. "China Has A 'Space Force.' What Are Its Lessons For The Pentagon?", *Defense One*, 2018, https://www.defenseone.com/ideas/2018/09/china-has-space-force-what-are-its-lessons-pentagon/151665/.

Liddell Hart, B. *Strategy*, New York: Plume, 1991.

Martindale, M. and D. Deptula. "Organizing Spacepower: Conditions For Creating A US Space Force", *Mitchell Institute Policy Papers* 16 (2018).

Military and Security Developments Involving the People's Republic of China 2020 Annual Report to Congress, Defense Intelligence Agency, Washington

Morrow, J. "The First World War, 1914-1919", in *A History of Air Warfare*, 1st ed. Dulles: Potomac Books, 2010.

Myers, N. "The Russian Aerospace Force", *Security Forum* 2, no. 1 (2018): 91-103.

Nguyen, H. "Russia's Continuing Work On Space Forces", *Orbis* 37, no. 3 (1993): 413-423, doi:10.1016/0030-4387(93)90154-5.

National Defense Authorization Act for Fiscal Year 2020, vol. 952-954 (Washington, 2019).

Oberg, J. *Space Power Theory*, Colorado Springs: US Air Force Academy, 1999.

"Our Navy's Mission: How The Surface Forces Fit In", *Public.Navy.Mil*, https://www.public.navy.mil/surfor/swmag/Pages/Our-Navy%27s-Mission-How-the-surface-forces-fit-in.aspx.

"President Donald J. Trump Is Unveiling An America First National Space Strategy | The White House", *The White House*, 2018, https://www.whitehouse.gov/briefings-statements/president-donald-j-trump-unveiling-america-first-national-space-strategy/.

"President Trump Establishing America's Space Force", *The White House*, 2019, https://www.whitehouse.gov/briefings-statements/president-trump-establishing-americas-space-force/

Pollpeter, K. "Space, the New Domain: Space Operations and Chinese Military Reforms", *Journal of Strategic Studies* (2016) 39, no. 5-6, 709-727, DOI: 10.1080/01402390.2016.1219946

Pollpeter, K., E. Anderson, J. Wilson and F. Yang. *China Dream, Space Dream; China's Progress in Space Technologies and Implications for the United States*, Washington: U.S.-China Economic and Security Review Commission, 2015.

Rathbun, B. *Reasoning of State: Realists, Romantics and Rationality in International Relations*, 1st ed. Cambridge: Cambridge University Press, 2019.

"Russia Conducts Space-Based Anti-Satellite Weapons Test", *United States Space Command*, 2020, https://www.spacecom.mil/MEDIA/NEWS-ARTICLES/Article/2285098/russia-conducts-space-based-anti-satellite-weapons-test/.

Shabbir, Z., A. Sarosh and M. Nayyer. "Space Technology Applications For Maritime Intelligence, Surveillance, And Reconnaissance", *Astropolitics* 17, no. 2 (2019): 104-126, doi:10.1080/14777622.2019.1636634.

Shuang, G. "Foreign Ministry Spokesperson Geng Shuang's Regular Press Conference On December 18, 2019", *Sydney.Chineseconsulate.org*, 2019, http://sydney.chineseconsulate.org/eng/fyrth/t1725764.htm.

Smith, J. "Air Force Culture And Cohesion Building An Air And Space Force For The Twenty-First Century", *Airpower Journal* Fall (1998): 40-53.

Space Force, "Mark and Mallory Go To Washington", Season 1 Episode 3, Directed by Tom Marshall. Written by Shepard Boucher. Netflix, 29 May 2020

Speake, J. "The Wrong Kind of Wonder: Ben Jonson and Cornelis Drebbel." *The Review of English Studies*, 66, no.273 (2015): 60-70.

Talmon, S. "The South China Sea Arbitration and the Finality of 'Final' Awards", *Journal of International Dispute Settlement* 8, no. 2 (2017): 388-401.

"The Army's Vision And Strategy | The United States Army", *The Army's Vision and Strategy | The United States Army*, https://www.army.mil/about/.

"The Future Of Russian Space Strategy", *Stratfor*, 2013, https://worldview.stratfor.com/article/future-russian-space-strategy.

"Territorial Disputes In The South China Sea | Global Conflict Tracker", *Global Conflict Tracker*, 2020, https://www.cfr.org/global-conflict-tracker/conflict/territorial-disputes-south-china-sea.

"The Military Doctrine of the Russian Federation", *Посольство России В Великобритании*, 2014, https://www.rusemb.org.uk/press/2029.

Thomsen, J. "Pentagon: Mattis Originally Opposed Creating Space Force Over Budgetary Concerns", *The Hill*, 2018, https://thehill.com/policy/defense/budget-appropriations/401283-pentagon-mattis-originally-opposed-creating-space.

Townsend, B. "Space Power And The Foundations Of An Independent Space Force", *Air & Space Power Journal* (2019): 11-24.

Unal, B and M. Boulègue. "Russia's Behaviour Risks Weaponizing Outer Space", *Chatham House — International Affairs Think Tank*, 2020, https://www.chathamhouse.org/2020/07/russias-behaviour-risks-weaponizing-outer-space.

U.S. Department of Defense, *Defense Space Strategy Summary*, Washington, 2020

U.S. House of Representatives, *Report of the Committee on Appropriations Together with Minority Views*, Washington: US House of Representatives, 2020, https://www.congress.gov/116/crpt/hrpt453/CRPT-116hrpt453.pdf.

U.S. Space Force, "Spacepower Doctrine for Space Forces", *US Department of Defense* (2020): Colorado.

Vasani, H. "How China Is Weaponizing Outer Space", *thediplomat.com*, 2017, https://thediplomat.com/2017/01/how-china-is-weaponizing-outer-space/.

Wolf, M. "The Origin Of Mattis' Call Sign, 'Chaos'", *Marine Corps Times*, 2017, https://www.marinecorpstimes.com/news/your-marine-corps/2017/09/21/mattis-explains-the-origin-of-the-call-sign-chaos/.

Zak, A. "Russia Approves Its 10-Year Space Strategy", *The Planetary Society*, 2016, https://www.planetary.org/articles/0323-russia-space-budget.

Zivitski, L. "China Wants To Dominate Space, And The US Must Take Countermeasures", *Defense News*, 2020, https://www.defensenews.com/opinion/commentary/2020/06/23/china-wants-to-dominate-space-and-the-us-must-take-countermeasures/.

16

FROM PEACEFUL USES TO WARFIGHTING

The Dangers of the New Military Era in Space

Jessica West

Introduction

The "peaceful use of outer space," while vague and ill-defined, is a foundational tenet of global space governance and cornerstone of human activity there. But so, too, is military use. These competing functions are rooted in the political dynamics that defined the aftermath of World War Two and the Cold War competition between the United States and the Soviet Union. Both views of outer space were on display in the aftermath of the launch of the Soviet Union's Sputnik satellite on 4 October 1957. Both were institutionalised in the principles of the Outer Space Treaty (OST), which veiled strong military interests and uses of outer space with the language of humanitarian values and a global community. However, as military use of outer space has expanded and military actors have proliferated, the growing emphasis on warfighting in outer space has threatened to make any peaceful use of space untenable.

This chapter explores how competing values developed and coexisted to produce a space environment that allowed the rapid expansion of both military and peaceful uses of outer space. Although such development was punctuated by periodic testing of weapons and dreams of a military high ground, the balancing act has, until recently, been maintained by restraining military activities and adhering to a commitment to peace. But the current emphasis on warfighting in space could cause this fragile edifice to collapse.

The United States Space Force (USSF) is only one manifestation of a shift in the military uses of outer space from passive – and passably peaceful – to aggressive. As a domain reserved for all humanity, outer space must be preserved from weaponisation and warfare. Progress toward warfighting in outer space threatens not only principles, but the stability of the operating environment in outer space. From the contamination of the space environment to the damaging of civilian and commercial systems, aggression in outer space threatens to disrupt global space-based services on which people all around the world depend. At the same time, such actions increase the potential for nuclear escalation. For many practical and ethical reasons, we must not allow space to be viewed as – or become – a warzone.

The Competing Values of the Space Age

The space age was shaped by the competing values and tensions of the time period that followed World War Two. The Cold War between the United States and the Soviet Union, a techno-military competition for global superiority, was marked by an unprecedented arms race and military spending. But the period was also characterized by an emerging sense of a global community and universalistic impulses, expressed through rising movements for peace and disarmament, human rights, and environmental protection.[1] These two driving forces, seemingly opposites, were entwined in what has been described as a race for global hearts and minds.[2] This overarching dynamic is a key to the duality that emerged in the development of space – as both a domain of intense military competition and as the source of a peaceful future for humanity. Both impulses were revealed in the aftermath of the launch of the world's first artificial satellite on 4 October 1957 and left a lasting imprint on the Outer Space Treaty and the relationship that humans have with outer space today.

---

1 Akira Iriye, *Global Community: The Role of International Organizations in the Making of the Contemporary World* (Berkeley: University of California Press, 2004), 101.
2 Odd Arne Westad, *The Global Cold War* (Cambridge: Cambridge University Press, 2005).

Sputnik: Hope and Fear

The Soviet Union's launch of the Sputnik satellite is a seminal moment in the history of outer space. Although known for the technological achievement of being the first artificial satellite to orbit Earth, the social and political implications are most significant. Sputnik opened a new, untouched domain to human activity: space was a *tabula rasa* on which any future could be written. The launch of Sputnik made concrete hypothetical political and ethical debates about how space should be used, by whom, and in whose interests. Out of such controversies arose two competing visions that continue to influence the world today: space as a domain of military might and space as the source of peace for all humans.

Sputnik was launched amid intense military competition between the Soviet Union and the United States that was marked by a nuclear-arms race and the development of new rocketry capabilities to support intercontinental ballistic missiles (ICBMs). The Sputnik satellite program, as part of a supposed development of a space launch vehicle, was approved by the Soviet government to test these missile technologies. Subsequent "civilian" space efforts in the Soviet Union also had parallel military interests.[3] Similarly, in the United States, the rocket and missile capabilities that formed the basis for space launch vehicles were developed under the auspices of the U.S. Army, Navy, and Air Force. Both the Army and Air Force had satellite programs to develop reconnaissance capabilities long before the launch of Sputnik.[4]

Some analysts feared that Sputnik was ushering in a new era of deadly weapons. In the United States, some public commentators and media compared the launch of Sputnik to the bombing of Pearl Harbor and the development of the atomic bomb.[5] There were fears that States would develop space platforms from which bombs could be dropped on Earth and that space would become an extension of military rivalries on Earth.[6] This fear drove what became a race between the United States and the

---

3 James Harford, *Korolev* (Hoboken: John Wiley and Sons Inc., 1997), 123; Deborah Cadbury, *Space Race* (New York: Harper Collins Publishers, 2006), 255.
4 Robert Divine, *The Sputnik Challenge* (Oxford: Oxford University Press, 1993), 11, 21, 25; James Clay Moltz, *The Politics of Space Security* (Stanford: Stanford University Press, 2008), 83-85.
5 "US Talk of New 'Pearl Harbor,'" *London Times*, 4 November 1957,10; Divine, *The Sputnik Challenge*, 37.
6 Robert K. Plumb, "New Space Conquest Can Now be Foreseen," *New York Times*, 6 October 1957, 197; "The Implications," *New York Times*, 6 October 1957, 193.

Soviet Union – what Lyndon B. Johnson called "a race for survival".[7] Even the subsequent mission to the Moon and human spaceflight were driven in large part by interests within the U.S. military that wanted control of what was perceived as the "high ground" of space.[8]

But other observers viewed Sputnik as a symbol of peace. Launched during the International Geophysical Year that supported advancements and cooperation in science, the Soviet satellite was an inspiration to people around the world who believed that space exploration marked a new era of peace for humankind.[9] Many believed that a "humanity united in space" could transcend cultural and political barriers.[10] All around the world, people listened to their radios for the message of beeps that Sputnik transmitted from the heavens.[11] Referring to "peaceful co-existence", the Soviet Union called Sputnik a "success for all mankind".[12]

After the launch of the U.S. Explorer 1 satellite in January 1958, U.S. President Dwight Eisenhower issued a call "to solve what I consider to be the most important problem which faces the world today". Noting that both the United States and the Soviet Union were using outer space to test military missiles, he proposed that "outer space should be used only for peaceful purposes".[13] The Soviets responded with a proposal that put space under the control of the United Nations and banned military uses of space and rockets.[14] In an address to the United Nations General Assembly in September 1958, the United States insisted that "as we reach beyond this planet, we should move truly as 'united nations'".[15]

The Committee on the Peaceful Uses of Outer Space (COPUOS) was subsequently established to acknowledge "the common interest of mankind in outer space" and "the common aim that outer space should be

---

7 Lyndon B. Johnson quoted in Walter McDougall, *The Heavens and the Earth: A Political History of the Space Age* (Baltimore: Johns Hopkins University Press, 1997), 8; Lyndon B. Johnson cited in Divine, *The Sputnik Challenge*, 79.
8 Divine, *The Sputnik Challenge*, 98, 108, 150; Howard E. McCurdy, *Space and the American Imagination* (Washington: Smithsonian Institution Press, 1997), 41.
9 Paul Kecskemeti, "Outer Space and World Peace," in Joseph Goldsen ed., *Outer Space and World Politics* (Westport: Frederick A. Praeger, Publisher, 1963), 28.
10 McDougall, *The Heavens and the Earth*, 9; "space sanctuary" perspective described in Moltz, *The Politics of Space Security*, 27-31.
11 "Chicago 'Hams' Active: Illinois State Policy and Others Hear Satellite's Signals," *New York Times*, 6 October 1957, 42; "Calcutta Hears Sphere's Signal," *New York Times*, 6 October 1957, 43; "Cairo Assails U.S. on 'Moon,'" *New York Times*, 6 October 1957, 44.
12 Kecskemeit, "Outer Space and World Peace," 28.
13 McDougall, *The Heavens and the Earth*, 179.
14 McDougall, *The Heavens and the Earth*, 179.
15 John Foster Dulles quoted in McDougal, *The Heavens and the Earth*, 184.

used for peaceful purposes", with a mandate to negotiate an international treaty that would lay out the broad principles of how outer space would be used and by whom.[16]

Freedom of Space and Military Use

In 1955, President Eisenhower first mentioned "freedom of space" as key to the U.S. program to develop photoreconnaissance satellites, which would require safe passage over sovereign territories and thus depended on the universal recognition of space as a global commons.[17] The U.S. strategy was to establish the principle of freedom of space without raising protest by first launching a civilian satellite. When Sputnik was launched first, the effect was still the same.[18] Sputnik's launch and its unimpeded and unprotested orbiting of Earth made real the principle of "freedom of space", establishing the right of free use of, and passage through, space. It also established a critical base for the development of military activities, first for the United States and Soviet Union, and later for other countries as well.

In the decade between the launch of Sputnik and the signing of the Outer Space Treaty, the development of space and missile technologies accelerated, unimpeded by international regulation, and often hidden from the world. The Soviets publicly denied having a military space program.[19] The United States deflected attention from its military activities through the establishment in 1959 of the National Aeronautics and Space Administration (NASA) – a large, civilian flagship program to promote "peaceful purposes for all mankind".[20] The continuation of military space projects for reconnaissance, communications, navigation, and weather were heavily funded, but classified.[21]

---

16 United Nations General Assembly First Committee, Resolution 1348 (XIII) "Question of the Peaceful Use of Outer Space," (13 December 1958).
17 M. J. Peterson, *International Regimes for the Final Frontier* (New York: State University of New York Press, 2005), 50.
18 Divine, *The Sputnik Challenge*, 11.
19 Peterson, *International Regimes for the Final Frontier*, 134; Kecskemeit, "Outer Space and World Peace," 31.
20 Government of the United States, "National Astronautics and Space Act of 1958," Sec. 102 (a) (1958).
21 Michael Sheehan, *The International Politics of Space* (Milton Park: Routledge, 2007), 43. The National Reconnaissance Office, an entirely new, classified program for photoreconnaissance was established alongside NASA. According to McDougall, programs for the development of new military space technology received more funding than NASA until both were eclipsed by the race to the moon, 191.

In principle, human activity in outer space was rooted in universalist principles of peace, equality, and community. But in practice, many developments were based on military interests and imperatives. This duality can be found in the language and silences of the Outer Space Treaty.

Stated Principles and Strategic Silence in the Outer Space Treaty

The OST was signed on 10 October 1967, creating a legal framework for human activities in outer space. Recognizing the exploration and use of outer space as "the province of all mankind", it enshrined many of the universal values and ideals expressed following the launch of Sputnik, including provisions for "peaceful purposes", equality, cooperation, and specific arms control measures.[22] The treaty was modeled on the Antarctic treaty,[23] a non-armament agreement that restricted weapons and military activities and established a global commons, but its subtext, found in a series of strategic silences, was shaped by the military interests of the two superpowers.

Competing values are revealed in the description of three core mandates of the treaty: the concept of peaceful purposes, the nature of sovereignty in outer space, and arms control considerations. In these instances, the treaty both constrains and facilitates military activities in outer space.

Peaceful Purposes Versus Military Functions

The principle of peaceful purposes is a pillar of the OST. But while some states – particularly developing countries – sought to define peaceful purposes in the preamble to the treaty, their efforts were resisted.[24] Instead, the concept remained not only undefined, but operationally restricted to the Moon and other celestial bodies. These omissions have suited the Soviet

---

22 United Nations, "Treaty on Principles Governing the Activities of States in the Exploration and Use of Outer Space, Including the Moon and Other Celestial Bodies," (1967); Jill Stuart, "Unbundling Sovereignty, Territory and the State in Outer Space," in Natalie Bormann and Michael Sheehan eds, *Securing Outer Space* (Milton Park: Routledge, 2009), 15.
23 A. G. Mezerik, ed. "Outer Space: UN, US, USSR." *International Review Service* Volume VI, No. 56 (1960), 21; William Chapman, "LBJ Seeks U.N. Treaty on Space," *Washington Post* (8 May 1966), A1; "Next Step in Space," *New York Times* (28 January 1967), 22; y Marder, A1.
24 Peterson, *International Regimes for the Final Frontier*, 139, 141.

Union and the United States and many other states since then that have developed military space programs.

During the Cold War, both superpowers promoted the peaceful use of space to retain the support of the international community for space programs – programs that supported military strategies. The United States equated "peaceful" with "non-aggressive use" and saw its military programs as legitimate because they were "non-violen".[25] Although the Soviet Union initially advanced an interpretation of "peaceful use" as "non-military use", in practice the Soviets also equated "peaceful" with "non-aggressive".[26] This common approach allowed both States to claim the moral high ground by supporting peaceful uses. As well, camouflaged military space programs were exempted from subsequent governance practices such as prior notification of launches and registration of spacecraft with the United Nations.[27] The "non-aggression" tune, set off by strategic silences, permitted the emergence of a space environment that was both peaceful and military.

Non-Appropriation Versus Freedom of Space and Ownership of Objects

The OST's principle of non-appropriation of outer space "by claim of sovereignty, by means of use or occupation, or by any other means", is intended to prevent the extension of colonial competition in space. The treaty formally established space as a *res communis* or global commons, "free for exploration and use by all States without discrimination of any kind, on a basis of equality and in accordance with international law".[28]

However, the principle of freedom of space facilitated rampant military use. As well, Article VIII describes the principle of national jurisdiction and control, by which States maintain ownership of objects launched into outer space. The combination of these two principles embeds a kind of sovereignty in space, not through ownership of space itself, but through the freedom to access and use space and to maintain control over those uses. Not only does this provision allow for the physical extension of the State into space and facilitate the rampant development of national

---

25 Kecskemeit, "Outer Space and World Peace", 34.
26 Robert D. Crane, "Soviet Attitude Toward International Space Law", *American Journal of International Law* (July 1962), 704.
27 Ivan A. Vlasic, "The Space Treaty: A Preliminary Evaluation", *California Law Review* (May 1967), 516; Carol Kilpatrick, "60 Nations Sign Pact Here on Outer Space", *Washington Post* (28 January 1967), A1.
28 *Outer Space Treaty*, Article I.

security systems in orbit, but it establishes the right of non-interference and reinforces the UN Charter right to self defence in outer space.

Non-Armament Versus Strategic Restraint

The OST is known as a non-armament treaty, designed to prevent an arms race in outer space. But, in practice, this focus on arms control is incomplete, and progress deadlocked. In the meantime, strategic military advances have been made, restrained only by bilateral and political agreements.

The United States first adopted the rhetoric of non-armament in 1957 when President Eisenhower proposed to the UN Disarmament Committee that space be used exclusively for peaceful purposes and that the development of missile technologies be subjected to international oversight.[29] However, the American proposal to relocate space discussions to the newly created COPUOS subsequently dropped non-armament from the agenda.[30]

The Soviet Union's initial proposal was for the complete demilitarisation of space, with a particular focus on dismantling of American overseas bases.[31] This extreme approach mired COPUOS in controversy and allowed the development of weapons programs to continue unfettered.

Both the United States and the Soviet Union recognized, however, that nuclear explosions could damage their ability to use outer space for national military purposes. In 1963, they agreed to the first restraint on nuclear testing, the Partial Test Ban Treaty.[32]

What the public feared most were bombardment satellites that would drop nuclear weapons from space. But, in reality, such satellites had little military value compared to inter-continental ballistic missiles.[33] Limited bi-lateral arms-control agreements were created to maintain the stability needed to operate essential military systems such as those used for missile

---

29 Philip C. Jessup and Howard T. Taubenfeld. "Outer Space, Antarctica, and the United Nations", *International Organization* (Summer 1959), 367.
30 Peterson, *International Regimes for the Final Frontier*, 135; United Nations General Assembly First Committee, Resolution 1348 (XIII) "Question of the Peaceful Use of Outer Space" (13 December 1958).
31 Crane, "Soviet Attitudes Toward International Space Law", 701-702.
32 United Nations, "Treaty Banning Nuclear Tests in the Atmosphere, in Outer Space, and Under Water" (1963).
33 Bob Preston et al., *Space Weapons: Earth Wars* (RAND, 2002), 11-12; McDougall, *The Heavens and the Earth*, 191; Thomas Schelling, "The Military Use of Outer Space: Bombardment Satellites", in Joseph Goldsen ed., *Outer Space in World Politics* (Frederick A. Praeger, Publisher, 1963), 97-113.

early warning and arms-control verification. A gentleman's agreement that banned nuclear weapons from space was reached in 1963.[34]

Meanwhile, technology development and testing programs continued for both Earth and space-based anti-satellite weapons using conventional weapons. However, as everyone recognized that "space had become too important for war",[35] restraint was exercised in deploying these weapons.

In the 1967 OST, these conflicting interests and understandings can be seen in its non-armament provisions. Article IV banned the use or orbiting of nuclear weapons and other weapons of mass destruction in outer space.[36] The deployment of conventional weapons in space is not banned.

Proliferation of military space systems by the United States and the Soviet Union continued, as did the development of weapons to attack and defend these systems. Such developments were only checked by voluntary mutual restraint and a continued deference to the peaceful uses of outer space.

Outer Space Today: Mingling Military and Peaceful Uses

Outer space today is very different from what it was 50 years ago. In many ways, competing military interests and global humanitarianism of the early space age have struck a productive balance. Military uses and users have proliferated, but the worst fears have not been realized. Space today resembles the global community envisaged in the OST.

A Global Community

No longer the exclusive domain of two superpowers, outer space is now home to more than 3,000 satellites, which are owned and operated by more than 70 states.[37] The vast majority of satellites are civil and commercial.

Access to and use of space are now much more than military imperatives. Space systems are critical to civilian life; they form a meta-capability that enables almost all essential services on Earth. Consider cell phone

---

34 Theresa Hitchens, "Rushing to Weaponize the Final Frontier", *Arms Control Today*, (September 2001).
35 Laura Grego, "A History of Anti-Satellite Programs", Union of Concerned Scientists (2003); Preston, 12; McDougall, 343; Moltz, 125; Peterson, 141.
36 "Outer Space Treaty", Article IV.
37 "Satellite Database", Union of Concerned Scientists, last updated August 1, 2020, https://www.ucsusa.org/resources/satellite-database.

connectivity, air traffic control, disaster warnings, agricultural production, electronic banking, shipping, power grids, and the Internet. Space is essential for monitoring climate change and tracking weather patterns.

The commercial space sector was growing rapidly. The value of the global space economy in 2019 is estimated at $366 billion, almost half of which relates to satellite services.[38] Satellite-enabled communications meet the daily needs of billions of users on Earth but the benefits of space extend far beyond the private sector.

The ability of satellites to detect wildfires and monitor hurricanes and cyclones makes them indispensable for disaster early warning. They are also critical for disaster response. This need is recognized by the International Charter on Space and Major Disasters, which provides satellite data to help manage disasters. Satellite-enabled communications are critical when other ways of communicating are lost. The Crisis Connectivity Charter is designed to make satellite-based communications more readily available during disasters to those providing humanitarian aid and to affected communities. Efforts to mobilize the power of space to support the United Nation's Sustainable Development Goals (SDGs) are growing.

These peaceful uses of space and the achievement of the values of a global community in space are entwined with military uses.[39] Many civilian space capabilities and applications – such as communications and Earth observation – originated with military users. The development of global navigation satellite services (GNSS) is a case in point. The United States Global Positioning System (GPS) – one of several GNSS systems – is the central nervous system of the U.S. military. It provides precision timing, navigation, and targeting capabilities to military units and weapons systems. But GPS also communicates with individual wayfinding and fitness apps, and supports global travel, financial systems, civil communications, and power grids. During a disaster, GNSS signals such as GPS are critical in establishing the precise location of those in need. GNSS is also key to using outer space to achieve the SDGs.[40]

---

38 "State of the Satellite Industry Report", Bryce Space and Technology, LLC, July 2020, https://sia.org/wp-content/uploads/2020/07/2020-SSIR-2-Pager-20200701.pdf.
39 Jessica West, "Protecting Humans on Earth from Warfare in Space", *Ploughshares Monitor* 41, 4 (Winter 2020), https://ploughshares.ca/pl_publications/protecting-humans-on-earth-from-war-in-space/.
40 United Nations Office for Outer Space Affairs, *European Global Navigation Satellite System and Copernicus: Supporting the Sustainable Development Goals* (Vienna: United Nations, 2018), https://www.unoosa.org/res/oosadoc/data/documents/2018/stspace/stspace71_0_html/st_space_71E.pdf.

Expanding the Military Base in Space

"Peaceful" military use of space has expanded dramatically. Today, both dedicated and dual-use civilian satellite systems are used by modern militaries all around the world. Of the more than 3,000 active satellites, the Union of Concerned Scientists identifies 501 as having military functions.

Almost the same number of satellites are listed as "government" and may have both civil and military functions. China identifies almost all its 103 satellites as "government". At least 29 states operate satellites with dedicated military or dual-use functions: Australia, Belgium, Brazil, Canada, Chile, China, Colombia, Denmark, Egypt, France, Germany, Greece, India, Iran, Israel, Italy, Japan, Luxembourg, Mexico, Qatar, Russia, South Africa, South Korea, Spain, Sweden, Turkey, United Arab Emirates, the United Kingdom, and the United States.[41]

At the same time, warfighting on Earth has become more automated and more remote. Operation Desert Storm, the combat phase of the first Gulf War in 1991, is popularly known as both the first "information war" and the first "space war", demonstrating the ability of militaries to deploy space-based assets to directly control and enable the use of force throughout the world. Today that ability has morphed into a seamless, constant, real-time process. Among the most advanced militaries, space-based data and services are now integrated into almost every military function, including precision targeting as well as command, control, communications, computers, intelligence, surveillance, and reconnaissance (C4ISR) capabilities. Space is now essential to enabling military activity across all other domains.[42]

Upending the Balance: Warfighting in Space

Today, outer space is populated with valuable and vulnerable military targets. Satellites are easily identified and travel with minimal protection in predictable orbits; they use the electromagnetic spectrum and computer networks to send and receive information; and they are difficult to move

---

41 Union of Concerned Scientists, "Satellite Database".
42 "Military Uses of Outer Space", Space Security Index, updated November 2020, http://spacesecurityindex.org/wp-content/uploads/2020/11/IssueGuide_Military-Uses.pdf.

out of harm's way. In response, anti-satellite capabilities on Earth and in space are expanding.[43]

Thinking about space from the perspective of warfighting is not new.[44] Some analysts argue that the rhetorical and operational support for "peaceful uses" of outer space has always been illusory.[45] A belief in outer space as the "ultimate high ground" has persisted, encouraging the contemplation of such new weapons as space lasers, fighter satellites, space-based interceptors, and space planes. None of these systems have so far been realized, because of real-life limits imposed by the laws of physics, financial barriers, and a lack of strategic necessity. But while warfighting *from* space has been seriously contemplated, warfighting *in* space has remained in check, governed by the delicate ethical balances embedded in the OST.

But the ability to balance military capabilities with peaceful uses is eroding as a new era of space activities emerges. In August 2019, the U.S. Space Command was re-established as a geographic combatant command. Constituted as the sixth branch of the U.S. military in December 2019, the U.S. Space Force is tasked with missions and operations in outer space. And in 2020 the United States published its intention to engage in offensive space operations.[46]

The United States is not alone in focusing on warfare in outer space. In recent years, both Russia and China have reorganised military units that incorporate space into more traditional warfighting functions. More recently, the United Kingdom, France, India, and Japan took steps to create new military units, commands, or departments that incorporate active military capabilities in space. In 2019, the North Atlantic Treaty Organization (NATO) declared space an "operational domain", underlining a need to protect civilian and military assets in space. These moves have been controversial, not least because they call into question the future of peaceful uses of outer space.

---

43 Brian Weeden and Victoria Samson, eds, "*Global Counterspace Capabilities: An Open Source Assessment*" (Secure World Foundation, April 2020), https://swfound.org/media/206970/swf_counterspace2020_electronic_final.pdf; Jessica West ed., *Space Security Index 2019* (Project Ploughshares, 2019), http://spacesecurityindex.org/ssi_editions/space-security-2019/.

44 Bleddyn Bowen, *War in Space: Strategy, Spacepower, Geopolitics* (Edinburgh: Edinburgh University Press, 2020).

45 Robin Dickey, "The Rise of Fall of Space Sanctuary in U.S. Policy", Aerospace Corporation (1 September 2020), https://aerospace.org/paper/rise-and-fall-space-sanctuary-us-policy.

46 United States Space Force, *Spacepower*, Space Capstone Publication (June 2020), https://www.spaceforce.mil/Portals/1/Space%20Capstone%20Publication_10%20Aug%202020.pdf.

According to an article in 2020 by Tom Ayres, the General Counsel of the Department of the U.S. Air Force, the U.S. Space Force tweeted that "the Outer Space Treaty does not limit how states organize military forces. It does mandate there will be no weapons of mass destruction or military bases on celestial bodies – neither of which are implied by the creation of the U.S. Space Force".[47] However, while it is reassuring to know that the United States intends to abide by the specific tenets of the OST, there is significant space between the letter of the treaty and the spirit of peaceful – at the very least non-aggressive – uses of space.

Implications for Civilians

Many military space systems are multiuse and support civilian functions, while many commercial systems have military clients. Some States operate only a few satellites, which must meet military, government, and civilian needs.

There is a significant risk that civilian and humanitarian space functions will be inhibited or lost if space becomes a warfighting domain. Civilian GPS signals are already a target of hostile forces, even during peacetime. For example, Russia has been accused of deliberately interfering with GPS signals in Finland and Norway, threatening the safety of passengers and crew on local airlines.[48] Such interference, while targeted and temporary, is still dangerous. A greater use of force against critical military systems could be devastating.

The outer-space environment challenges any attempt to target only military targets. Beyond the multiple uses of many satellite systems, the domain itself is multiuse. The portions nearest to Earth are crowded with military, civilian, and commercial satellites. There is no separate military zone. Warfighting in space is akin to urban warfare.

Environmental Risks

The risks to the global community posed by warfighting in space are compounded by the fragile and unprotected space environment. We have already witnessed the indiscriminate effects posed by using weapons in

---

[47] United States Space Force, Tweet, 27 January 2020. https://twitter.com/SpaceForceDoD/status/1221831921581223939.
[48] "Russia Suspected of Jamming GPS Signal in Finland", *BBC* (12 November 2018), https://www.bbc.com/news/world-europe-46178940.

space. In 1958, the United States secretly conducted the first nuclear tests in space, which culminated in the Starfish Prime test of a 1.4-megaton hydrogen bomb that disrupted radio transmissions in California and Australia, contaminated the Van Allen belts around Earth with additional radiation,[49] and disabled at least six satellites.[50]

Today, debris is the most pressing concern. Anything that is sent into space stays there. And when those objects break apart, the clouds of bits of debris that they create also stay there. These bits can then collide with other objects in space, creating a cascade of damage that not only harms other satellites, but makes surrounding orbits unusable.

While accidental collisions can and have occurred, the intentional destruction of objects is a key source of contamination.[51] China's anti-satellite test in 2008 created the largest debris field to date.[52] And all the pieces are still up there in space.

Escalation and Nuclear Catastrophe

More aggressive uses of outer space also threaten the military stability that emerged in the early years of the space age. Much of the restraint then was rooted in fear of nuclear war. Today, that fear could be realized. Effective command and control of nuclear weapons systems runs through space; both functions are threatened by the growing acceptance of space as a legitimate domain of warfighting. The space assets that control these functions could be deliberately targeted or accidentally hit during combat, setting off nuclear devastation.

The notion of extended deterrence heightens the risk because it operates on the principle that any suspected interference with nuclear capabilities will increase the chances of nuclear retaliation. This is alluded to in the U.S. 2018 Nuclear Posture Review, which for the first time since the end of World War Two extended the conditions under which the United States might use nuclear weapons to include "significant non-nuclear

---

49 The Van Allen belt is a radiation belt of charged particles around the Earth that damages electronics on spacecraft and human spaceflight vehicles that must travel through it.
50 Moltz, 67, 119.
51 Laura Grego, "Why we need to avoid more anti-satellite tests", *SpaceNews* (16 April 2019), https://spacenews.com/why-we-need-to-avoid-more-anti-satellite-tests/.
52 T.S. Kelso, "Chinese ASAT Test", CelesTrack (22 June 2012), https://celestrak.com/events/asat.php.

strategic attacks".⁵³ As recently as April 2020, Assistant Secretary of State for International Security and Nonproliferation Christopher Ford, stated that U.S. space-based command, control, and communications capabilities for both nuclear and non-nuclear forces fall under this condition.⁵⁴ But the United States has hundreds of military satellites and also relies on many foreign and commercial satellites for essential military services. It is far from clear which are considered strategic, or what constitutes an attack. The opportunity for misperceptions and unintended escalation of conflict in this situation is truly terrifying.

A military strategy that views space as an operational domain of warfighting is a recipe for an arms race in space, evidence of which can already be seen. While perhaps not a violation of the letter of the OST, it is a violation of the ethical principles that inspired it.

A Return to Restraint

It is imperative that diplomats, defence strategists, policymakers, and military operatives find a way to return to the restraint of the Cold War and the balance between military interests and the vision of a global community and humanitarian values that informed outer-space governance. Bilateral agreements provided protections for strategically sensitive capabilities such as Earth observation satellites and set out the means to verify nuclear and conventional arms control agreements. This approach can work again.

Operators must exercise extreme caution when contemplating any use of force. The international community also has a responsibility: to once again nurture norms that restrict the testing and use of anti-satellite weapons. At a minimum, military activities that inflict indiscriminate harm both on Earth and in space, such as the intentional creation of space debris and the provoking of nuclear escalation, must be banned. We must develop protections for critical civilian uses of space.

The universal and humanitarian values developed in the early years of the space age and enshrined in the OST reflect real aspirations

---

53 United States Department of Defense, *Nuclear Posture Review* (February 2018), VI. https://media.defense.gov/2018/Feb/02/2001872886/-1/-1/1/2018-NUCLEAR-POSTURE-REVIEW-FINAL-REPORT.PDF.
54 Christopher Ford, "Whither Arms Control in Outer Space? Space Threats, Space Hypocrisy, and the Hope of Space Norms", U.S. Department of State (6 April 2020), https://www.state.gov/whither-arms-control-in-outer-space-space-threats-space-hypocrisy-and-the-hope-of-space-norms/.

for a future in space that is different from the nationalism, inequality, and violence on Earth. All users of space should work toward the achieving of this aspiration.

References

Bowen, B. *War in Space: Strategy, Spacepower, Geopolitics*. Edinburgh: Edinburgh University Press, 2020.

Bryce Space and Technology, LLC. "State of the Satellite Industry Report." July 2020. https://sia.org/wp-content/uploads/2020/07/2020-SSIR-2-Pager-20200701.pdf.

Cadbury, D. *Space Race*. New York: Harper Collins Publishers, 2006.

"Cairo Assails U.S. on 'Moon.'" *New York Times*, 6 October 1957, 44.

"Calcutta Hears Sphere's Signal." *New York Times*, 6 October 1957, 43.

Chapman, W. "LBJ Seeks U.N. Treaty on Space." *Washington Post*, 8 May 1966, A1.

Crane, Robert D. "Soviet Attitude Toward International Space Law." *American Journal of International Law* (July 1962), 685-723.

"Chicago 'Hams' Active: Illinois State Policy and Others Hear Satellite's Signals." *New York Times*, 6 October 1957, 42.

Dickey, R. "The Rise of Fall of Space Sanctuary in U.S. Policy." Aerospace Corporation. 1 September 2020. https://aerospace.org/paper/rise-and-fall-space-sanctuary-us-policy.

Divine, R. A. *The Sputnik Challenge*. Oxford: Oxford University Press, 1993.

Ford, Christopher. "Whither Arms Control in Outer Space? Space Threats, Space Hypocrisy, and the Hope of Space Norms." U.S. Department of State. 6 April 2020. https://www.state.gov/whither-arms-control-in-outer-space-space-threats-space-hypocrisy-and-the-hope-of-space-norms/.

Government of the United States, "National Astronautics and Space Act of 1958." (1958).

Grego, L. "A History of Anti-Satellite Programs." Union of Concerned Scientists, 2003.

Grego, L. "Why we need to avoid more anti-satellite tests." *SpaceNews*. 16 April 2019. https://spacenews.com/why-we-need-to-avoid-more-anti-satellite-tests/.

Harford, J. *Korolev*. Hoboken: John Wiley and Sons Inc., 1997.

Hitchens, T. "Rushing to Weaponize the Final Frontier." *Arms Control Today* (September 2001).

Iriye, A. *Global Community: The Role of International Organizations in the Making of the Contemporary World*. Berkeley: University of California Press, 2004.

Jessup, P. C. and T. Howard. Taubenfeld. "Outer Space, Antarctica, and the United Nations." *International Organization* (Summer 1959), 363-379.

Kecskemeti, P. "Outer Space and World Peace," in Goldsen, J. ed., *Outer Space and World Politics*. Westport: Frederick A. Praeger, Publisher, 1963.

Kelso, T.S. "Chinese ASAT Test." CelesTrack, 22 June 2012. https://celestrak.com/events/asat.php.

Kilpatrick, C. "60 Nations Sign Pact Here on Outer Space." *Washington Post* (28 January 1967), A1.
McCurdy, H.E. *Space and the American Imagination*. Washington: Smithsonian Institution Press, 1997.
McDougall, W. *The Heavens and the Earth: A Political History of the Space Age*. Baltimore: Johns Hopkins University Press, 1997.
Mezerik, A.G. ed. "Outer Space: UN, US, USSR." *International Review Service* Volume VI, No. 56 (1960).
"Military Uses of Outer Space." Space Security Index, updated November 2020, http://spacesecurityindex.org/wp-content/uploads/2020/11/IssueGuide_Military-Uses.pdf.
Moltz, J. C. *The Politics of Space Security*. Stanford: Stanford University Press, 2008.
"Next Step in Space." *New York Times*, 28 January 1967, 22.
Plumb, R. K. "New Space Conquest Can Now be Foreseen." *New York Times*, 6 October 1957, 197.
Preston, B. et al., *Space Weapons: Earth Wars*. RAND, 2002.
"Russia Suspected of Jamming GPS Signal in Finland." *BBC*. 12 November 2018. https://www.bbc.com/news/world-europe-46178940.
Schelling, T. "The Military Use of Outer Space: Bombardment Satellites," in Goldsen, Joseph ed., *Outer Space in World Politics*. Frederick A. Praeger, Publisher, 1963, 97-113.
Sheehan, M. *The International Politics of Space*. Milton Park: Routledge, 2007.
Stuart, J. "Unbundling Sovereignty, Territory and the State in Outer Space," in Natalie Bormann and Michael Sheehan eds, *Securing Outer Space* (Milton Park: Routledge, 2009), 15.
"The Implications." *New York Times*, 6 October 1957, 193.
United Nations, "Treaty Banning Nuclear Tests in the Atmosphere, in Outer Space, and Under Water," (1963).
United Nations, "Treaty on Principles Governing the Activities of States in the Exploration and Use of Outer Space, Including the Moon and Other Celestial Bodies," 1967.
United Nations Office for Outer Space Affairs. *European Global Navigation Satellite System and Copernicus: Supporting the Sustainable Development Goals*. Vienna: United Nations, 2018. https://www.unoosa.org/res/oosadoc/data/documents/2018/stspace/stspace71_0_html/st_space_71E.pdf.
Union of Concerned Scientists. "Satellite Database." Last updated August 1, 2020. https://www.ucsusa.org/resources/satellite-database.
United States Department of Defense. *Nuclear Posture Review*. February 2018. https://media.defense.gov/2018/Feb/02/2001872886/-1/-1/1/2018-NUCLEAR-POSTURE-REVIEW-FINAL-REPORT.PDF.
United States Space Force. *Spacepower*. Space Capstone Publication. June 2020. https://www.spaceforce.mil/Portals/1/Space%20Capstone%20Publication_10%20Aug%202020.pdf.
"US Talk of New 'Pearl Harbor.'" *London Times*, 4 November 1957, 10.

Vlasic, I. A. "The Space Treaty: A Preliminary Evaluation." *California Law Review* (May 1967), 507-519.

Weeden, B. and V. Samson, eds. *"Global Counterspace Capabilities: An Open Source Assessment."* Secure World Foundation, April 2020. https://swfound.org/media/206970/swf_counterspace2020_electronic_final.pdf.

West, J. ed. *Space Security Index 2019.* Project Ploughshares, 2019. http://spacesecurityindex.org/ssi_editions/space-security-2019/.

West, J. "Protecting Humans on Earth from Warfare in Space." *Ploughshares Monitor* 41, 4, Winter 2020. https://ploughshares.ca/pl_publications/protecting-humans-on-earth-from-war-in-space/.

Westad, O. A. *The Global Cold War.* Cambridge: Cambridge University Press, 2005.

17

POST-TRAUMATIC STRESS AND MORAL INJURY IN EXTREME REMOTE WARFARE

Jayden Park[1]

This book has asked questions regarding the possible future militarisation of space and of the way international affairs may be affected if the current technological growth in the space domain continues. Militaries must consider the influence that these technological advancements have, not solely on a national political scale, but also on the individuals who will be interacting with this advanced technology. This chapter will investigate the issues raised by the fact that to protect their respective nations and national interests, militaries are at times asked to engage in warfare and kill, a fact which remains true despite any technological advancement that has previously occurred or is yet to occur. There is often a long chain of military decisions leading to the choice to take a life and it is the humans within this decision chain who ultimately accept responsibility for the results of any action or inaction. This chapter will investigate the potential for moral injury or other associated psychological trauma that individuals in this chain may experience when participating in warfare involving the emerging space domain. It will also discuss technological advancements through history, and the trends associated with the potential for psychological impact on combatants as the physical distance in warfare grew.

---

[1] The views expressed are those of the author and do not reflect the official policy or position of the Royal Australian Air Force, the Department of Defence, or the Australian Government.

Defining Moral Injury

Moral injury is commonly misunderstood, and the term is often incorrectly used interchangeably with other conditions such as Post-Traumatic Stress Disorder (PTSD). Psychologist Brett Litz describes moral injury as occurring when the individual perpetuates, fails to stop, or witnesses an act that infringes on the underlying moral framework of the individual.[2] This concept is deeply complex, partially due to the unique nature of an individual's psyche. An individual's unique perception of the world is developed throughout their lifetime and no two individuals share identical moral perceptions. The individualism of a person's moral framework is a factor as to why multiple individuals may have varied responses to similar or even identical situations or incidents.

In comparison, PTSD is defined in the American Psychiatric Association Diagnostic and Statistical Manual of Mental Disorders as a psychiatric disorder that may occur in people who have, directly or indirectly, experienced or witnessed a traumatic event.[3] Tom Frame suggests that moral injury can be independent of PTSD because "an individual can be morally injured without experiencing a traumatic event; and PTSD is not necessarily associated with any affront to moral principles or social conventions".[4]

Moral injury has been associated with military service throughout history. To highlight this association, clinical psychiatrist Jonathan Shay examines the stories of two Greek heroes in his books Achilles in Vietnam[5] and Odysseus in America.[6] Throughout these texts Shay uses the Greek heroes to show that wounds experienced by military personnel have existed throughout history. Litz suggests modern warfare has amplified the ethical challenges faced by combatants.[7] The lasting impacts of moral injury are not widely known, and more research is required to develop an understanding

---

[2] Brett Litz, et al., "Moral injury and moral repair in war veterans: A preliminary model and intervention strategy". *Clinical psychology review*, 29, no.8 (2009): 695–706.
[3] *Diagnostic and Statistical Manual of Mental Disorders: DSM-5*. Arlington, VA: American Psychiatric Association, 2013.
[4] Tom Frame, "Introduction" in *Moral Injury: Unseen Wounds in an Age of Barbarism*, ed. Tom Frame (Sydney: UNSW Press, 2015), 2.
[5] Jonathan Shay. *Achilles in Vietnam: Combat Trauma and the Undoing of Character* (New York: Simon and Schuster, 1994).
[6] Jonathan Shay, *Odysseus in America: Combat Trauma and the Trials of Homecoming* (New York: Simon and Schuster, 2002).
[7] Litz, et al., "Moral injury and moral repair in war veterans: A preliminary model and intervention strategy", 696.

of the impact it may have on individuals. For the purposes of this chapter, military engagements refer to both direct military actions and also support roles such as intelligence. Examples of intelligence gathering offered by the space domain include intercepting electromagnetic signals and optical monitoring.

Stepping Through History: Evaluating the Distance between Combatants

Throughout history, a correlation has been noted between the physical proximity of combatants and the psychological impacts associated with combat.[8] The physical distance between combatants has never been as large as it is in modern warfare, particularly when considering assets that utilise the space domain. By investigating the historical accounts of previous leaps in technology, such as the psychological detachment often associated with artillery combat and the emerging effects of combat through remotely piloted aircraft, the potential impacts of extreme remote warfare through the space domain can be considered. Due to the challenges that would be associated with attempting to make assumptions about what potential impacts are posed by extreme remote warfare, considerations must be given to previous historical leaps to validate any predictions made.

Throughout the book *On Killing*, Grossman describes the internal resistance to killing that exists in all humans; "the simple and demonstrable fact that there is within most men an intense resistance to killing their fellow man".[9] Grossman discusses the contributing factors to this resistance,[10] and also how modern militaries have sought to enable their personnel to overcome this psychological barrier.[11] He explains that this resistance has always existed in humans and is demonstrated throughout history. The factors influencing this inherent resistance have changed through the centuries, with Grossman identifying the relationship between an increasing physical distance between the perpetrator and the victim, and an associated reduction in the resistance to killing.[12] Military historian Gwynne Dyer

---

8 Dave Grossman, *On Killing: The Psychological Cost of Learning to Kill in War and Society* (rev. ed.) (Boston: Back Bay Books, 2009), 99–134.
9 Grossman, 4.
10 Grossman, 38.
11 Grossman, 252–263.
12 Grossman, 97–98.

specifically references the influence that technology has on increasing the physical distance between combatants and also the psychological impact that operating a crew operated weapon has on individuals. Gwynne states that "it is the same pressure that keeps machine gun crews firing – they are being observed by their fellows – but even more important is the intervention of distance and machinery between them and the enemy; they can pretend they are not killing human beings".[13]

It must be made clear that although the term moral injury has gained popularity in recent times, the concept is not unique to modern forms of warfare. It has always existed regardless of the physical distance between adversaries or the domain of warfare. By utilising historical accounts which detail the psychological impacts of killing, I will highlight the relationship between physical distance and the potential for associated psychological trauma which was identified by Grossman and Dyer. By analysing the factors influencing this relationship, a prediction can be made for the potential psychological trauma associated with utilising the space domain in warfare. Future research must be conducted to confirm that the historical trend continues.

The original and most primal form of killing is at the hand-to-hand range without weapons. When a human must look squarely into the eyes of an adversary and not use any tools to inflict damage, the instinctive resistance to killing is maximised.[14] Grossman notes that tools were the first step in increasing physical distance between combatants and that when humans first used tools to kill others they "gained psychological energy and leverage that was every bit as necessary in the killing process".[15]

Once hand-to-hand weapons grew to the size of spears this increased the physical distance between combatants allowing them to maintain a stand-off distance. Many ancient civilisations were able to take advantage of the stand-off distance offered by spears, seen in infantry formations such as the Ancient Greek phalanx.[16] This formation not only offered the physical advantages associated with tactics such as a shield wall but also offered the psychological advantage granted by the introduction of anonymity in combat. A single soldier in the phalanx could not be personally identified, but rather they were a single piece that formed a much larger group. This single, unidentifiable soldier also begins to feel the pressure of conformity

---

13 Gwynne Dyer, *War* (Toronto: Vintage Books Canada, 2005), 119.
14 Grossman, *On Killing*, 131.
15 Grossman, 132.
16 Grossman, 153–154.

granted through mob mentality and the human need to protect those around them.¹⁷

The introduction of the musket saw a new age of warfare begin. Musket and rifle infantry formations granted individuals a form of mob anonymity similar to that associated with spear formations, but individuals were also granted the ability to conceal their actions from those around them. This introduction of anonymity resulted in individual firers not being able to be singularly identified from the rest of the group. For example, in a typical Napoleonic era or American Civil War era formation, it would not be possible to identify who had fired the rounds that killed any individual enemy combatant. This granted individuals the option to not overcome their resistance to killing and either aim away and purposely miss the enemy, or not to fire at all. It must be noted that the musket and rifle range is the furthest distance of non-remote warfare in which the perpetrator could be reasonably expected to see the emotion on their target's face, which is an important psychological factor.

Dyer emphasises the role that physical distance has on the ability of a perpetrator to overcome internal psychological barriers to killing. He states that "[o]n the whole, however, distance is a sufficient buffer: gunners fire at grid references they cannot see; submarine crews fire torpedoes at 'ships' (and not, somehow, at the people in the ships); pilots launch their missiles at 'targets'".¹⁸ Dyer highlights the psychological leverage granted to operators who kill from long-range distances. This range describes warfare conducted beyond the capabilities of a non-specialised rifle, and situations where the perpetrator may no longer be targeting individuals but rather targeting larger groups, large assets, or areas of the battlefield. This subtle yet distinct difference means that operators do not necessarily need to consider the fact that they are killing individual human beings. The other distinct psychological advantage offered by using crew operated weapons is proposed by Grossman as the "diffusion of responsibility".¹⁹ The individual has now become a small part of the larger kill chain. The concept of the kill chain is used to describe the range of decisions and associated processes that must be completed up to the point of taking a life. Ship and associated naval warfare can be similarly described due to individual sailors effectively sharing in the responsibility of the destruction

---

17 Grossman, 154.
18 Dyer, *War*, 119.
19 Grossman, *On Killing*, 253.

of the enemy ships which they target. Air warfare can be similarly considered. Modern aircraft may only have a single pilot who is flying the aircraft and dropping ordinance, however, due to the large kill chain which has supported them up to the point of killing, they can similarly diffuse the responsibility of killing amongst others who have supported them. Grossman states that "death from twenty thousand feet is strangely impersonal and psychologically impotent".[20] Here Grossman identifies the reduced resistance to killing that the perpetrator feels due to the impact of long-range distances.

For the purposes of this chapter, remote warfare is defined as warfare conducted when an operator is engaging with an enemy whilst utilising technologically advanced weapons which do not require the user to be in the same physical location as the weapon itself. The operator is therefore typically utilising the technology using some form of communications link and is situated away from the immediate physical battlefield. Examples of remote warfare include the rise of Remotely Piloted Aircraft Systems (RPAS) which have enabled operators to engage enemies without the need to even leave their home country.

Extreme remote warfare, specifically through the space domain, refers to militaries conducting operations and engaging the enemy in space against other space assets, and the utilisation of space assets to influence and engage in warfare on Earth. By only discussing the psychological effects impacted by the physical and mechanical distance which has been offered by technological advancements, one could potentially make the dangerous assumption that remote warfare is psychologically the easiest form of killing to have occurred throughout history. It is vital to acknowledge the other psychological barriers which exist to the operator. Grossman proposes that other distances, such as cultural and moral, also influence the ability for an individual to see their target as a fellow human being.[21] These distances have fluctuated through history and have not solely increased or decreased consistently. It must be noted that these will not be consistent with all individuals, as some people are potentially more inclined to feel closer or more distant to an enemy based on many personal variables.

Cultural distance refers to the level of empathy that an operator may have on the target's appearance or customs. War-time propaganda has often attempted to maximise these distances through caricature style cartoons

---

20 Grossman, 210.
21 Grossman, 160.

or through mocking cultural traditions of the perceived enemy.[22] In the modern age of globalisation, cultural understanding is high compared to the gulfs of understanding seen during previous conflicts throughout history. This distance, therefore, contributes to the psychological barrier that an operator from an extreme remote distance may experience due to the humanisation of their target.

Moral distance refers to the ability for the operator to feel that their cause for war is morally justified and that their adversary is fighting for unjust or immoral reasons. Jus in Bello must be followed during extreme remote warfare, as in all forms of conflict, to ensure moral distance does not become a propaganda tool for adversaries.[23] It is conceivable that a refusal to uphold a high standard of moral beliefs could result in temporary tactical victories. These victories could come at the detriment of strategic goals and also vastly increase the psychological trauma on all involved. Rob Sutherland highlights this through the historical example of Bomber Command Aircrews fire-bombing the German city of Dresden.[24] Whilst the bombing of Dresden was a tactical success, the timing in the war and the question of necessity prompted the term moral injury to first be applied to the aircrews involved.[25]

A key factor that must be considered when discussing the historical evolution of distance in warfare is recent advancements in digital optics. Whilst extreme remote warfare is greatly increasing the physical distance, modern digital optics allow the operator to see detailed emotions on the face of their target, and intimately observe the devastating effects that their weapon systems have. This level of detail would have otherwise only been possible during hand-to-hand or "spear distance" combat. Expert in military drone operations, Peter Lee, describes these additional considerations for remote warfare and labelled it as the "visual intimacy" that aircraft such as the MQ-9 Reaper offer.[26] He describes the relationship of increasing physical separation whilst also increasing visual monitoring as the "distance paradox".[27] The rise of extreme remote warfare will only

---

22 Emily Robertson, "Atrocity Propaganda and Moral Injury" in *Moral Injury: Unseen Wounds in an Age of Barbarism*, ed. Tom Frame (Sydney: UNSW Press, 2015), 5.
23 Robertson, 8.
24 Rob Sutherland, "Is Moral Injury the Answer?" In *Moral Injury: Unseen Wounds in an Age of Barbarism*, ed. Tom Frame (Sydney: UNSW Press, 2015), 2–3.
25 Sutherland.
26 Peter Lee, The Distance Paradox: Reaper, the Human Dimension of Remote Warfare, and Future Challenges for the RAF. *Air Power Review* 21, no. 3 (2018), 12–14.
27 Lee, 3–6.

emphasise this paradox as assets in the space domain will allow operators to monitor their target for longer periods, through increasingly precise optics and other electromagnetic signal receivers.

Case Study

The potential impact on individuals who operate the advanced technology capable of remote warfare is illustrated by a real-world example of remote warfare enabled through RPAS as detailed in Peter Lee's Reaper Force.[28] In the book, he explores No. 39 Squadron of the Royal Air Force (RAF) and extensively interviews both the operators and the leadership team within the squadron. No. 39 Squadron operates the MQ-9 Reaper RPAS, developed by General Aeronautical Systems. They have operated this aircraft extensively throughout the Middle East region since the re-establishment of the unit on 1 January 2007 and are based in Creech U.S. Air Force Base, Nevada.[29] The MQ-9 can stay airborne for 12-20 hours dependent on its payload and typically operates at 20,000 feet.[30] This operational Squadron has had years of first-hand experience at engaging in remote warfare. The RPAS allows operators to remotely control the aircraft without the need to physically travel to and from the battlefield. They typically operate both out of reach and out of sight of their adversary.

Remote warfare enabled through an RPAS has many similar considerations that are transferrable to extreme remote technologies. These lessons learnt from the short period of experience with remote warfare, involving how to best prepare military personnel before operations and how to consider their mental health during operations, should be transferred as militaries introduce extreme remote warfare operations.

This exploration of No. 39 Squadron provides an insight into the emotions and perspective of the squadron whilst operating remote warfare technology and provides accounts of individual's experiences. One of these accounts is given from the perspective of the MQ-9 operators from Christmas Day in 2014 when the MQ-9 was operating in the vicinity of a friendly Iraqi Army Forward Operating Base (FOB). The Operators spotted an M113 Armoured Personnel Carrier (APC) in the nearby area which was

---

28 Peter Lee, *Reaper Force: Inside Britain's Drone Wars* (London: John Blake Publishing, 2018), 1–9.
29 Lee, 12.
30 Lee, 2.

acting suspiciously. This suspicion was compounded by the reports of a stolen Iraqi APC just days prior. The MQ-9 operators began conducting the appropriate procedures related to gaining approval to target the suspicious APC, ultimately ensuring that they could conduct a legally justifiable strike. The APC began driving aggressively toward the friendly Iraqi FOB and the operators had gained approval to strike. As the final processes of releasing an Air-to-Ground Missile (AGM) 114 Hellfire missile was underway, the legal approval to strike was withdrawn. Someone in the kill chain had requested that an additional check be performed on the identification of the APC. Once this identification had been performed and legal approval re-granted, the operators decided that the APC was too close to the FOB to release a Hellfire without posing a risk to the three Iraqi soldiers manning the entrance. One of the operators claimed that the approval had been re-granted fifteen seconds too late.[31] This lack of ability to strike forced the operators to sit and watch as the APC, which had been loaded with explosives, crashed into the FOB, detonating, and killing the three Iraqi soldiers. The operators watched as the bodies of those in the blast radius burned. After completing several other missions during the same shift, the operators all drove themselves home to celebrate what was left of Christmas Day with their families.

This example provides an insight into the stages of the kill chain that must be satisfied when conducting remote warfare. Due to a lack of appropriate legal authorisation, the operators failed to act in a situation that directly resulted in the deaths of allied personnel. This situation can be classified as having a high potential for moral injury, due to both the inability to act in a situation the operators felt they could have prevented, and the potential to feel that the organisation which they trusted had failed them. The operators understood the importance of the legal confirmation of targets but also understood that direct action could have prevented the loss of the lives of friendly forces. Shay[32] and Litz[33] have both proposed these as potential reasons for why moral injury occurs. Some of the considerations associated with unique features of remote warfare, such as the ability to monitor targets for long periods, may be amplified with the transition to extreme remote operations. Through critical analysis of lessons learnt from past operations, military organisations could be better prepared in the

---

31 Lee, 189.
32 Shay, *Achilles in Vietnam* and *Odysseus in America*.
33 Litz, "Moral Injury and Moral Repair in War Veterans"

future to handle morally ambiguous situations both in real-time and after the engagement when supporting operators. Better preparing operators and leaders for the moral dimension of warfare could result in quicker decisions with less potential for ethical ambiguity.

Hypothetical Case Study

A realistic example of a potential extreme remote engagement could involve observations of an adversary's patterns of life from an imaging satellite. Imagine a hypothetical satellite imagery analyst sitting in a room with no windows but numerous computer monitors which provide consistent monitoring over an area of interest. This analyst's role is to monitor the patterns of life of a potential target and provide reports to their Intelligence Officer who packages this information to be delivered to another joint military service or other friendly intelligence organisations. The analyst is allocated an individual of interest.

Over the course of a week, the analyst will become closer with this individual than almost anyone else in their own life. They will monitor all movement potentially provided from many space assets included optical monitoring, dependent on satellite availability, as well as interception of electromagnetic signals dependent on enemy emissions. This target goes about their life with no knowledge that in a distant land, the analyst is monitoring them, recording their movements in detail across numerous reports. The analyst is experienced at their job and has developed the organisational understanding to know that their reports, once compiled with large sums of other sources of intelligence, may contribute to direct military action being taken against the human they have been observing. After some time the analyst can confirm that the individual is alone multiple times a day. They confirm the times, locations, and surrounding environment, including infrastructure and other personnel who may be nearby. Along with other intelligence and appropriate military justifications, the analyst understands that their reporting will directly contribute to the military justification to conduct a direct-action engagement on the individual. The ability for operators to monitor their target in this way has never been achievable from other forms of surveillance. This is now only possible with vast arrays of satellites and other remote warfare technology.

The analyst may potentially develop some form of emotional connection to the individual they are monitoring. This connection may become extremely negative due to acts witnessed, or the knowledge of previous acts committed by the individual being tracked. The analyst may potentially see a humane side to this individual, as some level of empathy will most likely be generated when observing targets. Regardless, the fidelity offered by high-resolution cameras and the ability to track human patterns utilising extreme remote technology has never previously been achieved in warfare. The analyst may suffer a moral injury due to their actions or inactions, or through the feeling of betrayal by a trusted individual or organisation. It is clear that although the analyst is merely a single source of intelligence for the wider decision-making loop that occurs, questioning of the analyst's moral framework will most likely occur.

The Responsibilities of Future Leaders

Current and future leaders must have an understanding of the type of warfare they are engaging in, and the potential psychological costs typically associated with warfare. If a leader fails to understand the impacts posed to military personal before conducting warfare, the leader is failing to meet the moral responsibilities inherent with military command. With the increasing use of multiple intelligence sources, including space assets, military personnel will be expected to navigate morally ambiguous scenarios which occur at distances never previously experienced in warfare. These ambiguous scenarios must be navigated and internally justified both during the action itself and after the individuals are removed from the situation. A leader must ensure the moral ambiguity of situations is minimised and to minimise the ambiguity of actions taken by both themselves and those they lead. Once removed from an immediate situation, individuals must develop an understanding of their role in the ambiguous scenario. Completed tasks must be tied to an individual's beliefs and values to reinforce the necessity and moral implications of any actions taken or not taken.

Military personnel go forth and engage in acts of warfare on behalf of their State. Therefore, the State must accept responsibility for both the physical and psychological welfare of these trusted individuals.[34] Leaders within the military similarly accept this responsibility. This responsibility exists regardless of the domain of warfare, or the physical distance between

---

34 Nikki Coleman, "Moral Status and Reintegration" In *Moral Injury: Unseen Wounds in an Age of Barbarism*, ed. Tom Frame (Sydney: UNSW Press, 2015), 3.

combatants. Considering the moral dimension of wellbeing is as important for an Infantry Platoon Commander as it is for a Cyber Warfare Officer or any space domain job role, even those which may not exist yet. To not consider the lasting psychological impacts on individuals engaging in conflict would be a disregard of duty, regardless of rank, job role, or domain of warfare.

The future of extreme remote warfare must be met with an appreciation for the psychological barriers that exist within humans. To effectively engage an adversary using the space domain, similar to the other domains, the moral dimension of warfare must be considered. It is the individuals operating the advanced technologies who implicitly accept responsibility for what their State asks them to do. As more research is conducted into the impact of post-traumatic stress and moral injury, psychological barriers influencing military individuals must continue to be addressed.

References

Coleman, N. "Moral Status and Reintegration" in *Moral Injury: Unseen Wounds in an Age of Barbarism*, edited by Tom Frame. UNSW Press, 2015.

Diagnostic and Statistical Manual of Mental Disorders: DSM-5. Arlington, VA: American Psychiatric Association, 2013.

Dyer, G. *War*. Vintage Books Canada, 2005.

Frame, T. "Introduction" in *Moral Injury: Unseen Wounds in an Age of Barbarism*, edited by Tom Frame, UNSW Press, 2015.

Grossman, D. *On Killing: The Psychological Cost of Learning to Kill in War and Society* (rev. ed.) Back Bay Books, 2009.

Lee, P. "The Distance Paradox: Reaper, the Human Dimension of Remote Warfare, and Future Challenges for the RAF." *Air Power Review* 21, no. 3 (2018), 106–130.

Lee, P. *Reaper Force: Inside Britain's Drone Wars*. John Blake Publishing, 2018.

Litz, B., Stein N., Delaney E., Lebowitz L., Nash W., Silva C., and Maguen S. "Moral Injury and Moral Repair in War Veterans: A Preliminary Model and Intervention Strategy." *Clinical Psychology Review* 29, no. 8 (2009): 695–706.

Robertson, E. "Atrocity Propaganda and Moral Injury" in *Moral Injury: Unseen Wounds in an Age of Barbarism*, edited by Tom Frame. UNSW Press, 2015.

Shay, J. *Achilles in Vietnam: Combat Trauma and the Undoing of Character*. Simon and Schuster, 1994.

Shay, J. *Odysseus in America: Combat Trauma and the Trials of Homecoming*. Simon and Schuster, 2002.

Sutherland, R. "Is Moral Injury the Answer" in *Moral Injury: Unseen Wounds in an Age of Barbarism*, edited by Tom Frame. UNSW Press, 2015.

18

WHAT WE HAVE IS WHAT WE BRING THERE

Security in Space as Utopian Vision

Evie Kendal

Introduction

In the final week of October 2020, billionaire tech mogul and C.E.O. of private space company SpaceX, Elon Musk, made tabloid headlines for releasing a terms of service statement suggesting his proposed Mars colony would not recognise Earth-based laws or sovereignty. Instead, the document declares Mars a "free planet", claiming that any disputes arising would be settled according to the Mars colony's own "self-governing principles, established in good faith, at the time of Martian settlement".[1] While legal scholars have scoffed at this bold declaration and cited numerous international laws to the contrary,[2] this chapter will consider Musk's comments not in the light of what the law *is* regarding governing an independent Mars colony, but rather what it *should* be. Drawing from both real-world examples of enforcing law and order, and speculative fictions including representations

---

[1] Stacey Liberatore, "SpaceX Declares Independence: Elon Musk's Firm Says They Will Not Recognize Earth Law in Planned Mars Colony and Says 'Free Planet' Will Adopt 'Self-Governing Principles'", *Daily Mail*, October 31, 2020. https://www.dailymail.co.uk/sciencetech/article-8897601/Elon-Musks-SpaceX-says-not-recognize-Earth-laws-planned-Mars-colony.html.

[2] Cristian van Eijk, "Sorry, Elon: Mars is Not a Legal Vacuum – and It's Not Yours, Either", *Völkerrechtsblog: International Law & International Legal Thought*, November 5, 2020. https://voelkerrechtsblog.org/sorry-elon-mars-is-not-a-legal-vacuum-and-its-not-yours-either/; Mike Brown, "SpaceX Mars City: Legal Experts Respond to 'Gibberish' Free Planet Claim", *Inverse*, November 4, 2020. https://www.inverse.com/innovation/spacex-mars-city-legal.

of policing and military action in space, this chapter will argue that a Mars colony represents a unique opportunity to trial alternative legal systems that prioritise mental health and social work perspectives for promoting social harmony. For the purposes of this discussion this chapter will assume a single Mars colony with members from multiple countries on Earth, although many of the issues identified would also be relevant if multiple colonies were to form, possibly increasing the risk of inter-colony conflict.

In 2020 there were a number of global protests against police brutality, particularly acts targeting people of colour. In the U.S., Native Americans and black citizens have the highest rates of death caused by police officers, however, convictions for murder and manslaughter are rare.[3] Some sources cite as many as 1,134 young black men died in fatal police shootings in 2015, with zero convictions on record for the police officers involved.[4] In Australia, a Royal Commission was established in 1987 to investigate the disproportionate number of Indigenous deaths occurring in police custody, and while there have been improvements since the Commission's report and recommendations were released in 1991, by mid-2020 there had been at least 437 more Aboriginal and Torres Strait Islander deaths in custody.[5] In response to these and other damning figures, the *Black Lives Matter* movement is increasing in popularity, with some supporters arguing a more effective way of promoting peace in community is to consider criminal behaviour as a social problem, often exacerbated by mental health issues and disadvantage. Interventions against anti-social behaviour, it is purported, can best be managed through the creation of multi-disciplinary teams including health professionals and social workers, rather than solely having armed police officers as first responders. When traditional law enforcement is seen to be engaging in "predatory policing" of black communities and disproportionately arresting people of colour, this leads to a situation where crimes are not reported, and the police are seen as perpetuating injustice, rather than protecting the safety and rights of all

---

3 Shakira A. Kennedy, Folusho Otuyelu and Warren K. Graham, 'To Protect and Serve: Examining Race, Law Enforcement Culture and Social Work Practice' in eds. Sandra E. Weissinger and Dwayne A. Mack, *Law Enforcement in the Age of Black Lives Matter: Policing Black and Brown Bodies* (Lanham: Lexington Books, 2018), 153.
4 Kennedy et al., "To Protect and Serve".
5 Lorena Allam, Calla Wahlquist and Nick Evershed, "Aboriginal Deaths in Custody: Black Lives Matter Protests Referred to Our Count of 432. It's Now 437", *The Guardian*, June 9, 2020. https://www.theguardian.com/australia-news/2020/jun/09/black-lives-matter-protesters-referred-to-our-count-of-432-aboriginal-deaths-in-custody-its-now-437.

citizens.⁶ However, while many acknowledge the problems with current law enforcement methods, particularly as they relate to minority populations, there is disagreement regarding how to address these, with concerns that "softer" alternatives may not serve as effective deterrents against crime.⁷

The so-called "defund the police" campaign, which focuses on reducing investment in police forces and redirecting funds to crime prevention, education and community outreach,⁸ has faced opposition on the grounds that it cannot be proven such an approach would be safe to employ on a large scale. Spurred on by the shocking video footage showing the death of African-American man, George Floyd, at the hands of police in Minneapolis in May 2020, this movement claims the over-militarisation of police forces has created a situation where officers are not held accountable for brutality, included racially-motivated acts of violence.⁹ Supporters claim law enforcement systems need radical change, rather than continuing with incremental, superficial adjustment to current practices, with sociologist Alex S. Vitale noting diversity training and other policing reforms have heretofore shown minimal impact, as institutional pressures and prejudices remain.¹⁰ This highlights the unique opportunity designing a law enforcement system for a new Mars colony represents, as there are fewer obstacles to establishing a new form of policing in the absence of a status quo and existing institutional culture. Thus, the new Mars settlement could serve as a test site for exploring alternative police and military systems, including those proposed by the *Black Lives Matter* and other socially progressive movements.

There is not currently a Mars-based military or police force, so it is important to remember that whatever we may have in that regard in the future is what we have *chosen* to bring there. In other words, avoiding the injustices currently facing some populations on Earth from being replicated on another planet, and challenging the potentially negative influences of unfettered commercial interests imposing on legal systems, will require a conscious effort and careful consideration of ethical law enforcement practices. This chapter will first consider some of the sources of conflict we

---

6 Allum et al., "Aboriginal Deaths in Custody".
7 Stephen Rushin and Roger Michalski, "Police Funding", *Florida Law Review* 72, no. 2 (2020): 277.
8 James Purtill, "'Defund the Police': What it Means and How It Would Work", *ABC Triple J Hack*, June 9, 2020. https://www.abc.net.au/triplej/programs/hack/what-it-means-to-defund-the-police-and-how-it-would-work/12336014.
9 Purtill, "Defund the Police".
10 Alex S. Vitale, *The End of Policing* (London: Verso, 2017), 8.

can expect on a new human settlement, before exploring various alternative methods for promoting peace on the colony, using fictional representations as illustrations for how these may look in the future.

Conflict on the Colony

It is an indictment of our species that it is safe to assume if there are humans on Mars there will be conflict on Mars also. While the nature of this conflict is yet to be seen, the need for some form of legal system to protect colonists and space assets is certain. This is partly due to the intense stress new settlers will experience living off-world, which some astrosociologists and philosophers claim will increase the likelihood of interpersonal tension and aggression.[11] While it is expected those applying to emigrate to Mars will undergo some form of psychological screening, it is also anticipated that cultural, political and religious differences between colonists may lead to disagreements, especially when considering the lack of privacy that living in close quarters with a small group of people long-term is likely to entail.[12] Konrad Szocik *et al.* note that "effective law is a regulator of human conflicts", but that terrestrial legal systems are unlikely to be fit-for-purpose on a Martian settlement.[13] These authors claim that life on Mars for early colonists will be "characterized by high levels of stress and existential anxiety and uncertainty",[14] which may leave settlers vulnerable to exploitation by those with powerful interests who may wish to usurp authority if they perceive a power vacuum.[15]

In his guide for future space tourists and migrants, Neil F. Comins claims that despite the fact there will be a multitude of personal and cultural differences between colonists, each member of a proposed space exploration and settlement team will need to be able to demonstrate that they can coexist peacefully with others.[16] He notes group dynamics will be

---

11 David Lempert, "Living in Space: Cultural and Social Dynamics, Opportunities, and Challenges in Permanent Space Habitat", *Astropolitics* 9, no. 1 (2011): 84-111; Konrad Scozik, Kateryna Lysenko-Ryba, Sylwia Banaś, Sylwia Mazur, "Political and Legal Challenges in a Mars Colony", *Space Policy* 38 (2016): 27-9.
12 Neil F. Comins, *The Traveler's Guide to Space: For One-Way Settlers and Round-Trip Tourists* (New York: Columbia University Press, 2017), 109.
13 Scozik et al. "Political and Legal Challenges", 28.
14 Scozik et al. "Political and Legal Challenges".
15 Comins, *The Traveler's Guide*, 121.
16 Comins, 112.

influenced by differences in social expectations, affiliations, needs, tastes and interests, all of which can promote fascinating philosophical discussions while also fueling "initial discomfort, disagreements, resentment, animosity, and other tensions".[17] It is not difficult to predict how such challenges may lead to criminal outbursts for people living under extreme stress, both physically and psychologically. Comins notes the three main sources of stress for humans arise from their "interpersonal relationships, organizational activities, and interactions with the physical world", all three of which would be radically different on a Mars colony and significantly more dangerous than on Earth.[18] Consider, after an interpersonal conflict there may be no quiet place where two arguing colleagues can take refuge and calm down without risking running into each other, and the work of surviving the harsh Mars environment with limited resources is likely to occupy a lot of the colonists' time and mental energy. David Lempert further notes that views on privacy, human rights and acceptable uses of science and technology vary widely across individuals and cultures, even limited to those that currently represent major space powers, for example, China, Japan, the United States (U.S.), and Russia, to name a few.[19] As more countries and corporations extend their reach to the stars, such differences are likely to multiply, possibly leading to more conflicting values and tension.

As on Earth, a likely source of conflict on a Mars colony will relate to the use and ownership of limited natural resources. Private corporations and national organisations operating off-world will no doubt seek methods of protecting their space assets that require a robust legal system for enforcement. In other words, a space police force to protect against theft and other crimes involving property and personnel. The potential economic value of space exploration is vast, with scientists predicting asteroids rich in platinum, gold and titanium, among other minerals, could be worth billions or even trillions of dollars, assuming mining and transportation systems become logistically and financially feasible.[20] While the Outer Space Treaty (OST) of 1967 contains the principle of non-appropriation for space and

---

17 Comins, 119.
18 Comins, 122.
19 Lempert, "Living in Space", 96.
20 Namrata Goswami, "China in Space: Ambitions and Possible Conflict", *Strategic Studies Quarterly* 12, no. 1 (2018): 75.

celestial bodies,[21] it is less clear regarding the proper distribution of space-based resources, which may lead to property disputes. Space programs cost some governments billions of dollars annually, so it is reasonable to expect there is political appetite to recoup some of these funds through resource extraction or other means. The U.S. and Luxembourg have further encouraged private space industry by legislating in favour of private ownership of resources extracted through asteroid mining.[22] It is currently unclear whether other countries will follow suit, but regardless of national views and laws on the subject, it is likely individuals on a future Mars colony will have differences of opinion regarding the correct interpretation of the OST and other legal documents regarding space resources and property rights. And despite the claims made in Starlink's terms of service, private companies like SpaceX are bound by the Treaty since the U.S., as its State of incorporation, is a party to it and therefore required to supervise SpaceX's activities to ensure compliance with the Treaty and other relevant international laws.[23]

Another indication that legal conflict is likely to manifest on Mars is related to the increasing nationalism and territorialism seen among major space powers with regards to space security. It is perhaps not surprising that a lot of space-based fiction is focused on interplanetary war, with different factions fighting over limited resources and territory. Lempert predicts that the cultural groups that currently dominate Earth will also come to dominate in space colonisation efforts, "driven by the competition for resources off the planet and the substantial investment costs required in accessing them".[24] He further argues that international treaties have done little to protect diversity and sustainability on Earth, so are unlikely to be sufficient to protect colonists' rights and resources, particularly in the absence of systems of reliable oversight and enforcement.[25] James Clay Moltz suggests there has been a shift toward a more nationalistic U.S. military space policy following the terrorist attacks of 9/11 and increasing

---

21 United Nations Office for Outer Space Affairs (UNOOSA), *Treaty on Principles Governing the Activities of States in the Exploration and Use of Outer Space, Including the Moon and Other Celestial Bodies*, New York: United Nations, 1967. https://www.unoosa.org/oosa/en/ourwork/spacelaw/treaties/introouterspacetreaty.html.
22 Goswami, "China in Space", 75.
23 UNOOSA, *Treaty*, article VI.
24 Lempert, 'Living in Space', 90.
25 Lempert, 91.

suspicion regarding China's ambitions in space.[26] He claims that over the period 2001-2008, U.S. policies began to "reject negotiated forms of security" in favour of ones based on U.S. military strength.[27] The founding of the United States Space Force (USSF) as an independent armed service in 2019, and previous attempts to establish a U.S. Space Corp, continued this trend.[28] This led other countries to consider their own military assets in space and announce plans for equivalent forces,[29] while political scientists considered the ramifications of this increasing weaponisation of space for security, and whether such actions represented a violation of the OST.[30] Some have suggested so-called "middle range space powers", such as Australia and Canada, may have a vital role to play in mediating conflicts regarding space warfare and weaponisation between long-time rival countries, such as the U.S., Russia and China.[31]

Moltz notes that while the U.S. appears to be responding to a perceived security threat in China's increasing presence in space, the European Union has "readily embraced Beijing" and promoted space cooperation.[32] China's stated intentions for their space program include research, resource extraction, power generation, and long-term economic development, as well as a desire to develop effective planetary defense mechanisms against impact hazards and other threats.[33] According to Namrata Goswami, "China's space ambitions are unique and have the full backing of the Community Party of China", however, there are concerns these ambitions may include "territorial assertion" over regions and resources in space, particularly when China reaches them first.[34] There were also fears of hostile intent when China's *BX-1* pico-satellite came within 46 kilometres of the International Space Station in 2008.[35] In addition, the potential for military advantage in space to promote the same

---

26 James Clay Moltz, "Renewed U.S. Space Nationalism 2001-2008", in *The Politics of Space Security: Strategic Restraint and the Pursuit of National Interests* 2nd ed. (California: Stanford University Press, 2011), 260-1.
27 Moltz, "Renewed U.S. Space Nationalism 2001-2008", 283.
28 Brent Ziarnick, "The Coming Revolution in Military Space Professionalism", *Air & Space Power Journal* 32, no. 2 (2018): 12.
29 Chelsea Gohl, "Everyone Wants a Space Force – But Why?" *Space.com*, September 11, 2020. https://www.space.com/every-country-wants-space-force.html.
30 Joan Johnson-Freese and David Burbach, "The Outer Space Treaty and the Weaponization of Space", *Bulletin of the Atomic Scientists* 75, no. 4 (2019): 137-41.
31 Johnson-Freese, "The Outer Space Treaty".
32 Moltz, "Renewed U.S. Space Nationalism", 290.
33 Goswami, "China in Space", 80-5.
34 Goswami, 77.
35 Moltz, "Renewed U.S. Space Nationalism", 302.

in terrestrial warfare, for example, through the use of space weaponry and targeting enemy satellites and communication systems, has been the cause of concern for U.S. space policymakers over the last two decades of Chinese expansion in the space arena.[36] While the U.S. has been reluctant to engage China as an international space power, it has continued to engage strong collaborations with other nations, including Australia.[37] Moltz claims at the heart of this rejection of China as a full partner in space endeavours are underlying concerns about cultural differences between the two nations, particularly with regards to human rights.[38] This again highlights how national approaches to law and order, property rights and territory may cause friction in a future Mars colony. As Szocik *et al.* note, unless colonists and their spacecraft are considered to have "international status" different laws may be applied to different members, and it will be unclear what jurisdictional authority will prevail on the colony itself.[39]

A Martian settlement will represent a radical departure from human existence up until this point in history. As Comins notes, new methods of agriculture, mining, construction, transportation, communication, and all social activities will need to be developed to sustain life on the colony.[40] Given the various sources of interpersonal conflict, resource allocation dilemmas and territorial disputes likely to arise in a new Mars colony and the need to protect the welfare of vulnerable settlers, it is proposed a new method of policing will likewise need to be developed that will be adapted to the unique challenges of living off-world. As such an undertaking has not been attempted before, this is where science fiction and utopian studies can contribute to policy discussions regarding the future of human space exploration.

Space Security as Utopian Vision

The criticisms facing the *Black Lives Matter* and "defund the police" campaigns could also be levelled at any similar attempts to promote a more holistic approach to law enforcement on a Mars colony, prioritising

---

36 Dean Cheng, "China's Military Role in Space", *Strategic Studies Quarterly* 6, no. 1 (2012): 63.
37 John E. Hyten, "Challenges in Military Space Acquisition", *Hampton Roads International Security Quarterly* 11, no. 3 (2011): 48.
38 Moltz, "Renewed U.S. Space Nationalism", 276.
39 Szocik et al. "Political and Legal Challenges", 28.
40 Comins, *The Traveler's Guide*, 233.

mental health and social work perspectives over punitive measures and incarceration. However, one difference is a Mars colony could serve as a controlled environment for testing new methods of policing in a newly constructed society, where assumptions about the role of law enforcement may be more easily displaced. Representing human existence not as it is, but as it could be, is one of the major strengths of science fiction, as Dennis Livingston asserts:

> Science fiction is an important source for probing the interactions of science, technology, and society. First, it is a laboratory of the imagination, offering the author a framework for carrying out a series of thought-experiments on the consequences of current or foreseeable trends and events. …Second, many science fiction stories, though ostensibly set in the future or on other planets, are commentaries on our own society. With such tools as satire and metaphor to distance the reader from her own time and place, science fiction holds a mirror up to the present, reflecting in distorted or exaggerated form familiar human foibles. As with all good literature, the best science fiction provides us with new perceptions or questions to ask about the conventional wisdom.[41]

While we do not have real-world examples of successful or unsuccessful interplanetary police force structures, we do share a wealth of fictional explorations of the same. Science-fictional and utopian accounts of future human societies can help conceptualise different ways of organising structures we take for granted in our current reality. Through abstraction these stories can destabilise common beliefs about the ways things have to be, and in the process become powerfully persuasive tools for shaping what could be. What follows are some brief examples of possible policing structures for a Mars colony and some descriptions from science fiction and utopian narratives regarding what these might look like. Recognising such representations are often sensationalised for entertainment purposes, there are still many lessons that can be learned from engaging with fictional human societies, and the promises and pitfalls of alternatives methods of promoting law and order.

---

41 Dennis Livingston, "The Biology of Utopia: Science Fiction Perspectives on Ectogenesis", in eds. Helen B. Holmes, Betty B. Hoskins and Michael Gross, *The Custom-made Child? Women-centred Perspectives* (New Jersey: Humana Press Inc., 1981), 281.

Earth Cops 2.0: Different Laws for Different Regions?

The first method for establishing a police and military force on Mars is to replicate Earth systems and simply transplant these into space. As noted above, this is likely to lead to jurisdictional confusion, with different laws being applied to different colonists based on their country of terrestrial origin. While this is similar to the current situation on Earth, the confined space of an off-world colony might make such different applications of law untenable, especially if they relate to fundamental differences in the rights of some citizens, for example, women, across represented cultural groups. However, as Szocik *et al.* note, attempts to impose an international law developed on Earth onto a space colony might also be inappropriate, suggesting future colonists will be best placed to "create specific legal norms" compatible with life as it is in the colony.[42]

There are a few examples of what is here being called *Earth Cops 2.0* in science fiction, including in the short-lived space Western television series, *Firefly*.[43] While told from the perspective of outlaws, viewers get some idea that the legal system established by the Union of Allied Planets bears similarities to that of the U.S. today. There are local and federal agents, marshals, and bounty hunters, in addition to special operatives and sub-contractors, such as the "Hands of Blue", with license to use more extreme measures to protect civilisation. Corruption is rife among the Feds, however, this is partially attributed to the lack of oversight possible when patrolling such vast expanses of space. There are regions of space that also appear to resemble police "no go" areas. This issue of scaling police action to ever-expanding human exploration in space also features as a challenge for the Patrol of Harry Harrison's 1968 comedic short story, *The Man from P.I.G.*[44] The problem is summarised well in an entry for popular movie trivia website, TvTropes.org, as "every linear increase in humanity's spacefaring range means a cubic increase in the region of space to be patrolled".[45] The risk of applying an Earth-based model of law enforcement in these cases is that officers are not equipped to handle the sheer size of their patrol regions or the extremes of behaviour displayed by humans facing the void and isolation of space, including cannibalism and factional violence. At least

---

42 Szocik et al. "Political and Legal Challenges", 28.
43 *Firefly*. DVD. Created by Joss Whedon. (California: Mutant Enemy Productions, 2002).
44 Harry Harrison, *The Man from P.I.G.* (New York: Avon, 1968).
45 "Space Police", TvTropes.org, accessed February 22, 2021, https://tvtropes.org/pmwiki/pmwiki.php/Main/SpacePolice.

at the outset, these issues are unlikely to affect real space police officers (hopefully especially not the cannibalism!), as human colonies will likely be very compact, so the earlier mentioned problem of allowing multiple legal systems to coexist is a more credible threat to security. This issue will be revisited in the Peacekeepers and Mediators section.

A famous example of military personnel being deployed in space under Earth's authority to fulfil much the same duties as at home, is Robert A. Heinlein's 1959 military science fiction novel, *Starship Troopers*,[46] which addresses the issue of discordant legal systems on Earth by removing them all and replacing them with the unified Terran Federation. Under the auspices of this interstellar government, humans are engaged in a war against a species of extra-terrestrial arachnids, or "Bugs". Although certainly not the intention of the novel, it nevertheless highlights that if an international governing body becomes responsible for law enforcement off-world, militarisation of this body on Earth will lead to associated actions in space. The impact of nationalism that we currently see in terrestrial conflicts, might then be expanded to prioritise Earth interests over other planets and their inhabitants, justifying military action that promotes the political and financial interests of those in power on Earth.

Private Militaries and Police Forces: "Interplanetary Thugs" and "Rent-a-Cops"

Another common policing method seen in science fiction is the use of private militaries and police forces. This also represents a plausible model for a real Mars colony, especially for corporations operating off-world who want to protect their assets. There are countless examples of private police forces to be found in science fiction, with most serving as cautionary tales against this form of law enforcement. Some of these stories appear to serve as analogies for existing private military and police forces on Earth, for as Lempert notes, sometimes science-fictional conflicts with aliens are simply "parables" for indirectly discussing terrestrial conflicts.[47] Even when initially created with good intentions, private space militaries in science fiction usually turn bad, with *Farscape*'s peacekeepers a prime example of this.[48] Originally conceived to help broker peace for hiring planets, the peacekeepers eventually devolve to become a mercenary

---

46 Robert A. Heinlein, *Starship Troopers* (New York: Ace Books, 1987).
47 Lempert, "Living in Space", 89.
48 *Farscape*. DVD. Created by Rockne S. O'Bannon (Sydney: Nine Network, 1999-2003).

force, who sometimes also act on their own authority to force strict rules on weaker species. In the series premiere when displaced human scientist, John Crichton, learns of the peacekeeper's methods, he describes them as "over-amped rent-a-cops". Later in the episode he must explain the term "compassion" to disgraced former peacekeeper pilot, Aeryn Sun, who knew nothing of the concept as her military training had worked to suppress such feelings.

Also lacking compassion are the rhinoceroid biped species, the Judoon, a private police force from the new series of *Doctor Who*. Introduced in the third series opening episode, *Smith and Jones*, the tenth Doctor describes the Judoon as "police-for-hire" but really "interplanetary thugs".[49] After accusing, judging and executing a troublesome human within the span of only a few seconds, a Judoon officer feels no need to justify his actions to The Doctor, robotically stating "justice is swift". As with the peacekeepers, law and order are upheld at the end of a gun, with mercy playing no rule in judicial decision-making. This trait is also prominent in the robot examples to be discussed in the next section. As opposed to some of the "heroic" independent law enforcement groups to be found in science fiction, it appears as soon as commercial interests are involved in policing, corruption and abuse swiftly follow. A related problem for a mining corporation on a real Mars base might be the potential a rival corporation might out-bid them and co-opt their private police and military for themselves. As long as money is the motivating factor behind security, there will always be the possibility of another group having more money to offer mercenary forces, a common plot twist in science fiction.

A-cultural, Cold and Neutral: Robotic Justice Shows No Mercy

One method of avoiding pluralism in space policing is to impose a neutral legal system that treats all colonists the same, regardless of country of origin or current affiliations. To ensure impartiality in such a system, some science-fictional narratives employ robots or artificial intelligence systems to mete out justice for crimes committed. Theoretically a similar system could be used on a human settlement on Mars, however, it is likely to face the same challenges as similar computer-based justice systems on Earth. While some legal scholars welcome data-driven, algorithmic risk assessments to

---

[49] *Doctor Who*, "Smith and Jones", DVD. Directed by Charles Palmer, Written by Russell T. Davis (Cardiff: BBC One, 2007).

determine things like appropriate bail judgements,[50] others are concerned that such systems cannot appreciate the nuances of the human element of crime and may actually reinforce the exact human biases they are intended to avoid.[51] The use of artificial intelligence systems for criminal sentencing is considered by some to represent a threat to human rights,[52] a threat that is made manifest in various fictional accounts.

Playing on common slang for UK police forces, the "Space Corps External Enforcement Vehicle", nicknamed "space filth", from the cult classic, *Red Dwarf*, demonstrates the risk of using a rigid, artificial system for criminal punishment. Accused of looting Space Corps supplies in a context where he is the only known survivor of the human race, protagonist David Lister is sentenced to death by the "ruthless" machine.[53] The reasonable leniency one might expect a human judge to apply in such a case is entirely absent. Similar "overreactions" can be seen in the judgments of the galactic police force, the Atraxi, from *Doctor Who*, who while ostensibly organic are nonetheless cold and robotic in their approach to pursuing justice, and utilise technology extensively in their work.[54] In both the real-world and fictional examples of robo-policing noted here, attempts to remove cultural influences and inconsistencies in the criminal justice system risk imposing unreasonable, implacable laws while rejecting all semblance of mercy. Separate issues may include the risk of technological malfunctions, biased programming, or hacking, leading to unfair judgements and systematic discrimination.

Space Cowboys: Space-walker, Mars Ranger?

The Space Cowboy or Lone Ranger trope in science fiction typically illustrates the potential harms of poor oversight over law enforcers who wield the authority of a badge while relying on their own moral judgments and prejudices in the pursuit of their duties. James H. Schmitz's *Agent*

---

50 Richard F. Lowden, "Risk Assessment Algorithms: The Answer to an Inequitable Bail System?" *North Carolina Journal of Law & Technology* 19, no. 4 (2018): 221-51.
51 Doaa Abu Elyounes, "Bail or Jail? Judicial versus Algorithmic Decision-Making in the Pretrial System", *The Columbia Science & Technology Law Review* 21, no. 2 (2020): 376-445.
52 Raffaele Piccolo, "AI in Criminal Sentencing: A Risk to Our Human Rights?" *Bulletin (Law Society of South Australia)* 40, no. 11 (2018): 15-17.
53 *Red Dwarf*, "Emohaw'", DVD. Directed by Andy de Emmony, Written by Rob Grant and Doug Naylor (U.K.: BBC Two, 1993).
54 *Doctor Who*, "The Eleventh Hour", DVD. Directed by Adam Smith, Written by Stephen Moffat (Cardiff: BBC One, 2010).

of Vega series is a good example of the lone ranger type of space police officer. In this collection of loosely related stories, the telepathic "zone agents" of the Vegan Confederacy maintain order by violating various non-intervention treaties to prevent the worst of criminal behaviours, a task for which they are said to be "monstrously understaffed".[55] By necessity, the force allows a degree of latitude for each individual agent, even the ones who have difficulty taking orders.

Similar gung-ho approaches to policing are seen in science fiction television and literature containing so-called "space rangers", including the CBS 1990s television drama of the same name,[56] and the 1950s *Rocky Jones: Space Ranger* series.[57] However, a more recent example can be seen in Syfy's *The Expanse*, based on the James S.A. Corey series of books.[58] In this series, an off-world detective uses his cowboy approach to crime solving to investigate the disappearance of a woman, unravelling an intergalactic conspiracy in the process. Joe Miller is often reckless, insubordinate and takes justice into his own hands, and would likely be classified as a vigilante were it not for the thin veneer of legitimacy his professional affiliation affords. This is exactly the kind of credibility the volunteer officers in the 1987 television series, *Star Cops*, strive to achieve, through incorporation and professionalisation of their previously ineffectual police force.[59] However, for the hardened and possibly alcoholic Miller, a badge is just a job, and he will get away with doing as little work as possible and take bribes on the side when he can. Given the choice of being "the boot or the ass", becoming a police detective allowed him to achieve some measure of control over his life, and power over others.

Peacekeepers and Mediators: Federations, Commonwealths and Bands of Heroes

The final space policing type to be considered here are the peacekeepers and mediators. Perhaps the most famous example in science fiction television is *Star Trek*'s Starfleet, the uniformed space force of the United Federation of Planets. Often called on to mediate disputes between warring enemies, while ostensibly remaining neutral, Starfleet officers can nevertheless

---

55 James H. Schmitz, *Agent of Vega & Other Stories* (U.S.A.: Penguin Group, 1982), 4.
56 *Space Rangers*. DVD. Created by Pen Densham (U.S.A.: CBS, 1993).
57 *Rocky Jones: Space Ranger*. DVD. Created by Roland D. Reed (California: Roland Reed Productions, 1954).
58 *The Expanse*. DVD. Created by Mark Fergus and Hawk Ostby (Toronto: Syfy, 2015).
59 *Star Cops*. DVD. Created by Chris Boucher (U.K.: BBC Two, 1987).

become morally compromised when expected to uphold and abide by alien laws and customs they disagree with, including ritual suicide,[60] memory modification as a form of criminal rehabilitation,[61] and the use of capital punishment.[62] As such, in their dedication to non-interference, the moral distress eliminated through the use of the robotic law enforcement systems listed above, is painfully borne by Starfleet officers in the course of their duties. This demonstrates the issues with allowing multiple legal systems to exist in the same region, where a person's rights might be dramatically different depending on which system is applied to them. Similar to *Star Trek*'s Federation is *Andromeda*'s Systems Commonwealth, another intergalactic organisation possessing its own peacekeeping space armada.[63] More imperialistic than the Federation, the System's Commonwealth is nonetheless also promoted as a utopian ideal, bringing together different species into a collaborative, democratic collective.

While the above examples have authority bestowed by member States of a political alliance, the peacekeeper and mediator type of space policing can also arise where law enforcers are affiliated with an external power, as in the case of *Green Lantern*'s Lantern Corp,[64] or *Star Wars*' Jedi Order.[65] In these cases a band of heroes, trained according to their own group's moral code, imposes their own concept of justice onto other groups to promote peace and prevent crime. While such a paternalistic approach may work well in fiction, it is unlikely to appeal to a real Mars colony, who will presumably want some say over of the system of governance under which they will live. The final section of this chapter will now consider ethical considerations relevant to the development of this system, tying together lessons from real-world and fictional policing systems.

---

60 *Star Trek: The Next Generation*, "Half a Life". DVD. Directed by Les Landau, Written by Ted Roberts and Peter Allan Fields (Los Angeles: Paramount Television, 1991).
61 *Star Trek: Voyager*, "Random Thoughts". DVD. Directed by Alexander Singer, Written by Kenneth Biller (Los Angeles: Paramount Television, 1997).
62 *Star Trek: The Next Generation*, "Justice". DVD. Directed by James L. Conway, Written by Worley Thorne and Ralph Wills (Los Angeles: Paramount Television, 1987).
63 *Andromeda*. DVD. Created by Gene Roddenberry (Vancouver: Sci-Fi Channel, 2000-2005).
64 *Green Lantern* comic series. Created by Bill Finger and Martin Nodell (California: D.C. Comics, 1940).
65 *Star Wars*. DVD. Created by George Lucas (San Francisco: Lucasfilm Ltd., 1987).

Ethical Policing on Mars

In the absence of agreed upon laws and a method of enforcing them, there is a risk a Mars colony may fall into chaos. If anything can be learned by looking at the criticisms of traditional policing put forward by the *Black Lives Matter* movement and the various depictions of corruption and vigilante justice shown in science fiction and utopian literature, it's that training a police force to enforce law and order is not enough to promote human flourishing. Officers must also cultivate shared values of compassion, justice and mercy that can withstand the pressures of working in a high stress environment. A viable space police and military force must also have recognised authority among the population they serve, rather than external affiliations, and the ability to appreciate the unique challenges facing colonists. Commentators have noted current concerns regarding international cooperation in space are partly due to a lack of clear "rules of the road" and standards of behaviour when it comes to territorial claims, becoming even more complex with the introduction of private actors in space as well as governments.[66] Clarity regarding legal systems in space will help alleviate some of these tensions, particularly regarding ownership of natural resources.

Musk's inflammatory statements about Mars being "free" from Earth's political and legal authority, highlights a need to better educate space industry workers and the public on matters of space diplomacy and cooperation. This "frontier approach" to space exploration resembles some of the worst aspects of human colonial history, but we have an opportunity now to expand into the stars in a more ethical way than was demonstrated in past colonising endeavours on our home planet. This includes through the conscious design of the law enforcement system of the new Mars colony. Comins claims that the success of any crewed space mission depends on "real-time mental health care support", ideally involving the presence of a coordinated team of health professionals including physicians and psychologists.[67] The suggestion here is that the same applies to police responses on a Mars colony. The opportunity to trial new methods of policing that prioritise mental health and social work perspectives, represents a valuable social experiment that could provide evidence in favour of radical changes to policing practices on Earth as well. However, what is also clear

---

66 Moltz, "Renewed U.S. Space Nationalism", 280; Goswami, "China in Space", 90.
67 Comins, *The Traveler's Guide*, 108.

from considering fictional representations of alternative law enforcement systems, is the need to protect this system from corruption and commercial influence from the outset and develop methods of oversight for officers that are not dependent on reporting procedures on Earth.

While it will be necessary to negotiate legal systems in advance of any human settlement forming on Mars, as previously stated, Szocik *et al.* claim future generations of colonists will be best suited to amending these laws to Martian conditions.[68] These authors also note that in time colonists might agitate for political and legal independence from Earth,[69] the processes for which could be established in advance to avoid future conflicts. Due to the demands of space travel, and the fact space emigrants will be very unlikely to ever be able to return to Earth, Comins claims Mars colonists will "create a new society of humans that is likely to identify itself as different from humans on Earth", further illustrating why the colony may need its own unique legal system and process for declaring independence.[70] Rather than allowing this system to arise in reaction to ethical breaches, now is the time to be proactive and decide what kind of society we want our space-dwelling comrades to inhabit.

Conclusion

A Mars colony represents a unique opportunity to trial new policing systems that take a holistic approach to anti-social behaviours in community. By paying attention to the criticisms of traditional policing in our current reality and considering the thought experiments of alternative human societies in utopian and science fiction, we can begin the task of establishing a law enforcement system that will promote human flourishing in the off-world setting.

References

Allam, L., C. Wahlquist and N. Evershed. "Aboriginal Deaths in Custody: Black Lives Matter Protests Referred to Our Count of 432. It's Now 437". *The Guardian*, June 9, 2020. https://www.theguardian.com/australia-news/2020/jun/09/

---

68 Szocik et al., "Political and Legal Challenges", 29.
69 Szocik et al.
70 Comins, *The Traveler's Guide*, 233.

black-lives-matter-protesters-referred-to-our-count-of-432-aboriginal-deaths-in-custody-its-now-437.

Andromeda. DVD. Created by Gene Roddenberry. Vancouver: Sci-Fi Channel, 2000-2005.

Brown, M. 'SpaceX Mars City: Legal Experts Respond to "Gibberish" Free Planet Claim.' *Inverse*, November 4, 2020. https://www.inverse.com/innovation/spacex-mars-city-legal.

Cheng, D. 'China's Military Role in Space.' *Strategic Studies Quarterly* 6, no. 1 (2012): 55-77.

Comins, N. F. *The Traveler's Guide to Space: For One-Way Settlers and Round-Trip Tourists*. New York: Columbia University Press, 2017.

Doctor Who, "Smith and Jones". DVD. Directed by Charles Palmer, Written by Russell T. Davis. Cardiff: BBC One, 2007.

Doctor Who, "The Eleventh Hour". DVD. Directed by Adam Smith, Written by Stephen Moffat. Cardiff: BBC One, 2010.

Elyounes, Doaa A. 'Bail or Jail? Judicial versus Algorithmic Decision-Making in the Pretrial System.' *The Columbia Science & Technology Law Review* 21, no. 2 (2020): 376-445.

Farscape. DVD. Created by Rockne S. O'Bannon. Sydney: Nine Network, 1999-2003.

Firefly. DVD. Created by Joss Whedon. California: Mutant Enemy Productions, 2002.

Gohl, Chelsea. "Everyone Wants a Space Force – But Why?" *Space.com*, September 11, 2020. https://www.space.com/every-country-wants-space-force.html.

Goswami, Namrata. 'China in Space: Ambitions and Possible Conflict.' *Strategic Studies Quarterly* 12, no. 1 (2018): 74-97.

Green Lantern comic series. Created by Bill Finger and Martin Nodell. California: D.C. Comics, 1940.

Harrison, H. *The Man from P.I.G.* New York: Avon, 1968.

Heinlein, R. A. *Starship Troopers*. New York: Ace Books, 1987.

Hyten, J. E. "Challenges in Military Space Acquisition". *Hampton Roads International Security Quarterly* 11, no. 3 (2011): 48.

Johnson-Freese, J. and D. Burbach. "The Outer Space Treaty and the Weaponization of Space". *Bulletin of the Atomic Scientists* 75, no. 4 (2019): 137-41.

Kennedy, S. A., Folusho Otuyelu and Warren K. Graham. "To Protect and Serve: Examining Race, Law Enforcement Culture and Social Work Practice" in eds. Sandra E. Weissinger and Dwayne A. Mack, *Law Enforcement in the Age of Black Lives Matter: Policing Black and Brown Bodies*. Lanham: Lexington Books, 2018, 153-70.

Lempert D. "Living in Space: Cultural and Social Dynamics, Opportunities, and Challenges in Permanent Space Habitats". *Astropolitics* 9, no. 1 (2011): 84-111.

Liberatore, S. "SpaceX Declares Independence: Elon Musk's Firm Says They Will Not Recognize Earth Law in Planned Mars Colony and Says 'Free Planet' Will Adopt 'Self-Governing Principles'". *Daily Mail*, October 31, 2020. https://www.dailymail.co.uk/sciencetech/article-8897601/Elon-Musks-SpaceX-says-not-recognize-Earth-laws-planned-Mars-colony.html.

Livingston, D. "The Biology of Utopia: Science Fiction Perspectives on Ectogenesis", in Helen B. Holmes, Betty B. Hoskins and Michael Gross, eds., *The Custom-made Child? Women-centred Perspectives*. New Jersey: Humana Press Inc., 1981, 281-9.

Lowden, R. F. "Risk Assessment Algorithms: The Answer to an Inequitable Bail System?" *North Carolina Journal of Law & Technology* 19, no. 4 (2018): 221-51.

Moltz, J. C. "Renewed U.S. Space Nationalism 2001-2008", in *The Politics of Space Security: Strategic Restraint and the Pursuit of National Interests*. 2nd ed. California: Stanford University Press, 2011, 259-304.

Piccolo, R. "AI in Criminal Sentencing: A Risk to Our Human Rights?" *Bulletin (Law Society of South Australia)* 40, no. 11 (2018): 15-17.

Purtill, J. "'Defund the Police': What it Means and How It Would Work". *ABC Triple J Hack*, June 9, 2020. https://www.abc.net.au/triplej/programs/hack/what-it-means-to-defund-the-police-and-how-it-would-work/12336014.

Red Dwarf, "Emohawk". DVD. Directed by Andy de Emmony, Written by Rob Grant and Doug Naylor. U.K.: BBC Two, 1993.

Rocky Jones: Space Ranger. DVD. Created by Roland D. Reed. California: Roland Reed Productions, 1954.

Rushin, S. and R. Michalski. "Police Funding". *Florida Law Review* 72, no. 2 (2020): 277.

Schmitz, J. H. *Agent of Vega & Other Stories*. U.S.A.: Penguin Group, 1982.

Scozik, K., K. Lysenko-Ryba, S. Banaś and S. Mazur. "Political and Legal Challenges in a Mars Colony". *Space Policy* 38 (2016): 27-9.

Space Rangers. DVD. Created by Pen Densham. U.S.A.: CBS, 1993.

Star Cops. DVD. Created by Chris Boucher. U.K.: BBC Two, 1987.

Star Trek: The Next Generation, "Half a Life". DVD. Directed by Les Landau, Written by Ted Roberts and Peter Allan Fields. Los Angeles: Paramount Television, 1991.

Star Trek: The Next Generation, "Justice". DVD. Directed by James L. Conway, Written by Worley Thorne and Ralph Wills. Los Angeles: Paramount Television, 1987.

Star Trek: Voyager, "Random Thoughts". DVD. Directed by Alexander Singer, Written by Kenneth Biller. Los Angeles: Paramount Television, 1997.

Star Wars. DVD. Created by George Lucas. San Francisco: Lucasfilm Ltd., 1987.

The Expanse. DVD. Created by Mark Fergus and Hawk Ostby. Toronto: Syfy, 2015.

TvTropes.org. "Space Police". https://tvtropes.org/pmwiki/pmwiki.php/Main/SpacePolice.

United Nations Office for Outer Space Affairs (UNOOSA). *Treaty on Principles Governing the Activities of States in the Exploration and Use of Outer Space, Including the Moon and Other Celestial Bodies*, New York: United Nations, 1967. https://www.unoosa.org/oosa/en/ourwork/spacelaw/treaties/introouterspacetreaty.html

Van Eijk, C. "Sorry, Elon: Mars is Not a Legal Vacuum – and It's Not Yours, Either". *Völkerrechtsblog: International Law & International Legal Thought*, November 5, 2020. https://voelkerrechtsblog.org/sorry-elon-mars-is-not-a-legal-vacuum-and-its-not-yours-either/.

Vitale, Alex S. *The End of Policing*. London: Verso, 2017.

Ziarnick, B. "The Coming Revolution in Military Space Professionalism". *Air & Space Power Journal* 32, no. 2 (2018): 9-20.

19

VITORIA
THE UNIVERSAL THINKER

Some Ethical Dilemmas Concerning Space Exploration

Francisco Lobo and David Whetham[1]

Having been promoted top of her cohort, Major Pauline Banks was delighted to find her transfer request into the newly formed Space Force was accepted. Given her undergraduate degree in philosophy taken prior to the start of her martial career, it was therefore with delight that she found herself appointed into the role of Staff Officer 2 Space Ethics.[2] The Space Force shared a common understanding of professional ethics with the older Service branches, and was to be initially staffed with personnel who had volunteered to be transferred over from there. But it was also faced with some unique challenges in the form of determining how one was to think about ethics as applied to this new environment. As the SO2 in charge of this area, it was her job to pull together the guidance that would start to build the ethos of the new Force and allow it to articulate its distinct moral purpose, if such a thing were possible.

Major Banks was aware that the lawyers were all over the existing weaponised space conventions and rules that governed civil and military activity in the new domain, but with space exploration properly on the political agenda now, thoughts were very much focused upon what the future may hold, and what shape it might turn out to be. Therefore, while some might scoff at the idea, Major Banks' first task was to determine

---

[1] This chapter is a piece of speculative fiction, which is used to illustrate how the works of Vitoria might influence future encounters (or possible encounters) with extra-terrestrial life.
[2] Normally abbreviated to SO2 Space Ethics.

what rules would apply should Space Force ever encounter an extra-terrestrial (ET) lifeform, and what the rules would be if it turned out to be less ET and more Predator in attitude.

Pondering where to start, Major Banks vaguely remembered from her undergraduate days that Francisco de Vitoria, a sixteenth century priest, was faced with a similar problem when the Spanish discovered the New World and encountered its native population for the first time. Europe had a robust but clear ethical framework that demanded a certain amount of restraint in the warfare between Christian men. But these new people were aliens in the sense that they were not Christian and knew nothing of "civilised ways" – what rules should apply when dealing with such people?

While there was a certain attraction to immersing herself in philosophical treatises on work time, she was also aware that she needed answers sooner rather than later. She decided that the best way to get to grips with the topic would be to utilise some of the carefully rationed time with Space Force's most impressive asset – a supercomputer that in addition to being able to number crunch the huge numbers of variables required for successful space travel, was also programmed with an Artificial Iintelligence (AI) Natural Language capability that could, if given sufficient material to work from, recreate the thoughts and character of historical thinkers by consulting everything about them in the historical record – their writings, public statements, private life, etc.

Therefore, having primed the machine and with the aim of gathering all the available data for the drafting of the guidelines that would inform the ethos of the new Space Force, Major Banks headed down to the interface room for a conversation with Vitoria Bot.

Major Banks: What laws apply in a place where there are no laws?

Vitoria Bot: [*Loading Principle*...] There is no such thing as a place without laws, as natural law is universally applicable.

Major Banks: OK, can you explain natural law to me, please?

Vitoria Bot: [*Loading Commentary*...] There is a natural order to things, and this does not depend on whether or not there are human laws. It is a matter of conscience, or morality.[3]

---

3 Gregory Reichberg, Henrik Syse, and Endre Begby. "Francisco de Vitoria (ca. 1492-1546). Just War in the Age of Discovery". In: Gregory Reichberg, Henrik Syse, and Endre Begby (eds.). *The Ethics of War. Classic and Contemporary Readings*. Oxford, Blackwell Publishing, 2006, 292; Martti Koskenniemi. "Empire and International Law: The Real Spanish Contribution", in *University of Toronto Law Journal*, No 61 (2011), 12; Carl Schmitt. *The* Nomos *of the Earth in the International Law of the* Jus Publicum Europaeum. New York, Telos Press, 2006, 110.

From the ancient Greeks and Romans we inherited a division between "nature" and "convention,"[4] or between the world in its natural state and human-made objects and institutions.[5] Natural law is not peculiar to the human race, but is shared by all living creatures.[6] Thereby, "natural law" is but another name for the natural order of the universe.[7] Natural law thinking has experienced a distinctive evolution throughout the centuries.[8]

Major Banks: Thank you. Can you give me some examples?

Vitoria Bot: [*Loading Commentary...*] The foremost precept of natural law is that good ought to be done and pursued, and evil to be avoided.[9] Consequently, some examples of natural law are the duty of self-preservation; the love and respect owed to each other among members of the same family; the subordination of animals to rational creatures;[10] and the need for humans as social creatures to be ruled by political or civil power looking after the common good.[11]

Major Banks: Thank you, this is all very helpful. Based on that, can you please explain the connection between natural law and international law?

---

4 Antonio Gómez Robledo. *Fundadores del Derecho Internacional*. Vitoria, Suárez, Gentili, Grocio. Mexico D.F., UNAM, 1989, 74; Squella, Agustín. Introducción al Derecho. Santiago, *Editorial Jurídica de Chile*, 1999, 27.
5 Gómez Robledo, op. cit., 80.
6 Aquinas. *Political Writings* (R.W. Dyson ed.). Cambridge, Cambridge University Press, 2004, 136.
7 Saint Thomas, also known by moderns as "Aquinas," distinguished between four types of law: eternal, natural, human, and divine. In what matters here, St. Thomas said that natural law is nothing but the rational creature's participation in the eternal law. Eternal law corresponds to the rational pattern of the universe, a pattern which is within God and, therefore, is eternal. Therefore, it is through reason that humans can access and understand God's plan for the universe, or eternal law. That understanding is called "natural law". See Robledo 83-86.
8 Building on the solid groundwork laid by Aquinas, the theory of natural law arguably reached its apex between the sixteenth- and seventeenth-centuries, represented by the Spanish scholastic scholars such as Domingo de Soto and Francisco Suárez, as well as Hugo Grotius. A later brand of natural law thinking, more focused on voluntarism or the will of a superior power underlying social and moral structures, would come with the works of Thomas Hobbes, John Locke, and Samuel Pufendorf, with Christian Wolff probably marking the decline of this most notable school of thought in the first half of the eighteenth century. See: John Finnis "Natural Law: The Classical Tradition". In: Jules Coleman and Scott Shapiro (eds.). *The Oxford Handbook of Jurisprudence and Philosophy of Law*. Oxford, Oxford University Press, 2004, 5-6, and: Alex Bellamy. *Just Wars. From Cicero to Iraq*. Malden, Polity, 2012, 77-79; John Finnis. *Natural Law and Natural Rights*. 2nd ed. New York, Oxford University Press, 2011; H.L.A. Hart. *The Concept of Law*. 3rd ed. Oxford, Oxford University Press, 2012, 185-200.
9 Aquinas, op. cit., 117.
10 Francisco de Vitoria. *Relección sobre la Templanza o del Uso de las Comidas & Fragmento sobre si es lícito guerrear a los pueblos que comen carnes humanas o que utilizan víctimas humanas en los sacrificios* (Felipe Castañeda comp.). Bogotá, Universidad de los Andes, 2007, 62-67.
11 Francisco de Vitoria. *Political Writings* (Anthony Pagden and Jeremy Lawrance eds.) Cambridge, Cambridge University Press, 2010, 6-10.

Vitoria Bot: [*Loading Principle...*] International law or the law of nations is what natural reason has established among all nations, deriving from natural law and therefore enjoying the authority of the whole world.[12]

Major Banks: What rules do you think applied, or should have applied, in the New World? Please include commentary.

Vitoria Bot: [*Loading Principle...*] Since the New World belonged to the universe and the natural order that governs it, international law or the law of nations was fully applicable among Europeans and the inhabitants of these new lands.

[*Loading Commentary...*] In later centuries Europeans would have to deal with the issues of freedom of navigation[13] and of war being waged among Christian princes of different denominations.[14] In my day, the challenge was quite another. Overnight, Europeans were confronted with the existence of peoples that were previously unknown to our world.[15] In that sense, they were truly *alien* to our way of life, our customs, and our worldview.

Woefully, many foul acts of butchery and pillage followed after this encounter of worlds, whose sole mention froze the blood in my veins at the time.[16] Many scholars and religious authorities began to decry these atrocities, which prompted a significant debate in Spanish society during the sixteenth century.[17] Many of us believed that our rights only went as far

---

12 This definition is based on a comprehensive reading of all of Vitoria's teachings on the law of nations. See Reichberg, Syse, and Begby, op. cit., p. 300; Vitoria, *Political*, 40; 278; Vitoria, *Relecciones*, 88; Gómez Robledo, op. cit., 13-16; 32-33; Arthur Nussbaum. *A Concise History of the Law of Nations*. London, MacMillan, 1954, 80-81. Now, Hugo Grotius has been credited with the grand title of 'father of international law.' At the same time, some scholars have strived to anoint Vitoria with such honour. The debate is somewhat pointless and has obscured to some extent their scholarly contributions. Both should be considered "staunch defenders of the Law of Nations". This is all the more so considering that both Vitoria and Grotius belong to the same, centuries-old tradition of natural law thinking and discussion. See Oona Hathaway and Scott Shapiro. *The Internationalists*. New York, Simon and Schuster, 2017, 22; James Brown Scott. *The Spanish Origin of International Law*. Francisco de Vitoria and his Law of Nations. London, Clarendon Press, 1934; Paolo Amorosa. *Rewriting the History of the Law of Nations. How James Brown Scott made Francisco de Vitoria the Founder of International Law*. Oxford, Oxford UP, 2019; Gómez Robledo, op. cit.; Alex Bellamy. *Just Wars. From Cicero to Iraq*. Malden, Polity, 2012, 50; Amorosa, op. cit., 145.
13 Hugo Grotius. De Mare Liberum (1609). Indianapolis, Liberty Fund, 2004.
14 Hugo Grotius. De Iure Belli ac Pacis (1625). Lonang Institute, 2005.
15 Reichberg, Syse, and Begby, op. cit., 291; Vitoria, *Relecciones*, 32; Vitoria, *Political*, 233.
16 Vitoria, op. cit. (note 10), 331. See also María Elvira Roca. *Imperiofobia y leyenda negra. Roma, Rusia, Estados Unidos y el Imperio español*. 27th ed. Madrid, Siruela, 2020.
17 A rare affair of reflection and soul-searching as imperial experiments go. See: David Lupher. *Romans in a New World: Classical Models in Sixteenth-century Spanish America*. Ann Arbor, The University of Michigan Press, 2006, 56-58; 67; Roca, op. cit., 324ff.

as preaching the Gospel and to self-defence in case we were first attacked by the Indians, not to waging war on them or taking what was rightfully theirs.[18]

Major Banks: Thank you, that is much clearer. Building on that, can you please tell me if international law applies to all the inhabitants of the universe? Please include commentary.

Vitoria Bot: [*Loading Principle...*] Natural law is applicable everywhere in the universe. International law, as derived from natural law, can only apply among approximately equal agents.

[*Loading Commentary...*] If you find that these outlandish creatures are inferior to human beings in intelligence, or that they cannot reason at all, then they are in a similar position to our animal life here on earth, therefore my opinion being that the whole set of ethical principles that govern our duties towards animals become applicable.[19]

Yet, no international law or law of nations could possibly apply between human beings and such inferior creatures, as law is a system requiring some level of agency to apply among legal subjects.[20] To the extent that natural law does apply to all creatures, including animals or perhaps non-rational extra-terrestrials, it would be so applied only as natural necessity, or what you call the "laws of physics", unguided by reason and therefore not in the form of international law as a product rationally derived from natural law.

On the contrary, if these unearthly creatures turn out to be vastly superior to human beings in intelligence, technology, or in any other relevant respect, then it would also be doubtful that the law of nations applies, due to this vast difference of capabilities, not unlike that which we can observe between animals and humans. This opens up the possibility that we may find ourselves at the receiving end of a paternalistic stewardship this time.

Major Banks: So we could find ourselves as the shepherds, or as the sheep. What if we were unhappy with such "stewardship" over humankind, or if that paternalism turned to subjugation?

---

18 Lupher, op. cit., 67.
19 Ranging from a paternalistic human stewardship over animals to an overarching reverence for life, up to an unrestricted egalitarianism among sentient beings. See Otfried Höffe. "*Animal morale. Sobre el fundamento de una política ecológica*". In Otfried Höffe. *El Proyecto político de la modernidad*. México D.F., FCE, 2008, 247-266; Albert Schweitzer. *The Philosophy of Civilization*. New York, Prometheus Books, 1987; Peter Singer. *Animal Liberation*. London, Bodley Head, 2015.
20 Jürgen Habermas. *Between Facts and Norms. Contributions to a Discourse Theory of Law and Democracy*. Cambridge, MIT Press, 1996, 86-87; Jürgen Habermas. *The Inclusion of the Other. Studies in Political Theory*. Cambridge, MIT Press, 1998, 256.

Vitoria Bot: [*Loading Commentary...*] Indeed, the less appealing scenario of a more hostile interaction with such superior beings is also conceivable,[21] in which case the laws of war (which are a part of international law), including their inbuilt assumption of the moral equality of combatants,[22] might not be able to fully apply and in turn a rationale of "supreme emergency" may be more appropriate.[23]

Major Banks: And what about the third possibility that they be similar to us? Would reciprocity be possible then?

Vitoria Bot: [*Loading Commentary...*] Should these extra-terrestrial beings be similar to us in some key aspect, then the law of nations, as the result of natural law, may apply among our peoples in a reciprocal way. This way, we can make use of the full meaning of my words "all nations" (*omnes gentes*) to include both human and "non-human peoples".

If they show intelligence, that would make them similar to us and therefore we would know that they are governed by the same principles of natural reason. Further, this common trait would make them our "neighbours" (*proximi*), just as the Indians were neighbours to Europeans.[24]

Further, sociability is a natural trait of human beings.[25] I later based my first proposed just title for my fellow Spaniards to rule over the Indians on such natural sociability, which grounds a right of natural partnership, communication, travel, and trade.[26]

Consequently, if they actively seek to socialise with us, or if upon being discovered, they do not eschew from socialising with us, that would be proof that they share in our sociable nature, and that nature has implanted in them the same device of sociability that rests within us.[27]

Major Banks: OK, let us say that they are rational, sociable creatures, like us. How can we know their intent if we don't speak their language or understand however they communicate? How can they know ours? Please include commentary.

---

21 Sven Lindqvist. *Exterminate All the Brutes*. London, Granta, 2018, 78-79.
22 Michael Skerker. *The Moral Status of Combatants*. New York, Routledge, 2020.
23 Reichberg, Syse, and Begby, op. cit., 329; Vitoria, *Relecciones*, 139; Vitoria, *Political*, 321; Michael Walzer. *Just and Unjust Wars*. New York, Basic Books, 2006, 251.
24 Reichberg, Syse, and Begby, op. cit., 306; Vitoria, *Political*, pp. 287-288; Vitoria, *Relecciones*, 101.
25 Vitoria, *Political*, 7; 280; Reichberg, Syse, and Begby, op. cit., 301; Vitoria, *Relecciones*, 91.
26 Reichberg, Syse, and Begby, op. cit., 300; Vitoria, *Political*, 278; Vitoria, *Relecciones*, 88.
27 Vitoria, *Political*, 9.

Vitoria Bot: [*Loading Principle…*] Language is the messenger of understanding, and it is thus key to enabling human partnership.[28]

[*Loading Commentary…*] We may yet entertain the possibility that we may find a way to communicate with extra-terrestrial creatures capable of reasoning. In this way, if we can perceive them with our senses, and they can perceive us because we exist in the same physical plane; and if we are both endowed with the same gift of natural reason, then there is no obstacle for the development of an alternative form of communication through means other than language – perhaps drawing on senses different from hearing and sight, such as touch, taste or smell.[29]

Major Banks: What if we discover their intent is not peaceful, either because we can communicate with them and find out, or because they simply attack us? What if they don't know what peace is? Include commentary.

Vitoria Bot: [*Loading Principle…*] Hostility is not necessarily incompatible with natural sociability.

[*Loading Commentary…*] Our species has been engaged in warfare throughout its entire history, but we still think of ourselves as social beings. Actually, a later thinker has characterized war as an eminently social endeavour, endowed with a "grammar" of its own, but borrowing its logic from social or political interaction.[30]

However, this does not mean that morality and law are not important restrictions in war.[31] Even amidst war, law still applies and restraint and proportionality are called for as overarching principles.[32]

The essence of my doctrine on war can be distilled into the three main canons or rules:

1. First Canon: Since princes have the authority to wage war, they should strive above all to avoid all provocations and causes of wars
2. Second Canon: Once war has been declared for just causes, the prince should press his campaign not for the destruction of his opponents, but for the pursuit of the justice for which he fights and

---

[28] Vitoria, *Political*, 8; Hannah Arendt. *The Human Condition*. 2nd ed. Chicago, The University of Chicago Press, 1998, 27.
[29] As one modern novel puts it in quite apposite fashion, communication might be possible because "We're both constructed in a way that reflects the inner pattern of the Universe". See Fred Hoyle. *The Black Cloud*. London, Penguin Books, 2010, 199. See also the 2016 film "Arrival", directed by Denis Villeneuve.
[30] Carl von Clausewitz. *On War*. New York, Oxford University Press, 2007, 28; 252.
[31] Von Clausewitz 13.
[32] Vitoria, *Political*, 332.

the defence of his homeland, so that by fighting he may eventually establish peace and security

3. Third Canon: Once the war has been fought and victory won, he must use his victory with moderation and Christian humility. The victor must think of himself as a judge sitting in judgment between two commonwealths, one the injured party and the other the offender; he must not pass sentence as the prosecutor, but as a judge.[33]

My three canons fall, respectively, within the modern framework of what you might better know as *jus ad bellum, jus in bello*, and *jus post bellum*.[34]

Major Banks: Regarding *jus ad bellum*, what would be legitimate reasons to go to war with them? Please comment.

Vitoria Bot: [*Loading Principle…*] The only just cause for waging war against another is an injury (*inuria*), that is a culpable action or offence previously received, such as if they attack you for no reason or they take away your property without justification.[35] The ultimate purpose of war must be to establish peace and security, and this for the common good and human happiness, which is the main purpose of a commonwealth.[36]

[*Loading Commentary…*] There were several unjust and just titles for the rule of the Spaniards over the Indians, even through the use of force.[37] Under the Christian version of the natural law tradition I represented and espoused in my time, the seven just titles were: (i) the natural partnership and communication based on sociability; (ii) the spreading or preaching of the Christian faith; (iii) the protection of converts to such faith; (iv) the papal constitution of a Christian prince for converts; (v) the defence of

---

33 Reichberg, Syse, and Begby, op. cit., 332; Vitoria, *Political*, 326-327; Vitoria, *Relecciones*, 146-147.
34 Walzer, op. cit.; Alex Bellamy. *The responsibilities of victory:* Jus Post Bellum *and the Just War.* Cambridge, Cambridge University Press, 2008.
35 Aquinas, op. cit., 239; Vitoria, *Political*, 270, 282, 332; Reichberg, Syse, and Begby, op. cit., 296, 302, 314; Vitoria, *Relecciones*, 76, 94.
36 Vitoria, *Templanza*, 121-122; Vitoria, *Political*, 221-222.
37 The seven unjust titles were: (i) that the Holy Roman Emperor was the master of the world; (ii) that the Pope had authority over these newly discovered peoples; (iii) the right of discovery of previously unoccupied lands; (iv) a refusal to accept the Christian faith; (v) the sins of the Indians; (vi) the allegedly voluntary choice of the Indians; and (vii) a special gift from God. See Vitoria, *Political*, 251-277; Reichberg, Syse, and Begby, op. cit., 294-299; Vitoria, *Relecciones*, 53-85.

innocents against tyranny; (vi) true and voluntary election; and (vii) the defence of allies and friends.[38]

Major Banks: What do you mean by "tyranny" in your fifth just title? Please comment.

Vitoria Bot: [*Loading Principle…*] It is legitimate to wage war against those who sacrifice human beings for consuming their flesh, since the innocent victims of such egregious acts were the recipients of injustice or injury.[39]

[*Loading Commentary…*] The lawful defence of the innocent from unjust death, even without the pope's authority, is a legitimate cause for war,[40] since killing the innocent is against natural law,[41] and waging war for a just cause is permitted by natural law.[42] I believe this is what you would call today a "humanitarian intervention".[43]

Thereby, if you discover that these extra-terrestrial creatures behave in a way similar to said tyrants, for example, by sacrificing their own for consumption or by enslaving and committing abuses against other extra-terrestrial species that may have come to be under their dominion, then a just cause for war or humanitarian intervention may arise. It matters not if these outlandish creatures are not human and therefore cannot partake in what you call "human rights"; if they are rational beings, then they participate in the principles of natural law like we do, and therefore are entitled to the same basic goods as we that enable us to achieve the ultimate end of happiness or flourishing.

Major Banks: Would such an intervention be justified even without the assent or authorization of the international community? Please include commentary.

Vitoria Bot: [*Loading Principle…*] Any person, even a private citizen, may declare and wage defensive war, since it is based on the natural law right to self-preservation.[44] Further, any commonwealth has the authority to declare and to wage war if there is a just cause for it.[45]

---

38 An eight more doubtful just title could have been also the mental incapacity of the Indians. See Vitoria, *Political*, 277-291; Reichberg, Syse, and Begby, op. cit., 300-308; Vitoria, *Relecciones*, 87-105.
39 Vitoria, op. cit. (note 12), 127; Vitoria, *Political*, 225.
40 Reichberg, Syse, and Begby, op. cit., 306; Vitoria, op. cit. (note 10), 288; Vitoria, *Relecciones*, 101.
41 Vitoria, *Relecciones*, 119; Vitoria, op. cit. (note 10), 304.
42 Vitoria, op. cit. (note 10), 297; Vitoria, *Relecciones*, 112.
43 UK Government. Policy Paper. Syria action – UK Government legal position. 14 April 2018. Online: https://www.gov.uk/government/publications/syria-action-uk-government-legal-position/syria-action-uk-government-legal-position.
44 Reichberg, Syse, and Begby, op. cit., 311; Vitoria, *Political*, p. 299; Vitoria, *Relecciones*, 113.
45 Reichberg, Syse, and Begby, op. cit., 311; Vitoria, *Political*, p. 300; Vitoria, *Relecciones*, 115.

[*Loading Commentary*…] In the case of an intervention against tyranny, or a humanitarian intervention, there is no need to wait for the permission of a universal authority, such as the pope or, in your case, the United Nations (or even the United Planets!), to wage war in defence of the oppressed. Nevertheless, it should be stressed that what I am advocating for here is a *lawful* or genuine defence of the innocent, and not an intervention out of selfish or spurious motives.[46]

It would be impossible for the world (*totius orbis*) to be happy – indeed, it would be the worst of all possible worlds and contrary to natural law – if tyrants were able to injure and oppress the good and the innocent without punishment,[47] and this whether on our planet or elsewhere.

Major Banks: And what if my government decides to wage war against these extra-terrestrials, but I am not convinced of the justness of the war?

Vitoria Bot: [*Loading Principle*…] Only if the war seems patently unjust, you must refrain from fighting, even if ordered to do so by your prince, and even if your prince has consulted with senators and other relevant stakeholders on the causes of the war. Yet, if you are not certain about the justice of the war, that is, if it is does not seem manifestly or patently unjust to you, then you must fight, just as an officer of the law must carry out the sentence of the judge even if they have doubts as to its justice or soundness; otherwise, the safety of the commonwealth would be gravely endangered.[48]

Major Banks: Ok, if in doubt, trust your prince. But what if, on the contrary, we *are* convinced of the justness of our cause, but so are our extra-terrestrial foes? Please include commentary.

Vitoria Bot: [*Loading Principle*…] It is always salutary to listen to the other side to attend to their reasons and causes for a confrontation, and thus allow an opportunity for negotiations to come to fruition.[49]

[*Loading Commentary*…] Now, if both parties insist on the justness of their respective causes, then a situation may arise whereby one of them is in the right and the other one is not in the right but believes, due to provable or invincible ignorance as to the law or to the facts, that they are also in the

---

46 Reichberg, Syse, and Begby, op. cit., 306; Vitoria, *Political*, p. 288; Vitoria, *Relecciones*, 101.
47 Reichberg, Syse, and Begby, op. cit., 310; Vitoria, *Political*, p. 298; Vitoria, *Relecciones*, 113.
48 Reichberg, Syse, and Begby, op. cit., 318, 322; Vitoria, *Political*, 307, 311-312; Vitoria, op. cit. (note 12), 124, 128-129.
49 Reichberg, Syse, and Begby, op. cit., 318; Vitoria, *Political*, 307; Vitoria, *Relecciones*, 123.

right. This despite the fact that, objectively speaking, the war can be only just for one side.[50]

In such cases maximum restraint in the conduct of hostilities is advised, as it might turn out that the other side is, ultimately, in the right.[51]

Major Banks: Thank you for your time – I have much to think about. Please end session.

No sooner had Major Banks logged off the programme than she opened a dictation software to write down the main takeaways of the conversation, with the aim of drafting the guidelines that would inform the ethics of the new Space Force. Thus, she proceeded to dictate:

> *"Note to self: The following are to be considered the core principles of Vitoria's universal philosophy that may be updated to become applicable to twenty-first century space exploration:*
> - *Natural law applies everywhere in the universe, including its precepts on the natural sociability of human beings.*
> - *International law, considered as a result of natural law, may apply among humans and extra-terrestrials only if the latter are capable of reciprocity.*
> - *The only just cause for the use of force against extra-terrestrials would be an injury, whether against humans or other creatures. Yet, everything must be done to avoid war; and after war, the victor must be merciful and fair to restore peace. During war, maximum restraint and proportionality must be employed.*
> - *Finally, if convinced of an endeavour's unjustness, then you should not participate, but, if you are not sure, then benefit of the doubt should be granted."*

References

Amorosa, P. *Rewriting the History of the Law of Nations. How James Brown Scott made Francisco de Vitoria the Founder of International Law.* Oxford, Oxford University Press, 2019.

Aquinas. *Political Writings* (R.W. Dyson ed.). Cambridge, Cambridge University Press, 2004.

Arendt, H. *The Human Condition.* 2nd ed. Chicago, The University of Chicago Press, 1998.

Bellamy, A. *Just Wars. From Cicero to Iraq.* Malden, Polity, 2012.

---

50 Reichberg, Syse, and Begby, op. cit., 322; Vitoria, *Political*, 312-313; Vitoria, *Relecciones*, 130.
51 Bellamy, *Just Wars*, 53.

Bellamy, A. *The responsibilities of victory:* Jus Post Bellum *and the Just War*. Cambridge, Cambridge University Press, 2008.
Brown Scott, J. *The Spanish Origin of International Law*. Francisco de Vitoria and his Law of Nations. London, Clarendon Press, 1934.
Dworkin, R. *Law's Empire*. Oxford, Hart Publishing, 1998.
Finnis, J. "Natural Law: The Classical Tradition". In: Jules Coleman and Scott Shapiro (eds.). *The Oxford Handbook of Jurisprudence and Philosophy of Law*. Oxford, Oxford University Press, 2004, 1-39.
Finnis, J. *Natural Law and Natural Rights*. 2nd ed. New York, Oxford University Press, 2011.
Gómez Robledo, A. *Fundadores del Derecho Internacional*. Vitoria, Suárez, Gentili, Grocio. Mexico D.F., UNAM, 1989.
Grotius, H. De Iure Belli ac Pacis (1625). Lonang Institute, 2005.
Grotius, H. De Mare Liberum (1609). Indianapolis, Liberty Fund, 2004.
Habermas, J. *Between Facts and Norms. Contributions to a Discourse Theory of Law and Democracy*. Cambridge, MIT Press, 1996.
Habermas, J. *The Inclusion of the Other. Studies in Political Theory*. Cambridge, MIT Press, 1998.
Hart, H.L.A. *The Concept of Law*. 3rd ed. Oxford, Oxford University Press, 2012.
Hathaway, O. and Shapiro, S. *The Internationalists*. New York, Simon and Schuster, 2017.
Höffe, O. "*Animal morale*. Sobre el fundamento de una política ecológica". In Höffe, O. *El Proyecto politico de la modernidad*. México D.F., FCE, 2008, 247-266.
Hoyle, F. *The Black Cloud*. London, Penguin Books, 2010.
Johnson v. M'Intosh. Supreme Court of the United States, 1823. 21 U.S. (8 Wheat.) 543.
Kelsen, H. *Pure Theory of Law*. 2nd ed. Berkeley, University of California Press, 1967.
Koskenniemi, M. "Empire and International Law: The Real Spanish Contribution", in *University of Toronto Law Journal*, No 61 (2011), 1-36.
Lindqvist, S. *Exterminate All the Brutes*. London, Granta, 2018.
Lupher, D. *Romans in a New World: Classical Models in Sixteenth-century Spanish America*. Ann Arbor, The University of Michigan Press, 2006.
MacIntyre, A. *After Virtue*. 3rd ed. Notre Dame, University of Notre Dame Press, 2007.
Nussbaum, A. *A Concise History of the Law of Nations*. London, MacMillan, 1954.
O'Rahilly, A. "The Law of Nations", in *An Irish Quarterly Review*, Vol. 9 No 36 (1920), 579-596.
Radbruch, G. "Statutory Lawlessness and Supra-Statutory Law (1946)", in *Oxford Journal of Legal Studies*, Vol. 26 No 1 (2006), 1-11.
Rawls, J. *A Theory of Justice*. Cambridge, Harvard University Press, 2005.
Raz, J. *The Authority of Law*. Oxford, Oxford University Press, 1979.
Reichberg, G., Syse, H., and Begby, E. "Francisco de Vitoria (ca. 1492-1546). Just War in the Age of Discovery". In: Gregory Reichberg, Henrik Syse, and Endre Begby (eds.). *The Ethics of War. Classic and Contemporary Readings*. Oxford, Blackwell Publishing, 2006.
Roca, M. *Imperiofobia y leyenda negra. Roma, Rusia, Estados Unidos y el Imperio español*. 27th ed. Madrid, Siruela, 2020.

Schmitt, C. *The Nomos of the Earth in the International Law of the Jus Publicum Europaeum*. New York, Telos Press, 2006.
Schweitzer, A. *The Philosophy of Civilization*. New York, Prometheus Books, 1987.
Singer, P. *Animal Liberation*. London, Bodley Head, 2015.
Skerker, M. *The Moral Status of Combatants*. New York, Routledge, 2020.
Squella, A. *Introducción al Derecho*. Santiago, Editorial Jurídica de Chile, 1999.
UK Government. Policy Paper. Syria action – UK Government legal position. 14 April 2018. Online: https://www.gov.uk/government/publications/syria-action-uk-government-legal-position/syria-action-uk-government-legal-position.
Vitoria, F. *Political Writings* (Anthony Pagden and Jeremy Lawrance eds.) Cambridge, Cambridge University Press, 2010.
Vitoria, F. *Relección sobre la Templanza o del Uso de las Comidas & Fragmento sobre si es lícito guerrear a los pueblos que comen carnes humanas o que utilizan víctimas humanas en los sacrificios* (Felipe Castañeda comp.). Bogotá, Universidad de los Andes, 2007.
Vitoria, F. *Relecciones sobre los Indios y el Derecho de Guerra*. 3rd ed. Madrid, Espasa-Calpe, 1975.
von Clausewitz, C. *On War*. New York, Oxford University Press, 2007.
Walzer, M. *Just and Unjust Wars*. New York, Basic Books, 2006.

INDEX

9/11 attacks, 117, 122, 203, 304

Afghanistan, 147, 205–206
Africa, 174, 232–246
AFRL, *see* Air Force Research Lab
AI, *see* Artificial intelligence
Air Force Research Lab, 219–220, 224
Air traffic control systems, 118
Air-to-ground missile, 295
al-Qaeda, 203
Alderaan (from Star Wars film series), 54, 65–66
Analogy, 45, 47, 56, 59–63, 194
Antarctic Treaty, 181, 198, 274
Antarctica, 181, 198, 274
Anti-satellite testing, 24, 25, 31, 71, 74, 76, 82, 106, 120, 130, 165, 182–183, 186, 191, 202, 269, 276–277, 283
Anti-satellite weapons, 24, 71, 81, 86, 105–106, 118, 127, 130–131, 138, 162–163, 179, 183, 202, 252, 255, 263, 277, 279, 282, 283
Arms race, 1, 69, 72–73, 75, 77, 82, 85, 161–162, 183–185, 190–191, 243, 253, 270–271, 276, 283
Artemis Accords, 169–170
Artic, 29, 62

Artificial intelligence, 39–40, 310–311
ASAT, *see* anti-satellite
Asteriod/s, 25, 28, 41, 108, 145, 168, 303–304
Asymmetries, 57–58, 60, 65, 104, 127, 144, 206
Australia, 22, 101–104, 126, 161, 168–170, 174, 193, 250, 279, 282, 300, 305, 306
Australian Space Agency, 23, 168
Autonomous, 21, 25, 32–33, 43
Azerbaijan, 116

BeiDou, Chinese global navigation satellite system, 23
Belgium, 279
Biological warfare, 38, 64
Blinding, 29, 105
Blue Origin, 226
Boeing C-17 Globemaster III, 224
Brazil, 279
Bush, George W, 262
BX-1 pico satellite (china), 305

Cambodia, 202
Canada, 149, 163, 168, 185, 193, 279, 305
CD, *see* Conference on disarmament

Celestial bodies, 5, 44, 82, 116, 159–161, 165, 173, 179, 181–182, 249, 258, 274, 281, 304
Chemical sprayers, 202
Chemical warfare, 38, 64, 129, 202
Chile, 279
China, 8, 24, 55, 62, 68, 73, 76–77, 88, 98, 103, 105–107, 116, 118, 131, 133, 134, 136, 139, 162, 163, 171, 183–184, 188, 191, 193, 201, 209, 232–233, 249–258, 264, 279–280, 303–306, 314
Civilian and commercial uses of space, 5, 270, 281
Civilian infrastructure, 27, 103
Civilians, *see* non-combatants
Clausewitz, Carl von, 41, 220–222, 324
Climate change, 66, 89, 160, 242–243, 278
Code of ethics, 96, 110
Cold war, 1, 5, 55, 58, 62, 72, 77, 139, 161, 164, 178, 181, 193, 254, 259, 269, 270, 275, 283
Collateral damage, 18, 27, 29, 58, 64, 99, 131, 211
Colonies, 5, 31, 32, 42, 50, 56, 300, 309
Columbia, 101, 279
Combatants, 19, 27, 62, 65, 104, 145, 221, 238, 260
Committee on the Peaceful Uses of Outer Space, 128, 174, 181, 272
Common Good, 55, 88, 128, 320, 325
Communication, 2, 23, 25, 30, 31, 40, 56, 59, 62, 68–69, 71, 81, 95–97, 101–104, 116–118, 125–126, 130, 134–138, 160–161, 167, 171, 188, 204, 206, 211, 214, 257, 262, 278–279, 283, 292, 306, 323, 324, 326

Conference on disarmament, 69, 184, 276
Contractors, 57, 308
COPUOS, *see* Committee on the Peaceful Uses of Outer Space
COSPAS-SARSAT, 98, 100
Cost benefit analysis (CBA), 107–110
Costa Rica, 116
Counterspace capabilities, 68, 69–70, 72–75, 77–78, 104–107, 131, 179, 183, 194, 280
Crimea, 252
Cubesats, 204–206
Customary law, 27, 45, 49, 186, 189
Cyber warfare, 3, 25, 30–31, 59, 62–65, 73, 119, 124, 129–130, 132–133, 139, 162, 178, 187–188, 203, 209, 254, 261, 298
Cybersecurity Ventures, 203, 206

Dazzling, 31, 105, 162
Debris removal, 3, 73, 81, 86, 88, 89–93
Denial of service (DoS) attacks, 131–132, 135
Denmark, 279
DIA, *see* United States Defense Intelligence Agency
Direct-ascent anti-satellite weapons, 24, 29, 73, 105, 162
Directional fragmentation charge, 205
Disaster response, 72, 96, 100, 102, 124, 133, 143, 278
Doctor Who (tv series), 310–311
Drone warfare, 3, 60, 92, 132, 137, 293–294
Dual-use, 3, 73, 89–90, 95–111, 139, 189, 204, 279

Earth observation data, 71, 96, 100–101, 104, 278, 283
Earth-to-space kinetic weapons, 24, 25, 56
Earth-to-space non-kinetic weapons, 25
Earth-to-space warfare, 56
Education, 2, 95–96, 101–103, 135, 213, 301
Egypt, 15, 279
Eisenhower, Dwight, 272, 273, 276
Electromagnetic pulse, 31, 71, 119–121, 124, 130–132, 279
Electronic warfare, 68, 73, 76, 105, 120, 129, 132, 137, 254
Emergency beacon location, 95, 104
EMP, *see* Electromagnetic pulse
The Empire (Star Wars: A New Hope, film), 54, 57
Environment, 5, 45, 66, 72, 95, 99, 102, 104, 106, 107, 108–111, 115, 125, 128–129, 133, 147, 174, 189, 191, 244, 254, 269, 270, 281, 296, 303, 318
ESA, *see* European Space Agency
European Space Agency, 134, 204, 207
European Space Policy Institute, 136
European Union International Code of Conduct for Space Activities, 75, 95
The Expanse (tv series), 312
Extra-terrestrial, 31, 110, 148, 309, 318, 322–324, 326–328
Extreme-remote warfare, 5, 125, 287–298

Falun Gong, 136
Farscape (tv series)– 309
FCC, *see* Federal Communications Commission
Federal Communications Commission, 206–207
Finland, 74, 281
Firefly (tv series), 308
FOB, *see* Forward Operating Base
Food distribution, 2, 95, 211
Force multiplier, 68, 261
Forward Operating Base, 294–295
France, 15, 18, 72, 139, 163, 171, 185, 192, 250, 263, 279, 280

Galileo, ESA global navigation satellite system, 23
Geneva Conventions, 38, 54–66, 186, 188, 205
GEO, *see* Geostationary orbit
Georgia, 85
Geostationary orbit, 28, 107, 117
Germany, 279
Ghost Fleet (novel), 55–56, 81
Global Commons, 2, 169, 273–275
Global navigation satellite system, 98–100, 278
Global positioning system, 23, 26, 44, 70, 74, 98–100, 103–104, 115, 118, 137, 160, 189, 257, 262, 278, 281
GLONASS, Russian global navigation satellite system, 23
GNSS, *see* global navigation satellite system
GPS, *see* global positioning system
Greece, 10, 279
Ground stations, 26, 132, 162, 214
Gulf War (1991), 121, 161, 279

Hacking, 45, 68, 73, 75, 132–133, 311
Hague Conventions, 38
Haitian earthquake (2010), 100
Hamas, 213–214

Harvard Manual on International Law Applicable to Air and Missile Warfare (2003), 187–188
Health systems, 2, 95–96, 102–103, 143, 150, 154–155, 314
Hezbollah, 213–214
High powered microwaves, 25, 131, 202
Humanitarian assistance, 143, 179, 229, 278, 281, 326–327

ICBM, *see* Intercontinental ballistic missiles
ICJ, *see* International Court of Justice
IED, *see* Improvised explosive device
IHL, *see* international humanitarian law
Improvised explosive device, 205
India, 33, 68, 76, 72, 101, 105–106, 134, 136, 139, 162, 163, 174, 183, 188, 193, 206, 232, 249, 279–280
Institute for Applied Space Policy and Strategy (IASPS), 145
Institute of Medicine, 143–147, 150–55
InStrat Global Health Solutions, 101–102
Intercontinental ballistic missiles, 25, 120, 130, 271
International Charter on Space and Major Disasters, 278
International Committee of the Red Cross, 172, 188, 194
International Court of Justice, 46–47, 50, 171–172, 174, 186
International humanitarian law, 36, 38–39, 44, 47, 103, 167, 179, 194
International Law regarding Military Space Activities, 4

International law, 4, 8, 11, 12, 14, 15, 41, 43, 45, 46, 49, 76, 87, 117, 126, 159–174, 179–182, 184–198, 209, 257, 275, 308, 320–328
International Space Station, 39–40, 55, 86, 93, 118, 139, 203–204, 305
International Telecommunications Union, 167–168
International treaties, 10, 27, 82, 128, 273, 304
Internet, 2, 101–103, 138, 160, 161, 202–203, 278
Interplanetary United Nations, 42–43, 307
IOM, *see* Institute of Medicine
Iran, 73, 116, 134, 136–137, 139, 206, 209, 214–215, 250, 279
Iraq, 132, 147, 206, 214–215
Islamic State, 135
Israel, 15, 83, 134, 184, 213, 263–264, 279
ISS, *see* International Space Station
Italian-Turkish War (1911–12), 202
Italy, 279
ITU, *see* International Telecommunications Union

Jamming/jammers, 25–26, 29, 31, 68, 73–75, 103, 105, 132, 136–137, 162, 202, 253
Japan, 36, 72, 102, 105, 120, 134, 139, 250, 279–280, 303
JASSM, *see* Joint Air to Surface Standoff Missile
Joint Air to Surface Standoff Missile, 99
Joseph, Robert, 137, 201–202, 210, 215
Jus ad bellum, 10–18, 37–38, 64, 179, 180, 186–188, 191–192, 194–195, 215, 325
   Appropriate authority, 11, 14

INDEX   335

Intentions, 10–13, 16–17, 66, 173, 188, 191, 235, 241, 280, 309
Just cause, 10–13, 16–17, 37–38, 325–326, 328
Last resort, 10–17, 36–37
Probability of success, 10, 11, 15–17
Proportionality (*jus ad bellum*), 10, 11, 15–17, 64
Public declaration, 10, 11, 13
Jus ad vim, 37–38, 47–48
Jus in bello, 3, 10, 11, 37–38, 54–66, 179–188, 191–195, 206, 209, 215, 293, 325
 Discrimination, 3, 11, 18, 25, 38, 54, 60, 64–66, 126, 164, 206, 209, 215, 275
 Proportionality (*jus in bello*), 11, 18, 40, 47, 48, 54, 58, 60, 65–66, 102–105, 110, 126, 167, 188, 206, 209, 215, 324, 328
Jus post bellum, 10, 37–38, 325;
Just War Theory, 2–3, 8–18, 36–51

Kármán line, 21
Kerala Infrastructure and Technology for Education, 101
Kessler Syndrome, 56, 86, 106, 108, 120–121, 130, 212
Killer robot/s, 49
Kinetic attack, 4, 25–29, 201–215
Kinetic kill vehicle, 202, 205
Kinetic weapons, 24–25
KITE, *see* Kerala Infrastructure and Technology for Education
Kosmos-2542 satellite, 251
Kosmos-2543 satellite, 251
Kosovo, 252
Kuwait, 121
Kyoto Protocol, 242

Landmines, 38, 80–93
Laos, 202
Laser, 25, 30, 55, 91, 105, 129, 131, 171, 202, 280
Law enforcement, 5, 40–42, 300–301, 306–310, 313–315
Law of Armed Conflict, 10, 13, 38–39, 44, 47, 102, 160, 165, 167, 172, 179, 190, 191, 195, 197, 203
Law of the Sea, 61
Lebanon, 213–214
Legal positivism, 42, 48–49
LEO, *see* Low earth orbit
Liberation Tigers of Tamil Eelam (LTTE), 136
Limited Test Ban Treaty (1963), *see* Partial Test Ban Treaty
LOAC, *see* Law of Armed Conflict
Low earth orbit, 86, 106–108, 117–118, 120–121, 130, 138, 204
Lucas, George Jr. (philosopher), 62, 65
Luxembourg, 169, 279, 304

M*A*S*H (television series), 60
mala in se, 18, 37–38
Manual on International Law Applicable to Armed Conflicts at Sea, 187
Manual on the International Law Applicable to Military Operations in Outer Space (MILAMOS), 187–188
Mapping, 96, 100
Maritime law, 61, 137
Maritime warfare, 3, 61
Mars, 32, 40, 43, 145, 149, 262, 299–315
Mattis, General Jim, 256–257
McGill Institute of Air and Space Law, 188
Metz, Thaddeus, 237, 240–41

Mexico, 279
Middle East, 206, 294
Military Ethics, 8–18
Military operations, 2, 4, 80–81, 86, 92–93, 97, 104–107, 120, 143–156, 160, 179, 187
Millennium Development Goal, 242
Minerals, 56, 303
Moon Village Association, 172
Moon, 25, 28, 31, 32, 45, 56, 82, 137, 145, 161, 164, 165, 169, 170, 179, 183, 248, 249, 263, 272, 274
Moral Injury, 287–298
Musk, Elon, 226, 299, 314

NASA Extreme Environment Mission Operations (NEEMO), 149
NASA, *see* National Aeronautics and Space Administration
National Aeronautics and Space Administration, 76, 108, 130, 145, 146, 148–154, 169, 170, 203, 226, 273
National Air and Space Intelligence Center (NASIC), 134
National Oceanic and Atmospheric Administration (NOAA), 133
NATO, *see* North Atlantic Treaty Organization
Navigation, 2, 23, 44, 81, 95, 98, 99, 104, 118, 125, 126, 134, 137, 160, 204, 211, 261, 273, 278
Nigeria, 102
"No first placement" of weapons in outer space statement, 85, 184
Non-aggressive militarisation of space, 97, 275, 281
Non-combatants, 10, 18–19, 56, 61, 63–65, 69, 211

Non-state groups, 4, 10, 56–57, 62–63, 115, 118–122, 124–139, 194–195, 201, 215
North Atlantic Treaty Organization, 73, 172, 187, 280
North Korea, 116, 120–121, 131, 134, 209
Norway, 132, 281
Nuclear arms race, 1, 69, 178, 271
Nuclear Suppliers Group, 121
Nuclear weapons, 25, 36, 71, 82, 121, 183, 242, 258, 276, 277, 282

Ocean pollution, 87, 89
On-orbit/co-orbital ASAT satellites, 88, 106–108, 162, 171, 202
Operation Burnt Frost, 73
Operation Desert Storm, 279
Operation Starfish Prime, 71, 282
OST, *see* United Nations Outer Space Treaty
Ottawa Treaty Against Anti-Personnel Landmines, 49
Outer Space Treaty, *see* United Nations Outer Space Treaty

Palestine, 213–214
PAROS, *see* Prevention of an Arms Race in Outer Space
Partial (Limited) Test Ban Treaty (1963), 71, 74, 171, 276
Permanent Court of International Justice, 171
Peru, 101
Petrus, Simon, 232, 244
Pirates, 57, 62
Pitso, 235
PNT, *see* Position-navigation-timing
Position-navigation-timing, 23, 26, 30

Positive identification (PID), 206, 208
Post-Traumatic Stress Disorder, 6, 287–298
POW, *see* Prisoner of War
Power stations, 19, 95
Pre-emptive attack, 12–15
Predator drone, 132, 319
Presidential Executive Order, 169
Preventative attack, 13–15
Prevention of an Arms Race in Outer Space, 69, 72–73, 82, 85, 183–185, 191, 198
Prevention of the Placement of Weapons in Outer Space Treaty, 85, 128, 165, 171, 183, 184, 191, 192
Principle of beneficence, 146, 152
Principle of fairness, 154–155
Principle of fidelity, 154
Principle of non-maleficence/avoiding harm, 146, 151–152
Principle of respect for autonomy, 153, 155–156
Prisoner of War, 39, 64
Privateers, 57
Prohibited weapons, 18
PTSD, *see* Post-Traumatic Stress Disorder

Qatar, 279

R2P, *see* Responsibility to Protect
Raymond, General John, 202
"realist" position, 2, 39
Rebel Alliance (Star Wars: A New Hope), 54
Red Dwarf (tv series), 311
Remote warfare, 3
Remotely Piloted Aircraft Systems (RPAS), 292
Resilience, 58

Resources, 22, 28, 31, 57, 59, 61, 62, 71, 100, 101, 116, 134, 135, 147, 155, 164–165, 169–170, 174, 221, 254, 257, 303–305, 314
Responsibility to Protect, 12, 14
Reversibility, 58, 66
Robot, 2, 20–33, 36, 39, 40, 42–43, 49, 202, 310
Rocket Cargo Program, 219–229
Rocket Lab, 204
Rogue states, 4, 120–122, 124–139
Russia, 1, 24, 55, 56, 68, 69, 70, 74, 77, 88, 96, 98, 105, 106, 107, 116, 130–131, 134, 139, 162, 163, 171, 183–184, 191–192, 201–202, 209, 222, 232, 249–255, 263–264, 279–281, 303, 305

Safety zones, 76, 78
San Remo International Institute of Humanitarian Law, 187
Satellite Operations Centres, 214
Satellite, 1, 3, 20, 22–30, 36, 40, 44, 56, 58–59, 62, 68–73, 76, 81, 82, 83, 85–86, 88–93, 95–111, 115–123, 126–127, 130–139, 160–163, 167–168, 171, 174, 180, 183, 188–189, 192, 196, 201–214, 233, 249, 251–259, 261–264, 269–273, 276–284, 296, 305–306
Science fiction, 5, 144, 306–315
Second Intifada, 213
Secure World Foundation, 2, 68, 72–73, 105, 114, 188
Self-defence, 12–14, 38, 41, 45–47, 76, 173, 181, 188, 191, 192, 197, 276, 322
Soft kill, 83
South Africa, 168, 233, 279
South China Sea, 62, 103, 254–255
South Korea, 73, 134, 279

Space age, 6, 117, 130, 139, 163–164, 169, 178–179, 181–182, 270, 282–283
Space architecture, 56–57, 59, 64, 66, 255
Space debris, xv, 2, 25, 27, 40, 44, 56–59, 61, 66, 80–93, 108, 120, 127, 130, 162, 166, 171, 174, 205, 208, 212, 214, 283
Space faring nations, xv, 3–4, 25, 75, 92, 165, 192
Space Force (tv series), 5, 248
Space infrastructure, 22, 27, 31, 64, 260
Space Marine, 20, 21, 26, 29, 32
Space policy, 108–111, 135–136, 145, 304, 306
Space race, 1, 5, 77, 233
Space situational awareness (SSA), 205
Space terrorism, 3–4, 40, 57, 62, 65, 115–123, 126, 134–137, 202–205, 213–216
Space warfare, xv, 3–4, 24, 26–29, 54–61, 64–66, 81–82, 85, 134–135, 144, 160, 178–180, 189–192, 196–197, 231–234, 238, 240, 245, 257, 305
Space-enabled conflict, 4–5, 294
Space-to-earth kinetic weapons, 25
Space-to-earth non-kinetic weapons, 25
Space-to-earth warfare, 56
Space-to-space kinetic weapons, 24, 25
Space-to-space non-kinetic weapons, 25
Space-to-space warfare, 56
SpaceX, 204, 207, 226–227, 229, 256, 299, 304
Spain, xv, 279
Spoofing, 105, 132, 134, 137, 162

Sputnik 1, 1, 20, 21, 96, 181, 203, 250, 269, 271–274
Star Trek, 56, 312, 313
Star Wars (film series), 22, 54–55, 57, 313
Starlink, 304
Starship Troopers (film and novel), 22, 309
Stuxnet, 62
Suicide kinetic kill vehicle, 205
Swarm, 206–207
Sweden, 279

Taliban, 205–206
Tallinn Manual on International Law Applicable to Cyber Warfare, 187–188
Tardigrades, 263
Thailand, 116
Thresholds, 37, 54, 58–59, 63, 76, 86, 120, 180
Treaty on the prevention of the placements of weapons in outer space treaty (PPWT), 184, 192
Trump, Donald, 59, 88, 169, 249, 251, 262
Turkey, 171, 279

U.S.S.R., 1, 161, 178–179
Ubuntu, 235–246
Ukraine, 56, 85
UNIDIR, *see* United Nations Institute of Disarmament Research
Union of Concerned Scientists, 70, 162, 168, 188, 213, 251, 277, 279
Union of South American Nations (UNASUR), 184
United Arab Emirates, 279
United Kingdom, 107, 279–280
United Launch Alliance, 226

United Nations Charter, 173
　Article 2(4), 46
　Article 51, 12, 46
United Nations Convention on certain conventional weapons, 80, 84
United Nations General Assembly Resolution 72/250, 75
United Nations Guidelines on the Long-term Sustainability of Outer Space Activities (2019), 174
United Nations Institute of Disarmament Research, 2, 69, 71, 73
United Nations Moon Treaty (1979), 31
United Nations Outer Space Treaty (1967), 4, 25, 76, 81–82, 87, 90, 159, 161, 164–174, 178, 180–183, 186, 190, 269, 274, 276–277, 280–281, 283, 303–305
United Nations Security Council, 14
United Nations Treaty on the Non-Proliferation of Nuclear Weapons, 121, 242
United States Air Force Space Command, 27, 127, 137, 139, 255, 259–261, 262, 280
United States Air Force, 126–127, 144, 224, 249, 256–63
United States Defense Intelligence Agency, 201–202
United States Department of Defense, 144, 203, 249, 279
United States Marine Corps, 88, 261
United States National Security Strategy (2017), 59
United States of America (USA), 1, 4, 22, 46, 50, 85, 88, 116, 118, 120, 134, 161–2, 168–169, 171, 174, 279
United States Space Command, 27, 106, 127, 137, 252, 255, 259–262, 280
United States Space Force, xvi, 5, 22, 58, 144–145, 202, 219, 223–228, 248–264, 270, 305
United States Space Strategy, 250, 259, 262
UNOST, see United Nations Outer Space Treaty
UNSC, see United Nations Security Council
US Explorer Satellite 1, 272
USAF, see United States Air Force
USSF, see United States Space Force

Van Allen Belts, 282
Vanguard projects, 219–229
Vietnam, 147–148, 202–203, 288
Violent non-state groups, 4, 125, 134
Virtue ethics, 96, 109–111

Walzer, Michael, 8, 10, 12, 15, 36, 63, 323, 325
Weaponisation of space, 91, 97, 128–129, 135, 171, 190, 197, 197, 253, 270, 305
Weapons of mass destruction, 37, 82, 126, 128, 161–162, 165, 171, 183, 197, 258, 277, 281
"Wild west", 159, 164, 174
WMD, see Weapons of mass destruction
Woomera manual, 4, 87, 178–198
World War II, 36, 126, 186, 202, 269–270, 282
Wright Brothers, 202–203, 248, 263

We hope that you have enjoyed this book.
Please consider posting a review for future readers.

Thank you.

www.ingramcontent.com/pod-product-compliance
Ingram Content Group UK Ltd.
Pitfield, Milton Keynes, MK11 3LW, UK
UKHW021317180426
11947UKWH00015B/1273